D0935257

*The Yellow River*

# The
# *Yellow River*

**THE PROBLEM OF WATER IN MODERN CHINA**

David A. Pietz

 Harvard University Press

*Cambridge, Massachusetts*
*London, England   2015*

*Library of Congress Cataloging-in-Publication Data*
Pietz, David Allen.
   The Yellow River : the problem of water in modern China /
David A. Pietz.
        pages cm
   Includes bibliographical references and index.
   ISBN 978-0-674-05824-8 (alk. paper)
   1. Yellow River (China)   2. North China Plain (China)   3. Water—
Pollution—China.   4. Water—Purification—China.   5. Economic
development—Environmental aspects—China.   I. Title.
   TD424.4.C6P54 2015
   333.91'6209511—dc23        2014008684

*To Milton, Nancy, and Olivia*

# Contents

# Figures, Maps, and Tables

## Figures

## Maps

## Tables

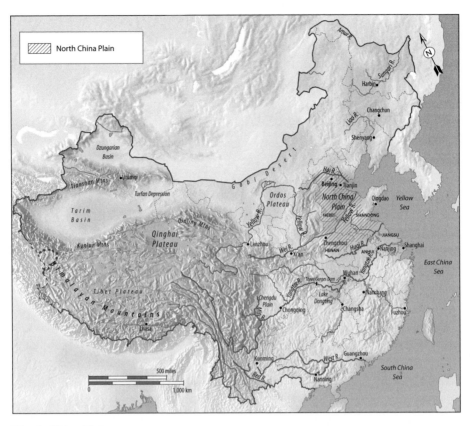

North China Plain

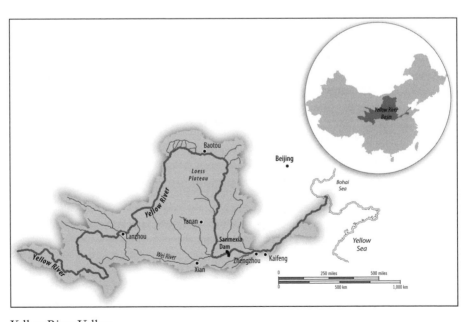

Yellow River Valley

*The Yellow River*

# Introduction

IN 1997 THE YELLOW RIVER dried up. The last remaining puddles and trickles of water soaked into the thick expanse of sediment on the riverbed or evaporated under the high, hot July sun of the North China Plain. The Yellow River had dried up on several occasions since the 1970s, but in 1997 the desiccation reached an unprecedented 400 miles inland from the river's mouth on the Bohai Sea. It is true that the inhabitants of the North China Plain have long struggled with water, at times with too much, but more persistently with too little. Indeed, historical records abound with chronicles of famine induced by drought in the Yellow River valley and the North China Plain. Yet the dry up in 1997 signaled something different—a particularly egregious set of water challenges embedded in the longer-run ecological setting of the North China Plain and, at the same time, conditioned by the forces of dynamic change in contemporary China.

The onset of a series of Yellow River dry-ups in the early 1970s was a relatively early indicator of water stress on the North China Plain. Beginning in 1978, this phenomenon became manifest, as demand for water resources skyrocketed during the remarkable economic growth of the post-Mao reform era. Rural reforms erased communal obligations by granting farmers long-term leases on their land and afforded them greater decisions over what to cultivate and where to market their goods. Shortly thereafter, the restructuring of the state-owned industrial sector and the introduction of market forces generated a manufacturing boom and a massive migration to urban centers, where rising incomes supported expanding patterns of consumption. A consequence of each of these reforms was the increased consumption and pollution of water. Per capita water availability plunged from 735 cubic meters ($m^3$) in 1952 to 302 $m^3$ in 2009. At the same time, in 2009 fully one-quarter of the water in

1

the Yellow River system did not meet government Grade V standards (i.e., the water was unfit for human consumption or agricultural use). China's consumption of water is projected to continue to increase, from 41 billion m$^3$ in 2009 to 46.2 billion m$^3$ in 2020. Further clouding tenuous future prospects, a recent joint publication by several government and research organizations in China forecasted a 27 percent decline in glacial volume on the Qinghai-Tibet Plateau and in the Himalayas. This region is the "water tower" of Asia and is the source of many of China's rivers, including the Yellow and Yangtze Rivers, as well as transboundary arteries, such as the Salween, Ganges, Mekong, Ayeryawady, Brahmaputra, and Indus Rivers.[1]

This book is not an exposé. Its purpose is not to catalog data in order to present a contemporary portrait of an ecological train wreck. Those sorts of analyses have been done.[2] At the most general level, the text explores continuity and change in the waterscape on the North China Plain. The objective is to explore China's contemporary water challenges from a historical perspective—to understand how cultural choices and biophysical processes shaped a historical trajectory that frames present and future constraints and opportunities. Particular emphasis is placed on the Maoist period (1949–1976), when unprecedented development of water resources occurred on the North China Plain. Although they sometimes resembled a "war on nature," Maoist-inspired developmental means and goals were shaped in important ways by long-run patterns as well as by global assumptions about the role of water in the pursuit of modernity. At the same time, the unprecedented capacity of the state and the Chinese Communist Party (CCP) to organize the exploitation of water on the North China Plain generated acute transformations of the landscape, which were the most direct legacy for contemporary China. Thus, in short, this book explores how China reached a state of acute water insecurity and what it might mean for China and the global community.

Flowing from these objectives are two fundamental arguments. First, China's contemporary water challenges are historically grounded. To be sure, China's post-Mao (1976–) economic boom has unleashed unprecedented urban growth, industrial expansion, and agricultural intensification, all with profound consequences for water resources. But patterns of supply and demand during the post-Mao market reforms have been grafted upon a physical, institutional, and symbolic landscape that was deeply affected by activities of the Maoist period. And both Mao and post-Mao waterscapes have been conditioned by longer-run historical realities and commitments. Second, these historical forces, or realities, will not disappear. As China necessarily confronts increasingly complex issues of resource allocation, the enduring legacies of the physical setting and cultural patterns will shape the parameters of available options, as its leaders formulate responses to a host of complex issues that impact national and international interests and constituencies.

The book's focus on the North China Plain is premised on several concerns. First, the region has long been among China's most ecologically vulnerable areas. At the same time, the region has been one of the most critical agricultural regions for sustaining the Chinese state and empire. Today, the region produces half of China's wheat and one-third of its maize (corn) and cotton. At the same time, nearly one-quarter of China's population resides on the North China Plain. The combination of agricultural importance, high population density, and limited water resources have historically induced successive Chinese governments to exert an appreciable effort at maintaining some semblance of ecological and social stability by expending considerable resources to prevent floods (e.g., by building dikes along the Yellow River), and by mitigating the social consequences of famines that result from floods and droughts (e.g., by transporting grain from the south via the Grand Canal to stock emergency granaries on the North China Plain).

Thus, as we begin to explore the state's unprecedented efforts to exploit China's water resources after 1949 and to examine the legacy of these efforts for contemporary China, we should remember that there has been a long history of state efforts to manage the ecological setting of the North China Plain. These historical endeavors left a legacy of practices and symbols that the post-1949 Chinese state rejected, adopted, or otherwise manipulated to sustain mobilization campaigns as the new state sought to wring every last drop of water from the North China Plain. After 1976, Party and state leaders continued to operate within the context of continuity and change of China's rich and celebrated history of manipulating water resources. But to this epic must be added an additional chapter, namely, the water development activities of the Maoist period, which had a profound effect on these resources.

Some have argued that China does not have a water problem; it has an *allocation* problem. This is but part of the challenge. Management choices necessary to increase overall agricultural and industrial efficiency—and to allocate water for the most productive uses—include increasing the price of water and other incentives that favor urban and industrial consumers. Such facile prescriptions, however, betray ignorance of the historically and culturally determined range of choices available to China's leadership. This leadership is acutely aware of the tensions between these legacies and the rational resource use necessary for modern economic growth. Political and academic actors in China are also keenly attuned to potential domestic and international consequences inherent in China's water dilemma.

China's political elites are also well aware of the potential for social instability that depleted and despoiled resources can engender. One need only heed the example of Soviet republics where environmental concerns and nascent environmental movements often formed the sharp edge of a wedge to broader reform movements. In China, the state and Party face a difficult balancing act—how to ensure agricultural productivity and social stability by allocating sufficient water supplies to rural areas, while at the same time providing water to industry and rural constituencies that are so vital to maintaining China's economic growth. The reemergence of China has generated considerable attention to global resource stocks of food, land, energy, air, and water. In 1995, public statements about China's capacity to feed itself in light of its growing population and its shrinking endowments of water and agricultural farmland by the American environmental activist Lester Brown aroused considerable attention, both within China and internationally. Brown posited that marginal economies in regions like sub-Saharan Africa would be priced out of international grain markets after China entered these markets to purchase grain to feed its people.[3] Indeed, the Party's continuing preoccupation with maintaining food self-sufficiency is one more factor complicating China's dilemma over water.

A range of historically informed practices, institutions, and resource endowments have shaped the state's view of a number of critical water issues: What are the domestic social, political, and economic consequences of water allocation decisions? What are the international consequences of exploiting the water resources of the Himalayas? What is the best way to legislate and enforce pollution control mechanisms? How will climate change affect China's water resources? How will China's resource management challenges shape its participation in international climate change negotiations? How will water concerns shape China's international relationships as it seeks nonmarket arrangements to ameliorate food insecurity?

The reemergence of China as a world power in an era of expanding global consumption and sustained population growth has impelled us to think in terms of "nontraditional drivers" of global relationships. Water and its fundamental relationship with economic growth, as well as social and political stability, occupy center stage in such considerations.[4] The management of water has a long, long history in China. And this history matters, as educated elites in China are keenly aware that present arrangements are very much influenced by precedent. Coming to grips with how resource constraints, like water shortages, are products of history and place not only allows us to better appreciate the range of choices available to China's leadership in addressing these challenges, but also affords us the opportunity to better comprehend the cognitive framework used by Chinese leaders in approaching the management of water resources.

The text is arranged largely chronologically, but several critical themes emerge in considering the historical foundations of China's contemporary water challenges. A conceptual framework that can help organize continuities and discontinuities in China's water management is suggested by the notion of a "technology complex." A technology complex refers to a specific kit of tools employed to achieve a particular goal. These tools include administrative organization, technological form, and cultural imaginary that can be used to legitimize the means and ends of water management projects. The goal of water management in China throughout the twentieth century and into the twenty-first, as in all industrial and industrializing regions of the world, has been the type of "multipurpose" water development for which the Tennessee Valley Authority (TVA) was the preeminent model. The TVA represented the promise of regional development powered by water. Industrial development based on hydroelectric generation, agricultural development through irrigation, expanded waterway transport, and valley-wide social and economic security through flood control—all were accomplished by storing water behind massive barricades. Created and sustained by the planning and administrative acumen of state or quasi-state bureaucracies, these high-modernist visions mapped, controlled, and exploited natural resources to create national wealth. In Europe and North America, this vision was premised on a technological complex that included advanced science and technology, capital-intensive inputs, rational bureaucratic organization, efficient labor inputs, and a sustaining mythos of industrial capitalism.

Both before and after 1949, China's political and technical elites subscribed to modernist goals of industrial growth and national wealth. With the victory of Mao and the CCP in 1949, the result of this quest was the building of a socialist society. After 1949, the means to achieve this end, China's technology complex, was grounded in a Soviet model of large-scale, technologically sophisticated, centrally controlled, and capital-intensive projects. For much of the post-1949 period, however, Mao was also smitten with enthusiasm for a technology complex that emphasized small-scale, local knowledge, local management, mass mobilization, and self-reliance. To what degree did this synthetic technology complex, referred to by Mao as "walking on two legs," suggest continuities and discontinuities with traditional management patterns, and what relationship did these programs have with those that followed during the post-Mao period? Although premised on high-modernist goals, was this hybrid technological complex uniquely Chinese? And, ultimately, did it generate unique consequences for China's water resources?

In many ways, the broader history of water on the North China Plain is a story of accommodation between imperial patterns of water management and the forces of modernity or, as the historian of technology Arnold Pacey described it, a "technological dialogue."[5] Accommodation implies continuity and change. The landscape, which was profoundly altered during the Maoist period, had already undergone a thorough "scape-lift" by the mid-twentieth century. By the time of Mao, there was no "wilderness" on the North China Plain. There was no distinction between natural and cultural. The landscape had been made and remade by inundations of the Yellow River that were largely consequences of human agency. The land-scape had also been shaped by state policies in the imperial period that sought to manage agro-ecological constraints with an extensive plumbing and replumbing of hydrologic systems on the North China Plain.

Engineering approaches designed to maintain a semblance of eco-logical balance have a long history in this region, but constraints on such projects included demographic pressures, limited land resources, and climatic forces (either too much water or too little water). Main-taining an ecological equilibrium to sustain agricultural communi-ties on the North China Plain was essential for a state that valorized the economic and moral order of these communities. Indeed, in im-perial China a mandate to rule was often premised on state capacity to "order the waters." In addition to controlling floods, such state concerns also led to the construction and maintenance of the massive Grand Canal system, completed during the Yuan dynasty (1271–1368) and designed to extract agricultural resources from the Yangtze River region to sustain the imperial capitals in North China. Such notions of ordering the water and of expropriating the resources of the south to sustain the north were extended by the state after 1949.

Transformative historical forces emerged in the twentieth cen-tury that critically informed water management practices after 1949. These include modern sciences like hydrology, industrial technologies, political ideologies, bureaucratic organization, inter-national technical collaboration, and nationalism. Beginning in the Republican period (1911–1949), China quickly adopted these components of a high-modernist technology complex through a va-riety of technical exchanges and by establishing technical training institutions in China. "Multipurpose" water development was the expression used to describe the ways in which this technology com-plex would allocate every drop of water to the cause of production. The efficient and rational harvesting of resources was intended as a means of strengthening the nation. Indeed, the construction of a nation and of a national identity through production implied the nationalization of nature. In other words, the landscape was etched (or perhaps as better stated for our digital world, "encrypted") with national identity. Beginning with Republican-era archaeological discoveries, including that of Peking Man, the North China Plain became celebrated as the birthplace of Chinese civilization and culture. This writing of the past on the landscape included the retelling of the creation myths of Yü the Great, who is credited in prehistoric times with draining the swampy expanses of the region by digging discrete channels that led water to the sea, thereby creating ecological conditions suitable for the rise of Chinese civilization. At the same time, the landscape of the North China Plain during the twentieth century was a symbol of the imperialist exploitation, of Japanese imperialist expansion in China beginning in 1931, of the U.S. "invasion" of Korea (1950), and of long-standing "feudal" social and political patterns. The post-1949 Chinese state combined all these factors with utopian images of a productive garden to suggest a second "creation story" of Chinese civilization precisely in that region where it all began millennia ago—the North China Plain. Epic narratives of the People's Victory Canal, the Sanmenxia Dam, and the Red Flag Canal represented the realization of the ancient saying: "When a prophet appears, the Yellow River will run clear." Mao was the modern Yü the Great. Conflation of landscape and national identity would be employed aggressively by the Party to legitimize its claim to rule and to sanction the reorganization of society in order to "make the mountains bow and the rivers yield."

The reconstruction of the waterscape on the North China Plain was accomplished through a negotiation (or accommodation) between received patterns and transformative forces, between the forces of continuity and discontinuity. Perhaps it would be better to think of China's water management on the North China Plain as a synthesis of these forces; the result was a technology complex that used unique means to reach familiar ends. What were the outcomes? Did this synthesis achieve a breakthrough in the notion of a traditional "hydraulic cycle" in which new states aggressively "ordered the waters" only to see the new ecological balance deteriorate over time? The last part of the book examines the outcomes and consequences of the transformations wrought in the Maoist period. How has present-day China been influenced by choices that were made during the Maoist period and the post-1978 reform period? The central argument presented here is that market reforms were overlaid upon a waterscape that had been extensively exploited during the Maoist period.

It is important to examine the responses by China's political elites to the domestic and global challenges regarding water in terms of options available to them as conditioned by historical forces. Continuity and discontinuity are clearly still at play. Engineering solutions to water management challenges continue to hold a powerful attraction for contemporary Chinese government leaders, as suggested by the state's determination to implement the massive South-to-North Water Diversion (Nanshui Beidiao) project, which will effectively replumb the drainage systems of central and northern China by diverting water from the Yangtze River system to the North China Plain to fuel agricultural, industrial, and urban expansion. At the same time, water allocation and water pollution challenges have themselves become critical forces impelling a reexamination of engineering approaches in favor of structural adjustments, such as demand management.

As one surveys long-term patterns of water use and transformation of water management on the North China Plain, what is truly unique in the contemporary era is the degree of China's integration into global networks. China's resource challenges are therefore global resource challenges. As Mark Elvin, a longtime observer of China's water management practices, notes, "In any future impasse that China faces, whether of resources, water, or food, [either] the world will . . . be part of China's solution, or China will become a large part of a global quandary."[6]

# 1  *On the Ecological Margins*

A QUICK GLANCE at selected data gives a compelling sense of China's water challenges. For example, over 20 percent of the world's population lives in China, sustained by only 9 percent of the world's arable land. Rendering these pressures more acute is the fact that China has only 6 percent of the world's supply of freshwater.[1] On a per capita basis, China's freshwater availability is 2,111 cubic meters ($m^3$) per year, compared to the global average of 6,466 $m^3$ per capita. But China gets by. Until very recently, the country has managed to feed itself despite imbalances in the population-land-water equation. Indeed, at the national level China has avoided large-scale water dependency. It has largely mitigated the importation of virtual water (i.e., water utilized to grow food and manufacture goods) by sustaining its production of agricultural and manufactured goods.

National data, however, obscure regional anomalies that suggest acute vulnerabilities. Differences between the water resources of North and South China, for example, are stunning. Annual river runoff on the North China Plain averages 6 percent of the national total. At the same time, the region accounts for 41 percent of total farmland in China. Viewed from another perspective, total water volume available per unit of farmland on the North China Plain is 15 percent of the national average, while the region supplies over 40 percent of the nation's grain.[2] The importance of the North China Plain is also illustrated by interregional transfer of agricultural goods in China. Nearly 10 percent of agricultural water use on the North China Plain is employed for food crops that are exported to South China (26 billion $m^3$/year). An additional water-equivalent (virtual water) of 14 billion $m^3$/year is sent from the North China Plain to other regions in North China, while 9 billion $m^3$/year is exported from the country.[3] In comparison, although the Yangtze River region is a net exporter of virtual water (both domestically

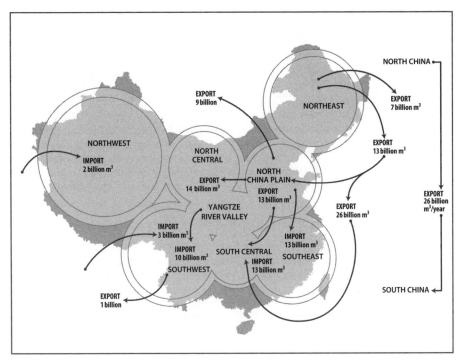

Regional virtual water transfers

and internationally), South China relies on North China to balance its total water demand. In order to sustain this transfer of virtual water to South China, large-scale diversions of water to the North China Plain from the Yangtze River valley, known as the South-to-North Water Diversion project, is well under way.

Thus, the North China Plain plays a critical role in China's economy. The region not only sustains 200 million inhabitants but also plays a pivotal role in the nation's food security, a central concern of the Chinese Communist Party (CCP) for decades. However, the strategic position of the North China Plain to China's economic, social, and political balance rests on a tenuous ecological foundation. The data cited above represent a snapshot, a moment frozen in time, of the region's contemporary hydrological balance. This profile obscures broader historical forces that have long rendered the North China Plain an ecologically vulnerable area. Contemporary realities and future trajectories lay along a historical continuum that, during any given period, is conditioned by short- and long-term natural and cultural forces. The most recent phase of this process, China's reform era, which began in 1978, has witnessed accelerated exploitation of water resources. Although many economists believe that the rate of economic growth will slow over the next decade or two, increased consumption in real terms will continue. Coupled with demographic growth, which is projected to continue until 2050, economic expansion will test the environmental capacity of the North China Plain to sustain region and country.

To better understand the economic, social, and political challenges of the North China Plain, it is necessary to explore the natural forces that, in part, shaped (and continue to shape) the ecological setting of the region. This landscape has also been heavily imprinted by humans for millennia. This long-running dialectic between nature and culture has forged a unique ecological setting on the North China Plain. Waterways, precipitation, temperature, and soils constitute the physical setting of the North China Plain, which we will examine on both its east-west and north-south axes. Virtually the entire area has been heavily influenced by the Yellow River, which traverses the plain from west to east. At the same time, the climate of the North China Plain is determined by weather patterns that largely flow from south to north (or vice versa).

## From West to East: The Transformative Powers of the Yellow River

The North China Plain is an alluvial plain occupying roughly 400,000 square kilometers (160,000 square miles), bounded to the north by the Hai River and to the south by the Huai River, with the Yellow River in between. The region is one of China's largest and most productive agricultural plains. The single-greatest force that has shaped the soils of the North China Plain, and that has been responsible for the creation of the plain itself, is the Yellow River. At one time, much of eastern China lay beneath the sea, with several large massifs rising from the sea to form islands. However, during the last great glacial period (ca. 18,000–15,000 BCE), sea levels declined, exposing large slabs of China's continental shelf. So great was sea-level regression that the continental coastline was located up to 1,000 kilometers (km) (650 miles) east of the present coastline and linked Japan and Taiwan to the mainland. With the end of glaciation, sea levels again rose. By the beginning of the Common Era, the oceans approximated modern levels.[4]

In addition to global climate fluctuation and tectonic movements, repeated inundation and silt deposition by the region's rivers (mainly the Yellow River) have contributed to the formation of the North China Plain. Serving as an artery connecting the far west with the littoral regions of the east, the Yellow River has for millennia redistributed surface soils from upstream to downstream, supplying the ecological basis for the steady growth of agricultural communities on the North China Plain. Gradual atmospheric warming further raised sea levels, but the impact of this phenomenon on China's eastern land mass was countered by the transport of massive loads of sediment to the sea by the Yellow River. Beginning at around 4000 BCE, the Yellow River delta gradually began to expand as the rate of sedimentation surpassed the rate of sea-level aggradation. But the alluvial soils brought from the west were not particularly fertile. In order to sustain cultivation of traditional crops like millet, wheat, and sorghum, the soil must be regularly seasoned with organic material (e.g., human and animal manure, pond growth). In addition, the waterways responsible for transporting sediment to the alluvial plain year after year were not stable and frequently changed course. To maintain a modicum of economic and social stability as human settlements increased, and as states emerged and empires expanded, rivers like the Yellow River required human intervention to keep them in their channels. In short, the geophysical attributes of the North China Plain are the result of an amalgam of natural and human forces.

From its source in Qinghai Province on the Tibetan Plateau, the Yellow River runs for 5,500 kilometers (3,400 miles) to the Bohai Sea. The river is the second longest in China after the Yangtze River and is the sixth longest in the world. Flowing eastward, the river descends what the famed Chinese geologist Li Siguang (J. S. Lee) described in 1939 as a "giant staircase" before reaching an alluvial plain that averages only 20 meters above sea level.[5] Nearly 40 percent of the Yellow River's flow originates on the Tibetan Plateau. Indeed, the entire plateau region, often referred to as the "water tower of Asia," is the source of most of China's major rivers, including the Yangtze and Yellow Rivers as well as the transnational rivers flowing from China into South and Southeast Asia. There has been considerable attention during the past decade on the potential impact of climate change on downstream water supplies. It is true that direct human impact in the source regions of Yellow River has increased in the past several decades, but climatic fluctuations have had the greatest single impact upon upstream hydrological conditions.[6]

In its upper course, the Yellow River is turbulent, with a rapid flow coursing through a series of steep gorges that have only recently been developed as dam sites. When the river meets the Gansu Corridor (part of the old Silk Road), the Yellow River enters its middle stream. At this point, the river current slows and heads northward to ultimately complete the Great Bend, an inverted horseshoe that goes north to east to south before abruptly turning eastward through its lower course on the North China Plain. It is in the middle course that the Yellow River derives its unique character. After the river completes the second leg of the Great Bend and heads southward, it enters the Loess Plateau. The region derives its name from the fine-grained material called "loess" that blankets most of Shaanxi Province and parts of western Shanxi, Henan, and Gansu Provinces. During his extensive travels in China during the 1920s, the American geographer George Cressey recorded his distinct impressions of the region:

> Sprinkled over the countryside as though by a giant flour sifter, a veneer of fine wind-blown silt blankets over a hundred thousand square miles of the northwestern provinces. Although this formation is described by the German word *löss*, derived from deposits along the Rhine, it would not be inappropriate for these far more extensive accumulations to be known by their Chinese name of hwang-tu; or yellow earth. The material consists of a very fine silt, yellowish brown in color, so fine that when rubbed between the fingers it disappears into the pores of the skin without noticeable gritty material.[7]

There is not universal agreement about the origins of loess soils, but there is consensus that the fine particles (which measure between 0.01 and 0.04 millimeters in diameter) blew in from the Gobi, Alashan, and Ordos Deserts sometime during the dry Pleistocene period. After accumulating for millennia, the loess has piled up to 80 meters (260 feet) deep in some locations in the Great Bend. One of the striking characteristics of consolidated loess is its tendency to cleave vertically to form sheer cliffs. Indeed, for centuries, local inhabitants have dug caves from the escarpments to fashion domiciles that are distinctive in their ability to retain a moderate temperature in an otherwise highly variable climate.[8] When exposed to the elements, the loess forms a thin but durable cement-like crust that provides some stability for these structures. However, the resilience of loess soils to extreme events is weak, and many caves have collapsed during earthquakes and torrential rains. It is precisely this relative instability that accounts for the erosion on the Loess Plateau sediment that drains into the Yellow River. Viewed from the air, the loess region resembles a treeless badlands region marked by deep and precipitous chasms. These chasms first appear when runoff from heavy rains begins to cut minor gullies into the loess. What starts as a gully is quickly transformed into a channel full of thick, ochre-colored torrent as runoff removes ever greater amounts of deposits. There is evidence that the loess highlands had significant forest cover at one time, but the expansion of human populations and their agricultural practices by 500 BCE rendered the region largely denuded. Indeed, texts from the Eastern Zhou dynasty (770–221 BCE) comment on the muddiness of the Yellow River, suggesting that deforestation was well under way by this time.

Upon completing the Great Bend, the Yellow River converges with the Wei River. This region is referred to as the Central Plain (Zhongyuan) and is "saturated with history; it saw the earliest Chinese civilization, the rise of the feudal Qin to continental domination, and the successive glories of the Han and Tang capitals at Chang-an."[9] Indeed, for much of China's modern history, the prevailing historical narrative identified North China as the birthplace of Chinese civilization. A powerful and oft-repeated variant on this theme equated the Yellow River with being the "mother of

Chinese civilization." The view that Chinese culture and civilization emanated solely in North China and eventually cast its influence spatially and temporally has been revised. Recent archaeological excavations support the notion of multiple nodes of early
Chinese civilization that collectively coalesced to comprise Chinese
culture. However, the historical connection between North China,
the Yellow River, and the genesis of Chinese civilization received
powerful sanction from nationalist actors, including the revolutionary CCP. This origin story was seized upon by the CCP, which had
its locus of power in North China. After 1949 Party elites constructed a narrative that simultaneously celebrated the Yellow River
as the mother of Chinese civilization while identifying the region
with a "feudal" and reactionary past in order to provide sanction for
ambitious but ecologically disruptive water management projects
on the Yellow River and the North China Plain.

After merging with the Wei River, the Yellow River turns sharply
at a 30-degree angle to enter the North China Plain. From here to
the river's mouth, the gradient of the downstream bed modulates,
causing the flow to slow. At this velocity, about one-third of the sediment picked up in the loess highlands settles on the bed, one-third
is deposited in the estuary region, and the remaining one-third is
flushed into the Bohai Sea.[10] The deposit of silt has been the determinative factor for the North China Plain. Sedimentation raised
the river until it poured over its banks and spread across the landscape, eventually carving a new channel to the sea. As early as the
Neolithic era (ca. 10,000–2000 BCE), the Yellow River changed its
course on multiple occasions. During the historical period (after
ca. 2000 BCE), the river continued to alter its course numerous times,
dropping sediment across the North China Plain as its lower reach
oscillated north and south across the Shandong Peninsula. As the
historian David Keightley observed, "These cataclysmic events
would have disrupted the lives of the Neolithic inhabitants as the
thick beds of Yellow River silt, which have concealed many Neolithic and Bronze Age sites from the eyes of modern archaeologists,
flattened out the landscape."[11]

Keightley also suggested that the memory of such inundations
informed the origins and subsequent reproduction of the myth of
Yü the Great (Da Yu; reputed ca. 2200–2100 BCE). Yü is claimed to

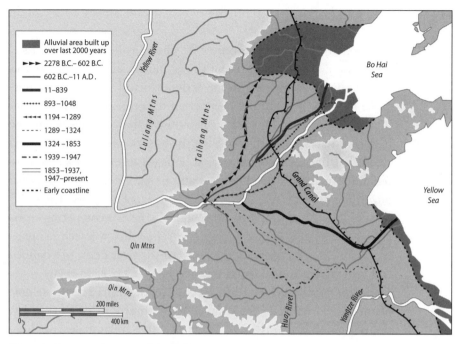

Historical course changes of the Yellow River

have carved out distinct river basins from the broad swamps that stretched across the landscape of the North China Plain. Esteemed as the founder of the first dynasty in China, the Xia dynasty, Yü the Great is celebrated for creating the ecological foundations for a stable and productive agricultural system and for establishing the precedent for sagacious rule—two foundational elements of imperial statecraft that would inform claims of ruling legitimacy well into the modern period. For these reasons, the Yellow River region has permeated the historical memory of China, and the mythic proportions of the river and region were appropriated by the hydraulic transformers of the twentieth century, particularly after 1949, when the dreams of such builders were at their most elaborate.

The moral dimensions of the Yellow River as a symbol of China have also had a material basis. Landscape-changing floods dispersed sediment across the plain and generated the "most important agricultural soils in China."[12] Although it needed substantial additions of organic manure and was susceptible to extreme climatic conditions, the region today holds over 40 percent of China's total farmland. It traditionally has been the center of wheat, millet, barley, and soy production, and, in more recent times, industrial crops such as cotton and hemp. Beginning in the early historical period, the importance of stable agricultural production to social and political order impelled the desire for greater hydraulic stability. Thus began the historical obsession with "ordering the waters" on the North China Plain. During the imperial period, villages formed associations to build and maintain flood control and irrigation systems. Imperial governments were focused primarily on large systems that stabilized Yellow River drainage. Although there were competing theories as to the best way to manage the Yellow River, a technological commitment to building dikes to restrict the river's flow to a defined bed remained a fundamental tenet of river control. In many ways, the history of the Yellow River on the North China Plain is an epic of the human capacity for controlling natural forces. To this day, the resolution remains in doubt, but these human endeavors have unquestionably transformed the landscape.

A dialectic developed on the North China Plain whereby dikes were built and progressively heightened to compensate for a river bed inexorably elevated by sediment deposits. In extended periods of weak state capacity, when political authority was insufficient to mobilize the human or material resources to maintain control infrastructure, the dikes gave way. Although attempts occasionally succeeded in repairing dikes to restore the river to its antediluvian channel, a "permanent" change of course periodically occurred. In such cases, the state, once again, assumed a commitment to keep the river in its new channel.

The commitment to maintain the flow of the Yellow River between dikes generated long periods of quiescence. The present is one such period, as there has not been a serious rupture of dikes in the past sixty years. The relative quietude of the river, however, belies the persistence of sedimentation. As river flows decreased in the several decades after 1949 (due to greater withdrawals for agricultural, industrial, and urban use), the rate of sedimentation increased. During the 1990s, roughly 90 percent of sediment carried from the Loess Plateau settled on the riverbed on the North China Plain. When the river dried up before reaching its outlet to the Bohai Sea, as happened

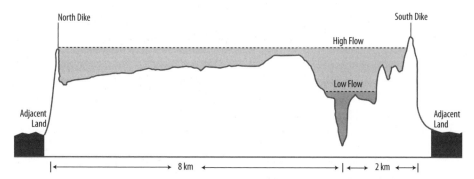

"Hanging" Yellow River bed. *Data source:* Adapted from http://www.iwmi.cgiar.org
/Publications/CABI_Publications/CA_CABI_Series/River_Basin_Trajectories
/Protected/Chap05%20Pietz.pdf, p. 114.

during the 1970s, 1980s, and 1990s, 100 percent of the sediment settled on the riverbed. According to one observer, in 1950 "Yellow
River flow could reach 6.5 acre-feet a second before it spilled over the
inner dikes. By 1990, that was down to 3.2 acre-feet, and the end of
the 1990s to a measly 1.6 acre-feet. . . . I looked at the marks (on Huayuankou dike) on the dike showing high water levels during past
flood in 1958 and 1996. The 1998 flood contained only a third as much
water as the 1958 flood, but it rose 3 feet higher on the dike wall."[13]

The aggradation of the riverbed generated the phenomenon of the
"hanging river" in the lower valley. Near Kaifeng, in present-day
Henan Province, the bed reaches 10 meters (33 feet) above the adjacent plain. This phenomenon has had two consequences. First, when
a dike rupture occurred, the floodwater, descending forcefully from
such an elevation, generated devastating floods. Second, the elevated
bed of the Yellow River could not absorb the inflow from the tributary streams in its lower reaches. One encounters the strange phenomenon of a riverbed forming a ridge that separates two other
drainage systems, the Huai River to the south and the Hai River to
the north. In effect, the lower Yellow River channel is an elevated
aqueduct. This complex hydrologic and hydraulic system has aggravated the challenges in regulating the drainage of the North China
Plain. As the Chinese geographer Xu Jiongxin stated, "Rivers
belong to the most active elements of a geo-system. There exist
complicated interactions and interrelationships between rivers and

other elements. A river flowing on an alluvial plain therefore can be regarded as a sub-system within the geographical system of the alluvial plan, which is coupled to other sub-systems such as landforms, soils, groundwater, vegetation, as well as to aquatic and agricultural ecosystems. Utilization and regulation of rivers by man will inevitably influence the state and stability of other sub-systems."[14]

## From South to North: Climate in the Making of the North China Plain

The second set of inputs that shape the ecology of the North China Plain is a function of climate, namely, precipitation and temperature. If one excludes the Tibetan Plateau and the northwest, continental China is a story of two worlds. The northern half of China lies in a temperate region, and the southern half in subtropical latitudes. Most of China lies in the monsoon region, a term defined "as a climatological phenomenon manifesting itself by a marked change of wind direction between summer and winter . . . producing noticeable effects on the weather and climate of the areas concerned."[15] The transition zone between south and north China runs 33–34°N, roughly following the Huai River. North China is predominantly influenced by continental climate patterns that tend to be cold and dry. During the summer, the south and north are subject to increased rainfall associated with the summer monsoon, but rainfall in the north does not approximate the larger amounts in the south.

On average, there is enough water on the North China Plain to sustain agriculture, but the margin is thin. Annual precipitation is 500–600 mm (20–24 inches)—below annual evaporation rates. Long stretches of hot and windy conditions prevail during the spring and summer seasons, and, compared to other regions of the world at similar latitudes, the North China Plain is, indeed, dry. Southern Europe, the southern United States, and much of Japan receive 2,000–3,000 mm of precipitation annually (80–120 inches). Combined with low annual river runoff (6 percent of national average), low precipitation on the North China Plain means a per capita annual runoff of only 431 m³. The generally accepted criterion for a water shortage area is below 1,700 m³ per capita. From a different statistical perspective, water volume per unit of farmland on the North China

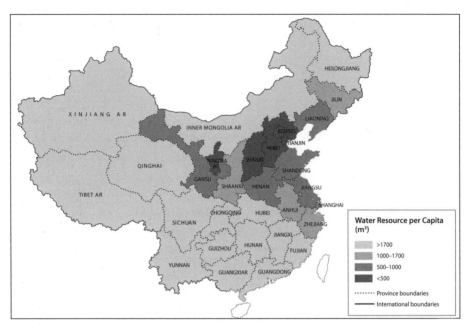

China's hydrologic divide

Plain is 15 percent of the national average.[16] Water availability combined with the porous and calcareous quality (high lime content) of loess soils prevent the cultivation of rice in most parts of the region. Wheat has been the principal crop in modern times, with millet, sorghum, beans, maize (corn), and cotton also cultivated in considerable quantities. Wheat is grown over the winter, but the relatively short growing season means that it cannot be double-cropped during the year. Instead, sweet potatoes, millet, and maize are grown during the summer season. In some regions of the North China Plain, double-cropping cannot be sustained, so a rotation of two crops in three years allows for a fallow period to regenerate soil fertility.

Although annual rainfall is indeed limited, precipitation on the North China Plain possesses unique temporal and spatial qualities. The vast majority of annual rainfall occurs in July and August, often in localized downpours. Agricultural communities thus face two major challenges: droughts and floods. The paucity of annual rainfall, particularly in the critical spring months, periodically results in wide-scale crop loss. In addition, the lack of regular rainfall contributes to the formation of saline soils. Hot spring and summer temperatures pull water and dissolved salts to the surface. The traditional solution to dry conditions has been irrigation, but the development of extensive irrigation on the North China Plain before the twentieth century was inhibited by surface water dynamics that were conditioned by the temporal irregularity of precipitation. Much of the expansion of irrigation occurred only after 1949, when the CCP organized massive irrigation development campaigns. In 1947 there were 16 million hectares (39.5 million acres) of irrigated farmland in China, but thirty years later this figure jumped to 45 million hectares. By 1987 one-fourth of all irrigated farmland in China was in the three North China Plain provinces of Henan, Hebei, and Shandong.[17] Breakneck exploitation of surface water resources for irrigation from the 1960s through the 1980s was simply not sustainable. The solution was to go underground. Massive pumping of underground reservoirs has sustained the viability of agriculture on the North China Plain in the face of near-total exploitation of surface waters.

When rain occurs, usually during July and August, when the southern monsoons hit North China with greatest strength, it often comes in a deluge. At such times, rivers and other channels surge with water. When dikes and other defense structures fail, floods wash over the landscape. During the late nineteenth and early twentieth centuries, floods occurred with frightening regularity as state capacity weakened during the waning years of the imperial system.

Given the thin ecological margin on the North China Plain, it is indeed remarkable that over the long run communities and states have adopted highly resilient agricultural, social, and cultural strategies. Adaptive strategies included an emphasis on managing water. The irony, of course, is that the historical success of these management choices have constrained the range of options that subsequent generations had, and continue to have, in managing the water resources of the North China Plain. For example, the choice of a particular technological complex that featured state-supported construction of dikes to constrain the Yellow River to a discrete channel has generated a technological "lock-in," or dependence, which, if abandoned, would generate unacceptable economic, social, and political costs engendered by regular floods. In other words, the entire

North China Plain is mortgaged. High population density and the remarkable productivity of regional agriculture have depended upon a continued commitment to managing the waters with dikes. Adopting an alternative means of managing the Yellow River (e.g., of allowing the river greater latitude to drain across the alluvial plain) would generate prohibitively high human and material costs. Technology choice created dependent human and material conditions. Ultimately, the state and local communities have had little choice but to continue to invest massive resources to maintain this human-built environment. Historically, there have been periods when this commitment has languished, resulting in disequilibria between nature and culture.

How will climate change impact the ecological setting on the North China Plain? Or, to phrase the question differently, will the state and local communities of the North China Plain continue to meet the challenges of more acute water constraints *within* the historical Chinese commitment to water management? Or will there be a fundamental transformation of the ecological, economic, and social foundations of the region? If climate change renders the ecological conditions of the North China Plain insufficient to reproduce economic and social patterns, what are the domestic and international consequences? We know that global climate change has impacted China in the past. The North China Plain has experienced a gradual cooling and drying of its environment over the past several millennia. From roughly 6000 to 1000 BCE, following the last *glacial maximus*, China experienced its wettest and warmest period. The expansion and development of Neolithic cultures was corollary to this warming climate.[18] Although an overall drying and cooling trend ensued, for several millennia rainfall and temperature on the North China Plain remained considerably greater than today. As the historian Chang Kwang-Chih has written, "In the river valleys and lowlands of North China, where now there is only loess land surface or bushes, there were dense forests and many bogs, marshes, and lakes, supporting such animals as tapir, elephant, water deer, elaphure [a species of deer], crocodile, and other species."[19] By the Bronze Age, however, archaeological and literary sources already indicated cooler and dryer conditions as "temperate-zone animals, like horses and cattle, increasingly replaced the subtropical fauna that are now found only to the south."[20] Southward migrations of fauna and human beings because of climate change or other pressures (like war or famine) are a key theme in the history of China. Cressey's observations of the North China Plain from the early twentieth century vividly contrast with the wet, warm, lush subtropical ecology of ten thousand years ago: "The region is characteristically brownish yellow and dusty. Houses, walls, and even roofs are of the same mud as the fields. The famous dust storms, so well known in Peiping [Beijing] but typical of the whole region, mantle crops and people with their yellowness. Agricultural implements and domestic utensils share this yellow dust; even the trees become yellow. The rivers, too, are yellow, as is the sea into which they flow. Even the tiles on the palace roofs in Peiping gleam with yellow, the Imperial color."[21]

The ecology of the North China Plain, then, is the expression of the long-term collaboration between natural forces and human activity. But the landscape has not been immune to the continuing forces of nature, whether by the transformative power of the Yellow River or the influence of climate. This relationship between the human-built and the natural worlds has made for a tenuous ecological balance on the North China Plain. Severe events like floods and droughts have been perhaps the most visible manifestation of a rupture in the relationship between the two worlds. However, these natural phenomena are, to some extent, social constructions. Floods and droughts would not occur, at least in the dramatic and obvious terms familiar to us, if human ingenuity did not impose its structures on the natural floodplain.

But floods and drought have occurred and have ravaged fields that were carefully cultivated on soils kept fertile by diligent human intervention. In the relative absence of farm animals (direct consumption of grains by humans was more efficient than converting into animal protein), organic sources of fertilizer have included mud, stalks, bean cake, and human excreta. Cressey observed in the 1920s that all wastes were gathered, "particularly . . . human waste or night soil, the collection and preparation of which is an important industry." Cressey added, "Cities receive considerable revenue for the concession, and in the early morning one may see long lines of wheelbarrows, carrier coolies, or canal boats engaged in transporting night soil. In the country districts many farmers whose land abuts on an important road construct comfort stations for the convenience of the passing traveler; and from the competition between adjoining farmers and the relative expense involved in erecting such shelters, one may judge of the very real value of the night soil in increasing the productivity of the land."[22]

The development of double-cropping dramatically increased the net cultivated area in China. Further intensification of farming is one of the few options available to increase agricultural output as virtually all potential agricultural land has been developed. Although statistics on arable land in China are notoriously unreliable, the consensus is that farmland has decreased in the past several decades as expanding cities and industrial plants gobble up agricultural land.[23] Thus, one of the few options that exist for increasing agricultural output is to invest in intensive efforts such as irrigation systems, chemical fertilizers, and the development of new seeds (including genetically modified seeds). But the margin will continue to be thin, and will perhaps become thinner, as extreme events like floods and drought increase because of climate change.

## Floods and Droughts

In the early twentieth century, travelers on the North China Plain like Cressey would have witnessed farmhouses built "on mounds raised five to ten feet above the general level of the plain. The traveler during the winter may wonder at the presence of boats from navigable waterways with not even a pond in sight. Should he return in the late summer of a flood year, he may see these same boats in use for harvesting crops partially submerged beneath a vast, but shallow, expanse of water."[24] Over the past two millennia, over 50 major, and countless smaller, floods have occurred in the Yellow River valley (virtually all on the North China Plain), and 1,500 dike breaks have generated over 20 course changes. In 1642 over 300,000 people died in Kaifeng during extreme flooding. In 1938 close to a million died when the Yellow River dike was destroyed by the Nationalist government to forestall a Japanese invasion from the north.[25]

As destructive as floods have been, droughts have historically been the greater problem on the North China Plain. Droughts have been caused by the weakness of the summer monsoon, the strength and position of the subtropical high-pressure area, spring snow depths on the Tibetan Plateau, and sea surface temperatures.[26] Several of these phenomena, in turn, were influenced by global events such as volcanic eruptions and El Niño–Southern Oscillation events (ENSO).[27] Data from tree rings, ice cores, and pollen samples reconstruct a history of drought. There is now general agreement that severe droughts occurred in the 1480s and 1490s, in the 1580s through the 1640s, from 1700 to about 1720, in the 1820s and 1830s, in the 1920s and 1930s, and the 1990s.[28]

Climate studies have added a new dimension to the historical study of dynastic change in China. Climate scientist Shen Caiming provides an explicit example:

> Many reasons have been cited for the collapse of the Ming Dynasty, including political corruption, economic breakdown, peasant rebellions, the invasion of the Manchu, and bad climatic conditions. The last is one of the direct reasons . . . China experienced a series of severe, extreme, and exceptional droughts from 1500 to 1600, which greatly worsened the economic condition of the late Ming Dynasty. In 1627, severe drought and ensuing famine in Shanxi Province triggered the longest peasant rebellion in Chinese history. . . . Eleven years later, the most extensive and intensive exceptional drought began. Following this exceptional drought, the Ming Dynasty collapsed in 1644. Between 1585 and 1645, the population of China seems to have declined by as much as 40%, partly due to economic distress, warfare, and the collapse of law and order, and partly due to droughts and floods and ensuing famine and disease.[29]

At the local and regional level, floods and droughts generated dislocation. The eminent anthropologist Fei Xiaotong (Fei Hsiao-t'ung) described conditions in the mid-twentieth century in the lower Huai River valley:

> With its water conservancy facilities falling into disrepair, northern Jiangsu was plagued by floods and famines in the

past. People fled to places south of the Changjiang [Yangtze] River to find food almost every year. Before the coming of floods, peasant families sealed the doors of their houses with mud and then fled their homes. In some localities like the Lixi- ahe area, whole villages were deserted during the flood sea- sons. Those who fled would never return again if they found employment and got established in other places. During my childhood, I had seen refugees from northern Jiangsu growing crops on newly formed land around Lake Taihu. When the wa- ter in the lake rose and submerged their crops, the refugees could only flee to cities to become coolies or beggars.[30]

It was these sorts of outcomes—namely, the threat to the agricul- tural stability of rural China and ultimately the threat to the entire political order—that explain why state and local elites exerted so much human and material capital on creating order in this tenuous environment. Traditional flood control regimes have more recently been joined by irrigation networks and massive dams in the search for ecological stability on the North China Plain. Failure to main- tain water systems, particularly flood control structures, would mean one of the most productive regions of China would revert to "fever-ridden and uninhabitable swamps."[31] The historical commit- ment to prevent this reversion to a natural state must be main- tained to avert a social and political catastrophe. But the develop- ment of this human-built world—with all its material artifacts to restrain flood waters, water the plain, fuel industry, and illuminate households—has, along with natural forces, become a factor in re- source scarcity. As part of the background of ecological marginal- ity, the forces of demographic growth, agricultural intensification, industrialization, and urban expansion have further pressured the ecological carrying capacity of the North China Plain. The same atmospheric and hydrological events as those that helped precipitate profound ecological change in the imperial era "would happen more easily now than during historical times."[32]

# 2  *Management and Mismanagement in the Imperial Period*

IN ADDITION to the biophysical forces already explored, the ecology of the North China Plain during the imperial period was deeply transformed by human action. During this long sweep of history, mutually supporting values and institutions evolved within which the manipulation of water was embedded. This system of values and institutional patterns encountered internal and external pressures, including demographic growth, commercialization of the economy, and imperialism, which collectively contributed to a breakdown of the imperial system and hydraulic structures by the late nineteenth century. During the twentieth century, new political, social, and cultural forces from the West were grafted upon Chinese values and practices to forge a new equilibrium between the human-built and natural worlds. This process would command the attention of a variety of Chinese state systems across the political spectrum during the twentieth century and would ultimately generate environmental outcomes both familiar and unprecedented.

## Water and the Rise of Chinese Civilization: The Formation of State Authority

As the geographer Vaclav Smil stated, "Few rivers had such a profound effect on a major civilization as the Huang Ho [Huanghe, or Yellow River] had on China."[1] Whether real or imagined, the river has been central to the historical memory of Chinese down to the present. The writing of history, which perpetuated this cultural mythos, is part of a historiographical tradition unparalleled in human history. The lion's share of artifacts from ancient China has been unearthed in the Yellow River valley, giving this region preeminence in the reconstruction of Chinese history. Oracle bone etchings, bronze inscriptions, and writing on bamboo, wood, and finally paper have all been discovered in the Yellow River valley. "Consequently, throughout Chinese history the Yellow River valley . . . and the Wei River valley of Shaanxi, were regarded by historians and common people alike as the cradle of Chinese civilization."[2] This was the literary context in which the myth of Yü the Great (Da Yü) was created and disseminated to subsequent generations. The myth's narrative focus was the regulation of the waters of the North China Plain, but the moral lessons served to connect the ordering of water and good governance for political elites up to and including the present.

Lying "at the root of every educated Chinese person's idea of the beginning of Chinese history," the story of Yü the Great is one among a set of "creation myths" that collectively laid the historical basis for Chinese civilization.[3] The greatest collator and transmitter of the lives and heroics of the "sage-kings" of antiquity was Sima Qian (ca. 145–86 BCE), who wrote the *Records of the Grand Historian (Shi ji)*, which incorporated oral traditions and earlier records by Mencius (372–289 BCE) to recreate the story of the Five Emperors, which began with the legend of the Yellow Emperor (Huang Di; presumed reign: 2696–2598 BCE). Each of the Five Emperors is credited by Sima Qian with foundational feats: institutions of rulership, the development of writing, the formulation of calendars, the separation of Heaven and Earth, and the moral foundations of ruling authority. The last of the legendary kings was Yü, and his great accomplishment was to direct the excavations of nine discrete waterways to channel the swampy waters of the North China Plain to the sea. Yü is also credited with recognizing that the beds carrying Yellow River water were already being raised by sedimentation. Accordingly, Yü adopted a method of constructing dikes along the riverbank to contain floodwaters. Constricting the flow of rivers by dikes, the normative approach to managing the rivers on the North China Plain throughout Chinese history, thus received historical sanction through the transmitted myth of Yü.

According to Chinese historiographical tradition, Yü's brilliance in controlling the floods and relieving his people established the ecological basis for agriculture. This accomplishment supplied the moral and material basis for Yü to establish China's first dynasty–the Xia (ca. 2070–1600 BCE).[4] At a number of levels, China's great flood myth and the epic narrative of reasserting social and ecological order are consistent with flood stories in a variety of cultures. Flood myths are often *re-creation* myths that are formulated *after* tales of human creation (the *original* creation myth). Such tales are sanctioned "for the existence of a particular body of practices that are central to the people who create the myth."[5] Two additional themes of these myths center on the flood as a form of punishment or repudiation of the immediate past and on water as the principal motif of the re-creation myths (i.e., as opposed to fire or other cataclysmic natural events). It is significant to note that, in these early myths, water is represented alternatively as a destructive or constructive force or as a symbol of chaos or order:

> Chinese flood myths exhibit all the features previously described in accounts of flood myths from other parts of the world. First, they were tales of the re-creation of the world that provided origin myths and thus justification for the major political institutions, particularly those associated with the role of the monarch or emperor and his servants. Second, they employed water as an image for the dissolution of all distinctions, and thus presented the taming of the flood as a process that recapitulated in the age of men, and, through human action, the creation of the world. Several versions of the myth also explicitly contrasted channeled water with the rampant waters of the flood that flowed properly and thus was beneficial. Third, those versions that touched on the origins of the flood attributed it to rebellion, and several versions identified the taming of the flood with the punishment of criminals.[6]

In the version that appears in the *Mencius*, the flood was a metaphor for the collapse of order during the Warring States period (475–221 BCE). In his telling, the flood represented the political and social disorder that resulted from bad rule or criminality by rulers in the past (represented in the myths by Yü's predecessors, Gao and Gun). "The Chinese myths insist that the flood began through rebellion, or some other criminality, and was ended through the successful actions of the ruler and his servants. The manner in which the flood was conquered in the tales of Yü, the bringing of water back into channels where its course could be guided for the benefit of humankind, was directly adapted from one of the major roles of the government in the period."[7] In the subsequent historical narrative of Sima Qian, the ordering of the waters by Yü, or the "second creation" of the Chinese world, was a "fashioning of an ordered physical space" that represented correct rule and sanctioned a political and social system premised on an "ordered and hierarchically divided human realm with the central role played by the family as the social unit."[8] From his perspective in the Han dynasty, Sima Qian was seeking moral sanction for central authority based on the Qin dynasty's (221–207 BCE) reunification of China after the disorder of the Warring States period. The world "re-created after the flood is often marked by the imposition of some new regime or institutions, and the purpose of many flood myths is to explain or justify such institutions."[9] Thus, Sima Qian's refashioning of the flood myth was a reflection of the Chinese world that was emerging from the chaos of the Warring States period. This same myth would be adopted and refashioned in a subsequent creation authored by the CCP after 1949.

The sanctioning power of myths, adapted and retold to legitimize political authority, was expressed in a host of water management projects throughout Chinese history. This ethos of "directing the waters" penetrated all levels of society. In his synthetic study of the rise of "Oriental despotisms," the eminent sinologist Karl Wittfogel observed the sophistication of water management schemes in China and concluded that such accomplishments could have been managed only by a state that had the absolute power to organize material and human resources necessary to build and maintain these projects. However, subsequent research on local society during China's imperial period has proved Wittfogel wrong. Although the state sponsored large-scale hydraulic projects, village-based organizations initiated and organized myriad projects at the local level.

"Historical" Yü the Great. (Collection of the
National Palace Museum, Taiwan)

But Wittfogel was not totally off base if we conclude that there
were unmistakable cultural patterns that valorized the manipulation of water patterns that pervaded the spatial and temporal dimensions of imperial statecraft in China.[10]

Contemporary Yü the Great. Robin McNeal, photographer. From Robin McNeal, "Constructing Myth in Modern China," *Journal of Asian Studies*, Vol. 71, no. 3 (August 2012), pp. 679–704. Copyright © 2012 The Association for Asian Studies, Inc. Reprinted with the permission of Cambridge University Press.

The recurring commitment to engage in water management in China was engendered by military conflict, economic transformations, and demographic expansion. Large projects were pursued by early Chinese states as an adjunct to the creation and sustenance of centralized authority. Maintaining such authority depended on the systematic extraction of agricultural wealth from a stable agrarian base. Thus, the manipulation of existing waterways and the creation of artificial arteries were deemed necessary to expand agricultural production through irrigation and to enhance military mobility. Chinese states had an important stake in the planning and implementation of large-scale projects in early China. The "Chinese imperial state was a meddlesome one, carefully looking after its own interests and, in keeping with cultural traditions, actively seeking to develop resources and rearrange nature so as to maximize tangible and taxable wealth."[11] In early China, this orientation of the state was expressed in a variety of large projects that contributed to the cultural ethos of "ordering the waters." In this way, the myth of Yü the Great served to perpetuate this form of statecraft for centuries in China.

The adoption of levees for flood control was one of the first organized expressions of water management on the North China Plain. Ever-larger protective dikes were part of the strategic culture (along with the technological and organizational know-how) of managing large public works projects. This strategy was also manifest in the walled boundaries between early Chinese kingdoms and, beginning in the Qin dynasty, in the construction of the Great Wall. Levees could also be utilized for offensive purposes during times of conflict. During the Warring States period, rupturing dikes to inundate the territory of a neighboring enemy was a well-developed practice. This strategy of war would be repeated on multiple occasions in Chinese history, most recently in the late 1930s, when Chiang Kai-shek ordered the destruction of the Yellow River dike in Henan Province to defend against Japanese invasion from the north. Thus, on the eve of China's imperial period, well-developed water management practices were adopted by a variety of states.

Early imperial-era water projects focused on the construction of canals for transportation and irrigation. Sima Qian and other chroniclers emphasized the moral dimension of rule that made these projects possible and indeed vital to social and political stability. Among China's early canals, the most celebrated was the Zhengguo Canal constructed by the Qin in 246 BCE at the end of the Warring States period. The story of the canal is part of every primary school lesson chronicling the Qin's rise as China's first empire. As recounted by Sima Qian, during the Warring States period, the prince of Han devised a ruse to sap the financial strength of the rival Qin by dispatching a river engineer named Zheng Guo to the Qin court. Zheng's task was to persuade the king of Qin to create a large irrigation district by constructing a great canal linking the Jing and Luo Rivers. Before the canal was completed, the king of Qin discovered the trick but was nevertheless convinced by Zheng of the benefits of the project. Instead of decreasing the financial strength of the Qin, the irrigation project enriched its coffers. Tax receipts from increased agricultural production supported the creation of a Qin army to unite China. Although sediment eventually clogged the Qin canal, the telling and retelling of the tale by historical chroniclers provided potent reminders of the historical imperatives of "ordering of the waters" to agricultural wealth and centralized rule.[12]

Although constructed well south of the North China Plain in today's Sichuan Province, a second remarkable irrigation system has informed the historical memory of every generation of China's educated elites for centuries. The Dujiangyan irrigation system was also built by the Qin during the Warring States period (in 265 BCE) under the leadership of Li Bing. Continuously in operation for two millennia, the vast system today hydrates 1.7 million acres of the Chengdu Plain. The project, completed by Li Bing's son, Li Erlang, largely retains its original form and function today. "Dig the channels deep and keep the spillways low" (*shen tao tan, qian zuo yan*) was the operative management approach attributed to Li Bing, an approach that contrasted with the tradition in the north that emphasized the construction of protective dikes.[13] Providing water control functions for over 2,000 years, the project has become a symbol of Chinese water management ingenuity. Li Bing and his son hold an eminent place in the pantheon of "water heroes" (*shui gong*) in China. As observed by Needham: "The Chinese were never content to regard notable works of great benefit to the people from a purely utilitarian point of view. With their characteristic ability to raise the highest secular to the level of the numinous they built . . . a magnificent temple . . . to commemorate Li Bing's heroic victory; and further back, in a scarcely less beautiful site . . . another one to that of his son Li Erh-Lang [Li Erlang]."[14] Serving as a link to the historical power of controlling the water, a temple dedicated to Yü the Great stands nearby.

The Zhengguo Canal and the Dujiangyang irrigation system are celebrated as exemplars of the capacity of centralized states to organize the human and material resources to manage large water projects. The Zhengguo Canal was but one of a set of early Chinese irrigation and transport canals that served to integrate regional polities and economies. Many years later, the famed Grand Canal (Da Yunhe), completed during the Yüan dynasty (1271–1368), was explicitly intended to integrate the political center in North China with the economic core of the lower Yangtze River valley. The official histories of subsequent dynasties consistently esteemed the brilliance of the water heroes in conceiving and directing projects that strengthened agricultural society and, by extension, the state. Imperial states viewed the management of water as a means of promoting agricultural production, thereby increasing the ability of the state to appropriate agricultural surplus to expand and sustain the empire and the state. The expansion of agriculture was due to many factors, including the development of agricultural markets and of an ethical ruling system that valorized agricultural pursuits. This interlocking rationale, based on material and moral factors, provided the underpinnings of state projects to manipulate water.

## Nature in Chinese Traditions

For the past several decades, a repeated refrain has emphasized traditional Chinese respect for nature, typically contrasted with the wanton disregard of the natural world exhibited by industrialized Western countries. The impulse for such comparisons, however, emanated from Western countries, not China, as connections were made between traditional Judeo-Christian attitudes toward the natural world and a continued human disregard for the natural environment that generated unsustainable environmental practices. An early articulation of this perspective was offered by the historian Lynn White Jr. in 1969. White argued that Judeo-Christian beliefs engendered a utilitarian attitude toward nature that promoted a thoroughgoing exploitation of the natural world. The destructive practices that such traditions promoted ultimately generated a scholarly and popular reaction in the West by the mid-twentieth century. One component of this reconsideration of Western ecological practices was the impulse by some to contrast traditions in

China that presumably reflected a moral and ethical system that venerated the natural world.[15] Yet fouled air, murky waters, and other ubiquitous images of environmental degradation in contemporary China seemingly run counter to these claims. Often these realties have been ascribed to an abandonment of traditional cultural attitudes as Chinese modernizers succumbed to the allure of post-Enlightenment notions of progress and twentieth-century modernist ideals of development.

The fact is that by the twentieth century the landscape in China had already been altered in a massive way. The "Chinese have . . . altered their environment, over three millennia, of close and intensive occupance, probably on a greater scale than has been the case in any other part of the world until the present century."[16] Deforestation, reengineering of watersheds, and manipulation of soils were all responses to China's demographic growth, empire building, and commercialization of the economy during the imperial era. The "Chinese agricultural landscape, and eventually almost every hectare of inner China, was thoroughly anthropogenic. People chose (sometimes unwittingly) which animals and plants lived. People governed (as best they could) the paths of waterways. Even the soil was a human construct."[17] In the realm of water, a "voluminous body of writing on the management of water systems indicates that the Chinese largely shared the Western attitude that nature should be managed for the benefit of humans."[18]

There are, however, strong ethical traditions in Chinese thought that have promoted the notion of harmony among Heaven, the natural world, and humans. But it is important to come to a nuanced view of what the "natural world" means in this context. This concept of nature did not stress a "wilderness ideal" that sought the maintenance of a pristine world devoid of human artifice. Even if we go back to the classic texts of Daoism, which seemingly prescribed human noninterference with nature, "there is no good reason to presuppose any necessary prehistoric balance with nature. The restraint preached by the environmental archaic wisdom found in certain Chinese classical texts is both familiar and in all likelihood commonly misunderstood; it was probably not a symptom of any ancient harmony, but, rather, of a rational reaction to an incipient but already visible ecological crisis."[19] Perhaps the ethos of human restraint did not reflect an impending "crisis" but, rather, an aspirational harmony with forces of nature that were in a delicate, if not precarious, balance in a landscape that had been thoroughly engineered to support human communities. In other words, the natural order envisaged here was one not untouched by humans but, rather, an order that included (perhaps even centered on) the stability of the landscape and the ability of the land to produce agricultural sustenance in the face of natural forces of climate and the marginal fertility of the soil. So, "it is not surprising that the society as a whole valued it [the land], or that the official belief in the rightness of the particular harmonious system of co-operation which had been worked out was confirmed. It was a demanding environment, especially in its water balance, from a farmer's point of view . . . it returned munificent rewards for the kind of sensitive attention which the Chinese lavished on it."[20]

Despite its rhetorical flourishes, the following account of a Westerner traveling on the North China Plain in the early twentieth century clearly illustrates the relationship between man and nature in China:

> The most significant element of the Chinese landscape is thus not the soil or vegetation or the climate, but the people. Everywhere there are human beings. In this old land one can scarcely find a spot unmodified by man and his activities. While life has been profoundly influenced by the environment, it is equally true that man has reshaped and modified nature and given it a human stamp. The Chinese landscape is a biophysical unity, knit together as intimately as a tree and the soil from which it grows. So deeply is man rooted in the earth that there is but one all-inclusive unity—not man and nature as separate phenomena but a single organic whole. The cheerful peasants at work in the fields are as much a part of nature as the very hills themselves. The Chinese landscape . . . is the product of long ages. Literally trillions of men and women have made their contribution to the contour of hill and valley and to the pattern of the fields. The very dust is alive with their heritage.[21]

Over the centuries, the received landscape, as originally manipulated by Yü the Great and further shaped by subsequent human action, was a challenge to maintain on the North China Plain. It is not surprising that an ethical canon developed that reflected these challenges. China's importation of development models from Japan and the West, beginning in the nineteenth century, introduced technologies and institutional arrangements that had profound environmental consequences. But twentieth-century high-modernist perspectives on nature, the state, national identity, and economic growth were not simply grafted onto an alien Chinese cultural and ecological landscape. Irrespective of their various ideological orientations, Chinese states of the twentieth century drew upon a long tradition of ecological intervention that could accommodate environmentally transformative structures like hydroelectric dams.

## Patterns of Environmental Change in Imperial China

Throughout the imperial period, sustaining the agricultural basis of the state and society required managing water resources. Periodic structural changes in the economy, climate change, demographic growth, and external pressures from non-Chinese groups have led to acute ecological challenges for the inhabitants of the North China Plain, including a breakdown in the hydraulic system toward the end of the imperial period. But for much of this period, the state employed variations of hydraulic management techniques established in the pre-imperial and early imperial periods. Traditions continued of managing water to support an expanding empire and a strong state structure, as well as the moral imperative to "order the waters" to buttress state authority. Much of the institutional and cultural innovation that occurred in China's history has been grounded in these traditions.

The growth of agriculture to sustain an expanding empire and a complex administrative bureaucracy depended on converting forest to tillable land and on exploiting water resources. Timber was also used in cooking, heating, smelters, and kilns and for constructing houses, boats, bridges, and pilings. Over the span of the imperial period, the vast majority of China's forests were felled. The removal of this forest cover resulted in the loss of topsoil through erosion, which was particularly severe in the hilly and mountainous regions after the demographic pressures of East and Central China increased migration to hill regions. The carrying capacity of these highland regions, however ephemeral, was considerably augmented by the introduction of New World food crops beginning in the sixteenth century. Upland migrants cleared land for the cultivation of maize or sweet potatoes following the introduction of these crops. As a result, the soil was exhausted after several agricultural seasons. Farmland was abandoned, and soils washed away. Deforestation in the loess highlands of the Yellow River valley had particular consequences for the Yellow River valley and the North China Plain. One estimate suggests that over the course of four thousand years, forest cover on the Loess Plateau declined from 53 percent to 8 percent.[22] The destabilized loess soils easily washed away, creating gullies of varying size that ultimately led to the Yellow River. Sediment was carried downstream and deposited on the riverbed as the speed of flow decreased in downstream regions, precipitating the never-ending battle to maintain hydraulic stability on the North China Plain.

From the later imperial period to the present, large levees have been constructed along the banks of major waterways like the Yellow River to contain spring and summer torrents. There has been a similar commitment to maintaining the artificially constrained waterways of smaller streams and channels, some of which had been developed for irrigation. Persistent sediment deposits in all these waterways required the periodic raising of levees. But the capacity of human engineering to maintain the Yellow River within these levees was limited. The river periodically ruptured its levees, and its water poured out over the raised bed to forge a new course to the sea. This pattern would be repeated as levees were constructed or augmented to ever-increasing heights. This "hydrological instability required the constant maintenance of large-scale engineering structures to achieve ecological stability."[23] This type of technological and managerial lock-in has continued to define water management on the North China Plain to the present day.

In addition to a constant emphasis on water management practices, another critically important aspect of China's imperial statecraft practices, and one that affected China's environmental history, was China's "medieval economic revolution," which occurred between the ninth and twelfth centuries, roughly coinciding with the Song dynasty (960–1279). Two major developments of this period with important consequences for China's natural environment were demographic growth and commercialization, two trends that would again accelerate in the late imperial period (ca. 1500–1911). From the beginning of the Common Era to the eighth century, China's population remained at around 50 million. The expansion of rice cultivation in Central and South China contributed to a doubling of the population by 1100. Agricultural prosperity and higher population density, in turn, encouraged greater commercialization. The extension of rice cultivation to the lower Yangtze River valley and to areas farther south meant the additional clearing of forest resources and the need to develop water resources for irrigation. At the same time, growing commercialization presented farmers with opportunities for regional specialization, as the extension of a market system allowed for a "national" exchange of goods. One example of the environmental impact of the development of China's markets was the demand for additional timber resources in Central and South China. The development of an efficient market for timber significantly depleted forests in the middle and upper river valleys.

Political transitions were also reflections of ecological change in imperial China. To be sure, military activity and its accompanying social and economic disruption had an environmental impact. Perhaps more importantly, dynastic transitions were often markers of a changed dynamic in China's frontier regions that separated the agricultural practices of mainly sedentary Chinese farmers from the pastoral practices of herding societies. The waxing and waning of the relative influence of these agricultural and social systems had significant ecological impacts. The extension of imperial power typically brought traditional patterns of deforestation followed by intensive cultivation. During periods of waning imperial influence, these regions would revert to more diversified groundcover as pastoral practices expanded. Throughout the imperial period, the degree and extent of anthropogenic ecological change reflected a "start-stop character," and there were periods in which human-created ecological change was reversed and natural forces prevailed. A variety of factors conditioned the pace and extent of ecological change in imperial China: the demographic and economic dynamics of dynastic transitions, the waxing and waning of frontier areas between pastoral and sedentary societies, and climate change. All these factors were mutually conditioning.

Although the character of the Chinese state was not uniquely defined by the imperatives of effective water management, as was elegantly argued by historians several generations ago, the state was nevertheless clearly involved in managing the ecological affairs of the empire in a variety of ways. In the case of water management, research has revealed a variety of arrangements whereby local, regional, and central administrative units independently or jointly managed projects. But the central state clearly had a critical function in large projects, particularly on the North China Plain, where the mandates of managing the Yellow River transcended the boundaries of bureaucratic and administrative constituencies. Ken Pomeranz sees a Chinese state interest in regulating ecological outcomes as an expression of maintaining the agricultural basis for state and society.[24]

Throughout the imperial period, the hydraulic system that sustained the human communities and their agricultural pursuits rested on an unstable ecological construct that could be undone by neglect. Thus, this was a "highly labile" environment, which required constant maintenance through regular investment of labor and materials (like timber and stone), and, above all, of the financial resources of the state and local communities. The state locked itself into maintaining large dike systems on the Yellow River, which, if neglected, could be destroyed in a summer torrent, as sedimentation steadily raised the riverbed. Such disasters did happen on a number of occasions during the imperial period, when the state lapsed in its commitment to control the waters on the North China Plain. An indication of the challenges faced in maintaining the great water systems of the imperial period is suggested by the history of the Zhengguo Canal. The source for the canal, the Jing River, rapidly bore down into the loess sediment below the irrigation intakes of the canal. Despite repeated lowering of the intakes, the system eventually collapsed by the late imperial period.[25]

## Water Management in the Imperial Period

As early as the onset of the Common Era, the state began to develop an abiding interest in maintaining the hydraulic stability of the Yellow River. This interest centered on encouraging irrigation development in the middle valley while committing to large-scale flood control efforts in the lower valley. Many large projects were administered with a mixture of central and local institutions and executed with labor-intensive methods. Even in smaller projects carried out under the leadership of local elites, the imperial state still provided moral encouragement and occasional technical assistance to help manage local waterways in order to advance rural prosperity and stability.[26] As the historian John McNeill argues:

> In comparative perspective the Chinese state . . . appears remarkable for its ecological role. The Chinese imperial state was a meddlesome one, carefully looking after its own interests and, in keeping with cultural traditions, actively seeking to develop resources and rearrange nature so as to maximize tangible and taxable wealth. Here, more than elsewhere, the state served (often unsuccessfully) as the guarantor of ecological stability. The state took primary responsibility for building and maintaining many big waterworks for flood control. . . . Its bureaucrats were taught to see a link between natural events and imperial politics, and to propitiate, placate, manage, and manipulate nature in the state's interest.[27]

During most of the imperial period, a focus of state activity on the North China Plain was control of the Yellow River. Although state management practices strongly supported the efforts of local organizations to develop irrigation systems, the principal objective of the state was to control flooding in the lower Yellow River valley. Routine dike maintenance was a requisite to maintain the integrity of the system, but during flood crises, renewed debate about control strategy inevitably emerged. The parameters of these debates, however, were largely bounded by assumptions about the importance of fixing the river channel through a system of dikes. This broad approach would be modified, but not fundamentally altered, by the introduction of modern hydraulic engineering and technologies in the twentieth century. Throughout the imperial period and down to the present, a succession of water managers became cultural heroes for rectifying the integrity of the flood control system in the lower Yellow River valley. In similar terms, late imperial-era water experts who recreated hydraulic stability, and hence could ensure social and economic order on the North China Plain, could consciously lay claim to a position among the pantheon of water heroes as they sought solutions to hydraulic breakdown on the North China Plain.

Here, it is worth returning to the legend of Yü the Great as recounted in the *Mencius* and the classical histories. Two great water heroes were appointed to relieve the massive floods that occurred at the time of the legendary emperor Yao. The first, Kun, spent nine years constructing dikes to contain the waters. Unfortunately for Kun, and perhaps as a signal of the importance of his task, Kun was exiled, killed, and cut to pieces by Yao after the former's efforts to tame the waters proved unsuccessful. Subsequently, Kun's son, Yü the Great was appointed to regulate the waters. After passing the door of his own house for years without entering (a sign of his unstinting efforts), Yü succeeded in dredging waterways for nine rivers and in permanently channeling the waters that had inundated the North China Plain to run into the sea. The recounting of these legends suggests that significant debates existed on how best to manage the Yellow River well before the imperial period began. The debates centered on the desirability of constructing a single diked channel versus multiple channels. The disagreement was an expression of a larger moral argument about the degree to which nature should be manipulated by humans. The "Confucian school" emphasized the need to control natural forces by constructing high dikes to constrict the flow. The other school (which we might call the "Daoist school") supported using the forces of nature to control water flow, by constructing low dikes farther apart to allow the river greater freedom to deposit its silt and to find its own course. Needham expands on this idea by noting that Confucians reflected "the forceful repression of Nature by the erection of convex 'masculine' ridges along the rivers was a case of what the Taoists [Daoists] called *wei* as opposed to *wu wei* (no action contrary to nature). The deepening of river beds by excavation of 'feminine' concavities was, on the contrary, a going along with nature [supported by Daoists]."[28]

The retelling of the legends of Yü and Kun during the imperial era suggests that both dredging and diking became orthodox river management doctrine. Although subsequent projects would adopt neither principle to the exclusion of the other, the subtleties in emphasis revealed a set of debates about how best to control the river. Charles Greer, an American geographer, argues that both schools of water management were based on the pre-Confucian and pre-Daoist precepts of controlling the river according to natural forces and of utilizing the river to control the river and that both of these approaches were reflected in the heroic efforts of Yü the Great.[29] Subsequent claims of the benefits of different technical approaches were really different claims for how best to achieve these ancient principles. The lynchpin for flood control, irrigation, and transport was the system of levees along the great Yellow, Huai, and Hai Rivers. Maintaining the integrity of the levees was a constant battle against sedimentation. During the two thousand years of China's imperial period, the Yellow River dikes experienced five major ruptures and subsequent course changes. Each occasion provided the opportunity for China's leaders to legitimize their political authority by reorganizing and renewing river management efforts.

During the Han dynasty (206 BCE–220), the state established many institutional practices to address the periodic ruptures of Yellow River dikes and to establish dike-maintenance regimes. The scale of such water management projects during the Han was greater than in any previous era. In the lower course of Yellow River, river management was of paramount concern, while irrigation development continued apace in the middle reaches. The strong Han central government also established the precedent of state direction for large projects as well as the recruitment of corvée labor to build such projects. In 11 CE, a series of ruptures of the restraining dikes along the lower Yellow River effected a dramatic course change that inundated large tracts of land on the North China Plain. Memorials to the Han imperial court advocated a variety of strategies to reassert control of the river, from restoring the beds of the nine rivers of the time of Yü the Great to creating an entirely new river bed, and from carving multiple drainage channels to allowing the river to meander freely across the alluvial plain.[30] Amid this hydraulic breakdown, the first two imperial-era river engineers to establish themselves as water heroes were Jia Zhang and Wang Jing. Attempting to alleviate the widespread flooding that followed the rupture of the Yellow River dikes, Jia advocated a three-pronged approach in 69. Jia concluded that restoring the Yellow River to its former bed was impossible because of the height of the current channel. He recommended: (1) reducing the flow of the river by drawing off water for irrigation in the middle section; (2) diverting sediment-free water from tributaries into the lower Yellow River valley; and (3) strengthening Yellow River dikes.[31] Jia's approach is generally identified with the Daoist tradition, for he sought both to dissipate the power of the river flow and to construct dikes some distance apart to allow the river to meander. His approach proved ineffective, but Jia's engineering principles were carefully studied by subsequent water managers.

Unlike the case of Jia, there is little historical material available to evaluate the precise methods of Wang Jing, the second and most famous of the first-century Han dynasty water heroes. Wang is reputed to have paid particular attention to strengthening dikes, dredging major tributaries, and building sluice gates to regulate flow in the main river channel. His general approach would likely have included maintaining strong dikes and restricting the Yellow River to a single channel. The greatest testament to the success of his engineering methods (whatever they were) is that the river did not experience a major course change for the next thousand years.[32]

From the beginning of the imperial period through the Song dynasty, the drainage systems of the Yellow River and the Huai River were integrated by a series of canals that linked the two regions. Further canal construction extended this exchange to the increasingly prosperous Yangtze River region. The integration of the major

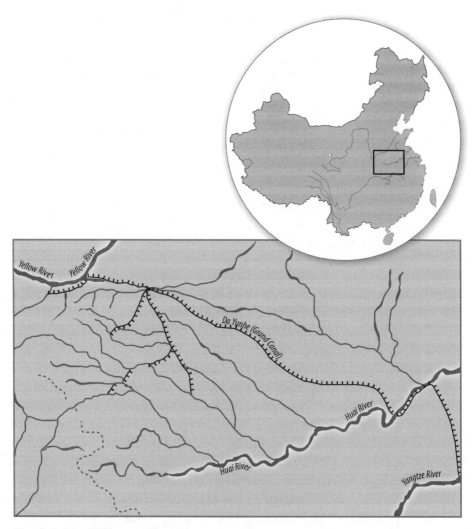

Canals linking Yellow and Huai River systems by 1000

drainage basins of the North China Plain generated long-term hydrologic transformations as silt from the Yellow River system clogged the canals and irrigation systems in the Huai valley. The full integration of the two river systems had reached its apogee when the whole system crumbled in the massive dike break of 1194. The rupture was the culmination of a series of breakages that were precipitated in part by direct human action.

During the politically fragmented and contentious Five Dynasties (907–960) and the later Song dynasty, control of waterways was a key strategic concern. During this period, water was viewed as one component of warfare—as it had been in the past and would be again in the future. For example, to resist the Jin army, which was advancing from the north, Song forces breached the Yellow River dikes. Later, the Jin army turned the tables by puncturing dikes near Kaifeng and sending Yellow River water southward to inundate Song forces. Deliberate destruction of the dikes meant that the waters inadvertently followed the Daoist ideal of a river "seeking its own course." The Yellow River drained through multiple channels, most leading to the Huai River. The continuing social and political commitment for the Chinese state to create and maintain stable hydraulic conditions on the North China Plain impelled the state to rectify this "ecological disorder." Attempts were made to regulate the drainage of the Yellow River, but between the twelfth and sixteenth centuries Yellow River waters followed a variety of courses both north and south of the Shandong highlands (on the Shandong Peninsula), although the majority of its waterflow during this period joined with the Huai River in what is now northern Jiangsu Province.[33]

To help defend the imperial system from nomadic and seminomadic peoples of the north, the imperial capital of the Yüan dynasty was located to the north in Beijing. It was during the Yüan that existing canals, between the Yangtze and Huai River valleys and between the Huai and Yellow River valleys, were augmented to form the Grand Canal. The Grand Canal sustained the capital and northern military garrisons with provisions from the Yangtze River valley. Thus, by the fourteenth century engineering projects had created a hydraulic system that fundamentally changed the drainage networks of the North China Plain. A new system of dikes on the Yellow River regulated all flow into the lower bed of the Huai River. At the same time, the monumental Grand Canal, running from south to north, bisected the drainage system of the entire region. The long-term struggle to maintain this new waterscape centered on silt. In short, the challenge was to engineer drainage of the Yellow and Huai Rivers in ways that did not threaten the integrity of the Grand Canal.

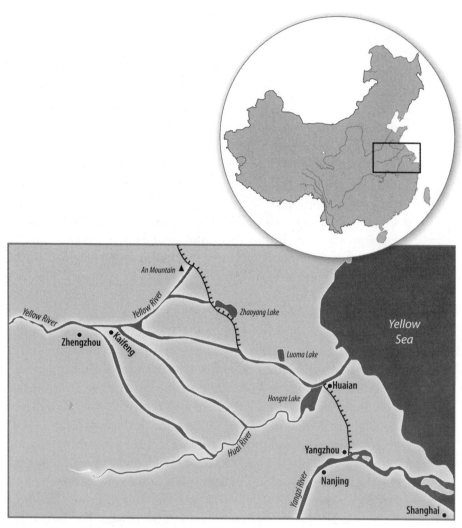

Yellow River drainage into the Huai River, ca. 1400

The increasingly complex task of managing the Yellow River–Huai River–Grand Canal matrix spawned the creation of centralized bureaucratic organizations. The water management strategies adopted during the Ming and Qing dynasties (1644–1911) were predicated on one goal: maintaining the transport of tribute grain to Beijing via the Grand Canal. This was precisely the responsibility that awaited Pan Jixun (1521–1595), later renowned as the greatest water hero in Chinese history, after he was appointed the Ming imperial commissioner of the Yellow River in 1565. The situation confronting Pan was formidable. Since the breakout of the Yellow River from its dikes in the late twelfth century, gradual progress had been made in building defensive bulwarks on the north bank of the former main stem of the Yellow River. This sent an increasing amount of Yellow River flow to the Huai River channel. By 1494 all Yellow River water was flowing into the Huai River through a dozen or so routes, with the main flow entering the Huai near Huaiyin in northern Jiangsu. Thus, Huaiyin was where the three main waterways of the region converged: the Yellow River, the Huai River, and the Grand Canal. Inexorably, silt did its thing. Silt gradually accumulated on the joint bed of the Yellow and Huai Rivers, ultimately preventing the Huai River from continuing in its original bed to the sea. As a result, Huai River water began to settle in depression areas west of Huaiyin, gradually forming a large retention basin that became known as Hongze Lake. Silt caused other problems. For example, at times of heavy Yellow River discharge, water would backflow into Hongze Lake, flooding massive tracts of agricultural land as well as the Ming Tombs west of the lake. Sediment deposits from the Yellow River heightened the bed of Hongze Lake during periods of Yellow River backflow, and eventually the Huai River could drain neither into Hongze Lake nor through its original bed. In addition to critical problems with Huai River drainage, the increased silting of the Grand Canal near Huaiyin raised the bed of the canal and resulted in frequent transportation blockages.[34]

Pan was ordered to solve this hydraulic mess. From the Ming government's point of view, there was no choice. For centuries, China's imperial states had staked the integrity of their empires, and their capacity to rule, on their ability to extract wealth from the lower Yangtze region. Pan understood what was at stake. The complexity of the task and the political pressure to maintain the hydraulic infrastructure resulted in Pan's being appointed to (and dismissed from) office on four different occasions between 1565 and 1592. Pan carefully studied past river control practices and concluded that the current approach (largely followed since the massive dike ruptures of 1194) was flawed. The method of "divert the Yellow, restore the Huai" (*fen Huang dao Huai*) focused on dredging multiple channels to channel Yellow River water into the Huai bed. The "divide the channel" strategy for the Yellow River was intended to alleviate dike breaks upstream from Huaiyin that could deprive the Grand Canal of the water necessary for a functioning transport system. Dissipating the power of the river (so that the river could not break its dikes) was one of the traditional approaches to river management. Pan opposed the "divide the channel" method, arguing that dividing and consequently slowing the flow only lessened the capacity of the waters to transport sediment. If engineers followed this scenario, the multiple channels carrying Yellow River waters would need constant dredging.[35] Based on his observations of the river's flow and its carrying capacity, Pan proposed building close-set dikes along the lower course of the Yellow River to restrict the spread of floodwater and to force the water to flow at an accelerated pace (which would flush silt out to sea).

The second important component to Pan's water management strategy centered on storing the clear waters of the Huai River in a catchment basin near the Huai's confluence with the Yellow River in northern Jiangsu. This basin, Hongze Lake, would release water into the Yellow/Huai channel to augment the silt-carrying capacity of the current, a strategy referred to as "storing the clear to scour the Yellow River" (*xu Qing shua Huang*). Pan's empirical observations about river current and silt led him to devise a system that was more comprehensive than earlier approaches. He drew on engineering principles that included the use of reservoirs, which informed a set of hydraulic practices that in many ways represented the greatest expression of river control up to his time, constrained as it was by social and economic organization and premodern technology. Pan's approach was ultimately accepted as orthodoxy. His dual approach (of restricting the flow and storing water) would also be followed in the Qing dynasty and became a fundamental tenet of water management during the twentieth century. This enduring legacy made him, after Yü the Great, the greatest of China's water heroes.[36]

It took Pan all four terms in office to realize his strategy, in part because of the fractured nature of the water management bureaucracy in Ming China and in part because of the persistent power of alternative visions for Yellow River management. During the Ming, the imperial commissioner of the Yellow River, the imperial commissioner of the Grand Canal, the minister of public works, and the provincial officials responsible for supplying labor all had a bureaucratic investment in managing the waters of the North China Plain. Although the goal of Ming policy was to protect grain transport, there was no single path to this goal. Different institutions advocated different solutions. Although Pan gained more administrative authority when he was concurrently appointed vice-minister of public works, provincial and central officials continued to promote different schemes. For example, the governor of Zhili Province sought to dig a different channel for the Yellow River, but the commissioner of the Grand Canal wanted to dig separate channels for the Huai and Yellow Rivers. Pan was dismissed from his second tenure as imperial commissioner after the minister of public works blamed him for inadequate grain shipments via the Grand Canal. In short, Pan was in and out of office, depending on which bureaucratic constituency could claim sufficient bureaucratic clout to pursue competing plans. This "fragmented bureaucracy" was an early harbinger of patterns repeated during the Mao and post-Mao eras.[37]

The Qing achieved an unprecedented degree of central control over Yellow River management as it established the Yellow River Administration (YRA) headed by a director general and staffed by thirty centrally appointed officials. The YRA was charged with overseeing the hydraulic integrity of the lower Yellow River. The efficient transportation of grain tribute via the Grand Canal continued to be of prime importance to Qing rulers. Indeed, the YRA was effectively an administrative adjunct of the Grain Transport Administration. Near the beginning of the Qing, the Kangxi emperor (reign: 1661–1722) appointed Jin Fu (1633–1692) as director general of the YRA. Jin extended the strategies of Pan Jixun as he raised and constricted the apertures of the Yellow River dikes to increase the sediment-flushing velocity of the river's flow. Jin's approach was based primarily on the idea of "storing the clean waters of the Huai to combat the silt of the Yellow River." Like the efforts of his distinguished predecessor, Pan Jixun, Jin Fu's success in creating a new ecological balance on the North China Plain was short-lived. The effective management of silt remained an elusive dream.[38]

By the advent of China's late imperial period (ca. 1500), successive imperial governments had committed themselves to economic, social, and political institutions on the North China Plain that required "controlling the waters" through perpetual investment in maintenance. The mandate to protect the lifeline of the imperial system, the Grand Canal, "committed the Chinese state to a conflict with the [Yellow] River that it could not afford to lose, and could only momentarily win."[39] There were two basic alternatives regarding technological and managerial approaches available to the Chinese state: "dividing the flow of the Yellow River" and "utilizing a single flow to scour [the riverbed]." In the Ming dynasty, Pan Jixun proposed the construction of retention basins in upstream segments of the Yellow River in order to regulate flows. This would have been a significant innovation in river management, but for reasons that are not entirely clear, this plan did not materialize. The sort of administrative control that would have been necessary for such a project (both horizontal and vertical control) would not appear in China until after 1949, but the technological approach articulated by Pan Jixun was largely consistent with the tenets of Yü the Great as incorporated into the canon of imperial statecraft. The necessity for Chinese authorities to "order the waters" continued to be a straitjacket on the state down to the present day.

Beyond formulating general goals, adjudicating disputes, articulating an idealized local society, and providing the moral encouragement to manage water in service to that ideal, imperial governments were not directly involved in managing small-scale projects. Yet much of the anthropogenic landscape of the North China Plain was (and is) the aggregate of the outcomes of innumerable water projects, large and small. The ecological foundation of the economic and social commitments of the imperial system required the state to engage in a perpetual struggle with the hydraulic forces and climate of the North China Plain. The constant effort to engineer and reengineer the landscape was a symbiosis of the small and the large, of the central and the local. The massive water management projects that made heroes of Li Bing, Wang Jing, and Pan Jixun were "made possible by Chinese administrative virtuosity in planning, organizing, conscripting, taxing, and coercing. These reshaped the Chinese physical environment, and committed a large part of the Chinese economy to a paradoxical relationship with water that was startlingly productive yet relentlessly costly to maintain, protective yet intermittently terrifyingly hazardous, and above all, one from which it could not, and has not yet to date been able to, extricate itself."[40] By the late imperial period, it seemed that practitioners of water management had won the old debate over the best means of "controlling the waters." The authors of this late imperial water management orthodoxy, such as Pan Jixun and Jin Fu, appear to have taken an approach rather similar to that of modern engineers. These Ming and Qing dynasty water officials "believed that by using dikes and sluice gates, by dredging, and by digging diversion channels, planners could determine the flow of even the largest rivers of the empire. Although they did not use concrete, they devised plans that required massive amounts of timber, bamboo, earth, and human labor in order to reshape nature radically. The builders of the Three Gorges Dam and other giant dams of today have followed their example."[41]

## Environmental Breakdown in Late Imperial China (1500–1911)

The challenges of maintaining the engineered landscape of the North China Plain were enormous. Throughout the imperial era, central states and local communities employed an array of strategies that, for significant spans of time, created a stasis between anthropogenic and natural forces. For example, local agricultural knowledge grew and local practices adapted to the vagaries of soil, hydrology, and climate. Villages also adopted organizational techniques based on collective responsibility that diffused the risks and burdens of managing scarce resources. To be sure, there was conflict inherent in this regime, as the interests of adjacent polities may have differed. Resource rights and allocations could be bitterly disputed. As a partner in promoting the health and stability of rural society, the state sought to mediate these disputes through its bureaucratic representatives. But beyond this judicial function, local societies were supported in the broader project of "ordering the waters" by an imperial state that attempted to impose ecological order on natural forces that were beyond the capacity of local institutions to control. Managing the rivers was one such endeavor. Other elements of imperial statecraft served to buttress the mutually supporting system of state and local society. Promotion of alternative agricultural practices (what in modern parlance we might call "agricultural extension"), technical partnerships between representatives of the bureaucracy and local elites and a granary reserve system to smooth out price fluctuations during periods of famine were all attempts by the central government to assist local communities in adapting to environmental stress. The continued existential task was to adapt to and further manipulate this high-risk environment. It was a challenging cycle. Despite the success that local and central institutions achieved during the imperial period, periodic floods, famines, and droughts signaled ecological breakdown. Such crises required additional adjustments to reimpose hydraulic stability.

One of these periodic breakdowns in the ecological balance of the North China Plain took place in the nineteenth century and lasted throughout much of the twentieth century. Some of the forces impelling this breakdown were familiar to political elites who had absorbed traditional notions of cyclical political change in China. For example, a sclerotic imperial government, increased corruption, and deleterious climate change (probably understood by observers at the time as ominous natural events, such as floods and drought) were all well-worn markers of social and political instability. At the same time, the same observers of the late imperial period witnessed historical forces that could well have been considered novel (or seen as traditional forces that had reached a new intensity). These forces included unprecedented commercial development, unprecedented demographic growth, and unprecedented global forces, including trade and technology transfer. The combination of internal and external dynamics, Chinese traditions, and foreign novelty profoundly shaped virtually every element of China's late-imperial experience. Collectively, these forces generated a rupture of the ecological balance on the North China Plain. An important part of the mission of every Chinese state in the post-imperial period has been to address this breakdown. Would the reimposition of hydraulic stability be guided by traditions in water management? Or would the seeds of China's stability be found in one of the forces that impelled breakdown, namely, China's integration into an international system dominated by the West?

## Forces of Ecological Change in Late Imperial China

During the late imperial period, a convergence of demographic, climatic, economic, and political factors led to unprecedented ecological transformations in China. By the beginning of the Qing dynasty in 1644, "China entered a new phase of definite and irreversible change," and over the next "two centuries the Chinese natural and economic environment had changed . . . beyond recognition."[42] During this period, China's population tripled. At the same time, food crops from the New World entered China, while foreign demand for tea and silk skyrocketed. Pressures to increase food production and commodity production combined to increase exploitation of China's forest, land, and water resources. To sustain an expanding empire and burgeoning population and to reproduce the material basis of social stability, China was walking an ecological tightrope. Consequences of these trends included denuded hills, flooded river valleys, and desertified border regions. Compared to other periods of profound change (e.g., China's so-called medieval economic revolution during approximately the eleventh to twelfth centuries), the pace and degree of ecological transformations in the late imperial period were unprecedented and surpassed only in the post-1949 era.

A critically important cause of these transformations was demographic growth. The population of China increased from roughly 100 million in 1500 to between 320 million and 350 million by 1800. There had been periods of population growth prior to the late imperial period, but checks such as epidemics and war had restrained growth rates.[43] Although the reasons for a sudden surge in population during the late imperial period are not entirely clear, the outcomes of this growth were. Cultivated land per person in China decreased from 1.4 hectares/person in the Han dynasty to 0.8 hectares per person in the late imperial period.[44] As a consequence, Chinese farmers began to exploit every parcel of land available. In addition, the empire doubled in size during the late imperial period, and much of this area was ecologically vulnerable. Internal migration to these ecologically fragile border regions relieved some of the demographic pressures on the North China Plain and lower and middle Yangtze River valley regions.

As China's population-to-land ratio declined during the eighteenth century, the Qing state was well aware of the mounting threat to its ability to promote the material basis of social and political stability. According to the historian Peter Perdue, "[The] emperor and his advisers grew increasingly concerned about the uncontrollable movements of millions of peasants, laborers, and refugees, but they recognized that it was impossible to force them back to the villages. In the pre-industrial empire, commerce could absorb only a fraction of the labor force. The clearance of new land had been the classic answer of previous dynasties, but expansion of the land area alone was insufficient. As the population to land ratio increased, the promotion of greater yields from existing lands became the primary means of providing for the people's livelihood."[45] Faced with these constraints, the state promoted a variety of measures to maximize yields from existing farmland:

> They [the rulers] focused most of their attention on the control of water and land. Absent inputs from new technologies, agriculturalists had to increase intensification of farming method—increasing output per acre . . . additional sources of fertilizer and development of new sources of irrigation were part of a suite of responses by local society and the state to address the worsening land-population calculus. The state and its clients embraced an ethic of agrarian developmentalism, dedicated to maximizing yields from China's limited supplied of arable land, increasing total land area as much as possible, and channeling water sources to supply the fields.[46]

Faced with an obvious potential for social instability, the Chinese state made a "sort of national priority" of increasing agricultural productivity in the eighteenth century.[47] Such statecraft explains the extraordinary success of the Qing dynasty to feed its people through the eighteenth century, but the environmental consequences of this success led to constraints as the land–population ratio continued to decline.

One of the outcomes of feeding a rapidly increasing population was deforestation and the directly related outcomes of erosion and sedimentation. Deforestation was particularly severe in hilly and mountainous regions where migrants opened up new land for cultivation. Often, upland areas were cleared, and then, as fertility was exhausted, they were abandoned after only a few agricultural seasons. In addition, New World food crops that were introduced to China in the sixteenth century, such as maize, peanuts, and sweet potatoes, could be grown on what had previously been considered marginal land. The introduction of such supplementary calorie sources increased the ability of China's land to support population increases, but the cultivation of these new crops hastened erosion. The increasing number of floods on the North China Plain in the late imperial period served as proxy for measuring the increasing rates of sedimentation coming from the loess region. Records indicate flooding every 1.89 years between 1645 and 1855.[48]

The increased commercial focus of China's rural economy was also a major driver of environmental stress during the late imperial period. This robust development of market relations in the imperial period has been a topic of considerable attention by historians of China. Commercialization led to a "radical simplification of the natural ecological order."[49] China's loss of ecological diversity in the late imperial period was a consequence of agricultural specialization, whereby large swaths of ground were given over to commodity production. During this period, other factors that led to specialized market production included an expanding population, a warmer climate in the eighteenth century, and integration into global silk and tea markets. Other studies, focused on the Yangtze River, have shown similar forces with familiar results: deforestation, erosion, monocropping, and heightened risk of flooding. This research has shown that state policies, formed in reaction to the growing population–land dilemma, promoted intensified agriculture and land reclamation. Otherwise resistant to aggressive land development efforts by local landowners, the late imperial state was forced to cede significant autonomy to local elites.[50] Such trends compromised the utility of water control structures that had, to that point, mitigated flood threats from the Yangtze River and its tributaries. By the nineteenth century, floods had become endemic in the region. It is not entirely clear to what degree similar processes were at work on the North China Plain. Patterns of landownership and social organization differed between the two regions. Small-scale, locally managed irrigation facilities were largely absent from the North China Plain. Still, there was increasing commercialization of agriculture in North China that may have engendered similar patterns in response to the deterioration of the land–population ratio.[51]

Climate change was an important contextual element of environmental transitions on the North China Plain during the late imperial period. Over the past two decades, there has been considerable work on reconstructing temperature and precipitation data for the North China Plain. As global climate change has emerged as a regional and national issue, China has directed research funds to exploring historical climate change in China. A variety of studies suggest that much of China, and indeed most of East Asia, followed the general pattern of global climate change over the past two thousand years. For example, average annual mean temperatures from 1951 to 1980 were 0.17° C higher than from the beginning of the Common Era to 510. The Medieval Warm Period ensued in 510 and extended to the mid-fourteenth century. This was followed by the so-called Little Ice Age, which lasted from the mid-fourteenth century to the late nineteenth century (during which mean annual temperatures declined by 0.10° C). Within the Little Ice Age period, however, there was also a relatively mild period in the eighteenth century.[52]

The North China Plain has been characterized by distinctive temperature oscillations (rapid warm-cold and cold-warm transitions). For example, in a short span of ninety years between the late fifth and the sixth centuries, temperatures decreased by an average of 1.3° C. Similarly quick transitions occurred between the mid-thirteenth and mid-fourteenth centuries, when temperatures decreased by 1.4° C, and between the late nineteenth and early twentieth centuries, when temperatures increased by 1.0° C. Perhaps even more striking is the temperature increase of 1.5° that occurred between the mid-nineteenth and late twentieth centuries.[53]

Coupling precipitation patterns with temperature data, many studies have concluded that the climate of the North China Plain followed a warm-humid/cool-dry climate pattern over the past two millennia.[54] Consistent with these correlations of cold-dry periods, most of the Little Ice Age was relatively dry, with periods of acute aridity. Indeed, the most arid period in the past two thousand years, as recorded in stalagmite reconstructions, occurred during the first half of the seventeenth century. But China has also seen other periods of significant drought, for example, 850–940 and 1350–1380. In contrast, sustained wet intervals preceded the Little Ice Age (during the periods 190–300, 920–1010, and 1090–1140), and similar conditions prevailed in the early twentieth century.[55]

There has been a temptation to rely on climate change to explain the rise and fall of polities in China; but regardless of the historical veracity of such views, there seems to be little question that fluctuations of temperature and precipitation have had profound effects on local and regional conditions. First, historical climate dynamics have coincided with shifting control of China's border areas between sedentary and pastoral communities of North and Northwest China. Generally speaking, during relatively cool and dry eras, nomadic herders moved southward, looking for better conditions to sustain their animals and their livelihoods. However, during warmer and wetter periods, Han Chinese farmers reestablished settlements to the north. This linkage has led some scholars to posit a connection between climate change and political transitions over the long run of Chinese history. The geographer David Zhang has perhaps argued most forcefully for the determinative power of climate change on the course of Chinese dynastic history. He contends that the frequency of warfare in eastern China correlates with Northern Hemisphere temperature oscillations and that virtually all peak periods of warfare and dynastic change have occurred during a cooling phase. "The reduction of thermal energy during cooling phases significantly shrank agricultural production. Such ecological stress interacted with population pressure and China's unique historic and geographic setting to bring about the high frequencies of warfare over the last millennium."[56] Indeed, a juxtaposition of data for temperature, precipitation, and political turbulence in China over the past millennia suggests that such correlations are not easily dismissed.[57]

As we have seen, by the late imperial period a variety of forces were working to upset China's ecological balance. The "biological old regime" had been stretched to its limit by population growth and by structural changes to China's agricultural economy. The brilliant success of China's agricultural practices, namely, high returns from high labor inputs, had come at a price. As pressures grew to feed China's growing population, agricultural capacity was maximized by the clearance of virtually all hilly and mountainous terrain. Double- and triple-cropping spread, and New World food crops filled marginal farmland. These areas were particularly vulnerable to floods and drought, but there was also a limit to further development of available water resources, given the available hydrological technology.[58]

Beyond this, intensification of agriculture fostered the growth of economic and social forces that threatened the stability of water control structures (if such structures were not carefully tended by local political organizations or the imperial bureaucracy). The situation at the end of the imperial period is consistent with the notion of the "hydraulic cycle," which described successive eras in the imperial period when robust state sponsorship of water control was followed by neglect and ultimately by a breakdown in water management.[59] Implicit in such a "cyclical" interpretation of hydraulic breakdown is the idea of "technological lock-in." The stability of China's late-imperial economy and society was dependent on the success of a particular hydraulic technological complex, which had been replicated throughout the late imperial period. The "hydrological instability of man-made water systems made the burden of maintenance perpetual."[60] During this period, China still lacked technological innovations that would be the focus of twentieth-century water management, and any disruptions in the heavy investment of material and labor that were necessary to maintain the viability of the water system inevitably led to deterioration. With the weakening of the Qing dynasty and the withdrawal of state patronage for Yellow River management in the nineteenth century, water management systems atrophied and ecologies deteriorated on the North China Plain.

Did China exceed its carrying capacity by the nineteenth century? Did it encounter a Malthusian crisis that ultimately led to China's modern environmental crisis? Political scientists, historians, anthropologists, demographers, and economists have all weighed in on the debates surrounding China's modern political transitions. For example, Esther Boserup has argued against the notion that Malthusian "positive checks" exist as a consequence of population's exceeding resources. On the contrary, she claims that population growth has been an inducement to increased agricultural output. Other paradigms, informed by Marxism, by liberal economics, or by agrarian studies, have advanced their respective arguments. However, to all such analytical and methodological frameworks for understanding China's late imperial crisis we must now add an ecological perspective. The dynamics of this approach are shaped by considerations of demographic growth, commercial development, and climate change.[61]

The causative factors employed by the theoretical models outlined above (which explain political, social, and economic crisis in one region of the world, e.g., Europe) may be quite inadequate to explore realities in China. Nevertheless, there were synchronistic social reactions to global environmental phenomenon in regions with differing cultural, social, political, and religious systems. Violence was one adaptive response to resource scarcity conditioned by climate change. Precipitating such responses were the pressures of a growing population upon the limited carrying capacity of land. The agricultural output of this land was, in turn, conditioned by climate. Climate oscillations impacted the length of the growing season (e.g., whether there would be two crops or one in any given year) and also governed the availability of water for irrigation. Of course, the flip side of the climate equation, namely warm-wet periods, could induce excessive precipitation and could lead to wide-spread inundation.

The role of the state has also been a central theme in analyses of environmental change in a variety of national and regional contexts. Similarly, maintaining social and political order in China's ecologically fragile regions was a continuing challenge to China's political elites during the imperial period. It was simply not possible for China to abandon the ecological status quo and, at the same time, to maintain the social, economic, and political structures upon which the imperial system rested. Imperial rulers were acutely aware of these challenges and strove to bolster the resilience of agriculturalists on the North China Plain. In effect, the state sought to manage risk to these agriculturalists (and, by extension, its own risk) by sponsoring hydraulic engineering schemes that provided favorable conditions for irrigation and transportation, as well as by establishing a social safety net in the form of a state granary system that served to smooth price fluctuations in times of agricultural privation. As long as the state maintained the capacity to effectively administer these functions, and as long as it maintained its own legitimacy, China's central government achieved what can only be considered extraordinary success in ensuring social and political order over much of the imperial period.

An influential and persuasive neo-Marxist interpretation of Chinese history links ecological destruction to the development of capitalist modes of production, particularly in the agricultural sector, where agribusiness practices supplanted subsistence production patterns.[62] Although they recognize that similar outcomes occurred in regions of China that experienced the impact of capitalist modes of production, scholars of China's late imperial environmental transformations rightly emphasize the presence and impact of the imperial state in the rural economy. The active role of the state was reflected in its propagation of new agricultural techniques and its provision of infrastructure to help increase agricultural production. Indeed, from a broader perspective, state policies toward the agricultural sector were long aimed at guaranteeing subsistence for farmers and generating a surplus to sustain the imperial structure. Will argues that "increasing agricultural productivity in the face of a population that was rapidly growing upon a severely limited land base had become a sort of national [Qing] priority."[63] That the state successfully utilized these policies to promote sufficient agricultural productivity to feed virtually every Chinese speaks to the strength of the imperial state. Thus, in considering the collection of variables that impacted the management of water, we must always include state capacity. In addition, at the end of the imperial period, along with population growth, diminishing returns on agricultural labor, and climate change, we should remember that the deterioration of state administrative capacity had a profound impact on the waterscape of the North China Plain.

## The 1855 Yellow River Course Change

Water management practices on the lower Yellow River reached their zenith in the early eighteenth century, during Jin Fu's tenure as general director of the Yellow River Administration. Given the imperial mandate to protect grain shipments from the south via the Grand Canal, Jin faced the fundamental problem that had been confronted by all Yellow River managers since the completion of the Grand Canal: how to prevent the Yellow River from overflowing its dikes and inundating the canal. Jin renewed Pan Jixun's strategy of constricting the flow to combat silt (shu shui gong sha), and he ordered the building of higher dikes along the Yellow River, along the Grand Canal, and around Hongze Lake. Over the next 250 years, this control system underwent a further process of articulation, driven primarily by the hydrologic peculiarities of the Yellow River. The result was a hydraulic system that could be maintained successfully only at the cost of an ever-greater share of the state's fiscal and administrative resources.[64]

Following the success of large-scale hydrological engineering projects in the early Qing, Jin created a permanent paid workforce to maintain these installations. The burgeoning YRA workforce was an early manifestation of the financial burden that the Qing assumed as it shouldered imperial commitments to maintain the delicate water matrix on the North China Plain. Indeed, by the early nineteenth century, the YRA bureaucracy was burning through Qing financial resources at a moment when imperial coffers were weakening. A century earlier, the Qing treasury had had an ample surplus, but by 1800 it was struggling with insolvency. Furthermore, the cost of sending 3 million to 4 million piculs (1.8 million to 2.4 million kilograms [kg], or 1,984 to 2,646 short tons) of rice to Beijing from the lower Yangtze River region via the Grand Canal was becoming absurdly expensive. One picul (60 kg) of tribute rice was estimated to cost four to five times the market price. Official appropriations for the YRA by the early nineteenth century were 4.5 million taels (1 tael = 37 grams) of silver per year. This amount, however, did not include dike repair. Regular expenditures on the YRA alone constituted more that 10 percent of Qing revenue. Faced with increasing financial demands on many fronts, the Qing simply could not keep up with the increasing demands of water management.[65]

A commitment to the technological solution of higher dikes coupled with the ineffectiveness of the YRA led to control bottlenecks. A series of serious dike breakages occurred in 1841, 1842, and 1843, and others occurred in 1851, 1852, and 1853. By the time of this second set of dike failures in the 1850s, imperial commitment to North China's hydraulic system was vacillating. In part, this was due to other urgent threats to the Qing state, including the Taiping Rebellion (1850–1864) and the growing European presence in South China. A buildup of silt in the estuary and the downstream stretches of the Yellow River backed up the entire flow and increased pressure on dikes.[66]

The most serious dike break came in 1855. Heavy rains in the upper- and midstream riparian regions generated massive runoff into the Yellow River. On June 19, the Yellow River overtopped dikes at Tongwagxiang in Henan Province. Within a short period, the roiling waters dissolved the loose soil of the northern dikes, carving a breach 5 kilometers wide and, within a day, leaving the original downstream channel completely dry.[67] Water released by the breach poured into the adjacent landscape, which lay 7 to 10 meters below the elevated riverbed (the "hanging" river). Under the weight of such a huge inflow of water, flooding spread across thirty counties in Henan, Jiangsu, and Shandong provinces. Eventually, water flowed eastward to the sea in two large swaths, to the north and to the south of the Shandong highlands.[68]

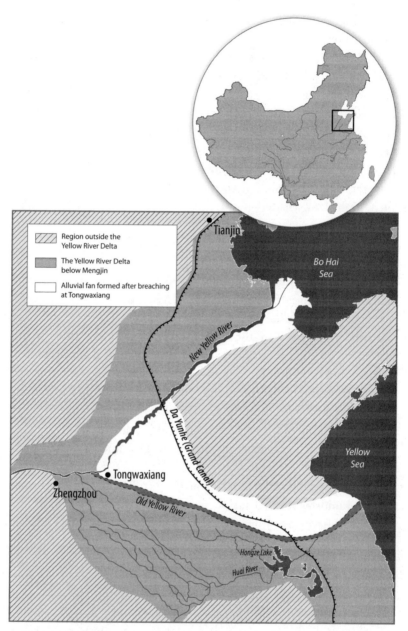

Yellow River drainage after 1855 dike break

The Yellow River remained uncontrolled for thirty years. The Qing dynasty hesitated to commit itself to renewed efforts to regulate the flow of the river. During these decades, vigorous debates took place among regional officials over how to reimpose hydraulic control. One school of thought advocated plugging the gap at Tongwaxiang and returning the flow to the former riverbed. This perspective was advocated by Shandong provincial interests, now burdened almost solely with managing the lower reaches of the Yellow River. A second school of thought advocated making the current northeasterly course permanent by constructing dikes and other control structures. This approach was championed by interests in Anhui and Jiangsu Provinces, which for the moment did not have to worry about the lower course of the Yellow River. Despite the urgency reflected in these petitions to the court, the Qing government prevaricated. As noted above, the imperial court had other concerns centered on domestic rebellion and treaty relations with the West. In the absence of alternative control strategies, the Qing government was forced to recommit itself to a system of control that relied on levees and related control structures. Qing authorities also had little choice but to accept the new course of the Yellow River. As was vehemently argued in petitions supporting the new course, the expense of restoring the river to its pre-1855 bed would have been prohibitive. The sixth major course change of the Yellow River was honored by the imperial court.[69]

The longer-term consequences of the Yellow River's 1855 course change profoundly affected the North China Plain. First, the shift prompted a retreat from state management of the rivers of the North China Plain. The course change debilitated tribute transport on the Grand Canal. An attempt to revive canal shipments was made, following a failed experiment with ocean shipping, but the days of the tribute system were numbered, as tribute obligations formerly in grain were increasingly converted to silver payments. Because regulation of the transport system was no longer necessary after the shift of the Yellow River, the YRA was abolished in 1861. The dismantling of the old institutional structure responsible for central water management on the North China Plain was followed by the abolition of the grain tribute system in 1904. This withdrawal of central administrative sponsorship of water management projects was one component of a larger shift in statecraft in the mid-nineteenth century, as competition from imperialist powers posed an existential threat to the dynasty. Intent on creating a modern industrial sector that could strengthen the dynasty's prospects for survival, Qing policy shifted investment from the interior to the coastal regions, where modern industrial bases could more efficiently be established. Beginning in the mid-nineteenth century, the long-held priority of "reproductive statecraft" that subsidized the well-being of ecologically marginal areas of the empire, such as the North China Plain, was gradually abandoned by the imperial state. This change in statecraft essentially left local interests on the North China Plain to their own devices. The late imperial devolution of water management to local and provincial purview meant that broader planning for controlling the water in the region would be plagued by a lack of consensus and would pit inter- and intraprovincial interests against one another.[70]

The second longer-term consequence of the 1855 Yellow River course change was a deterioration of the ecological foundations of the North China Plain. It is precisely beginning in this period that increasing numbers of Western missionaries visiting the North China Plain would call the Yellow River "China's Sorrow" and report to their constituencies back home that North China was a "land of famine." As summed up by the historian Mark Elvin:

> The same skill in water control that had contributed so greatly to the development of the Chinese economy . . . slowly fashioned a straitjacket that in the end hindered any easy reinvention of the economic structure. Neither water nor suitable terrain was available for further profitable hydraulic expansion. . . . Deadliest of all, hydrological systems kept twisting free from the grip of human would-be mastery, drying out, silting up, flooding over, or changing their channels. . . . No other society reshaped its hydraulic landscape with such sustained energy as did the Chinese, nor on such a scale, but the dialectic of long-

term interaction with the environment transformed what had
been a one-time strength into a source of weakness.[71]

In other words, by the late nineteenth century, the North China
Plain had entered a period of environmental breakdown that lasted
well into the twentieth century.

# 3 Transforming the Land of Famine

THE 1855 CHANGE of the Yellow River's course signaled the onset of ecological breakdown on the North China Plain. The longer-term environmental changes that led to the dramatic event were the consequence of demographic growth, commercial development, and the inexorable silting of the river. These historical dynamics were also shaped by internal rebellion, foreign pressure, and a general decline in Qing administrative acumen. Pierre-Étienne Will has argued that there was a discernible historical pattern that connected good governance and the "ordering of the rivers" in imperial-era China.[1] One of the first expressions of this connection between good governance and effective management of water was likely the legend of Yü the Great. From the perspective of the dynastic cycle (the rise and fall of dynasty), floods, droughts, and famine were seen as symptoms of ineffective rule.[2] But did the hydraulic cycle (and dynastic cycle for that matter) proceed to its next phase of "recovery" after 1855 based on the technology complex of the late imperial period? Or were there new patterns in the state, society, and technology matrix that were distinct from those of the late imperial period?

The story of twentieth-century water management on the North China Plain is one of continuities and discontinuities. On the one hand, there were agro-ecological continuities: the challenge of feeding a growing population and the challenge of river silting. There were institutional continuities as well: faith in the efficacy of administrative centralization and enduring management frameworks. In addition, the notion of "ordering the waters" continued to be a source of ruling legitimacy.

However, new forces impinged on the hydraulic cycle that suggest critical discontinuities in how water was managed on the North China Plain in the twentieth century. These forces might be organized under the rubric of "the internationalization of China."[3] These transformative dynamics included Western hydraulic science and technology, new forms of bureaucratic organization, and political and ideological innovations. Equally important was the development of a national Chinese identity embedded in the landscape of the Yellow River and the North China Plain and premised on a recent history of oppression by external and internal forces that led to social and political disintegration. This combination of continuities and discontinuities generated a new legitimizing discourse to manage water. Indeed, the manner in which enduring agro-ecological and institutional patterns were creatively assimilated by these forces of change suggests a unique Chinese approach to managing resources. Ultimately, questions arise as to the outcomes of this new approach to managing water on the North China Plain. Was the hydraulic cycle broken? Were long-term ecological constraints alleviated by new ideas, technologies, and organization? As evidenced by acute water challenges of recent decades in China, the answer to both questions is no.

## China's Sorrow and the Land of Famine

After the 1855 flood, the Yellow River and the North China Plain became both the symbol and the reality of China's modern dilemma. At the same time, floods, droughts, and famines became powerful symbols for nationalist and revolutionary impulses. Evoking China's exploitation by internal and external foes, these images of disaster would become part of a legitimizing discourse to manipulate the waters in order to rebuild the nation's wealth and power. The years between the flood of 1855 and the birth of the People's Republic of China in 1949 witnessed a profound transition from national lamentation about floods, droughts, and famine to a new national vision of resource development in service to the rebirth of Chinese civilization.

However, these visions remained beyond the horizon in the decades immediately following 1855. Although one could argue that the state, as well as generations of agriculturalists on the North China Plain, had developed practices to manage ecological risk, there was a sense that cycles of flood, drought, and famine in the late imperial (ca. 1500–1911) and Republican (1911–1949) periods was something quite different. To be sure, Western narrative and photographic images transmitted home by missionaries, consular officials, adventurers, and commercial agents amplified the impact of suffering in China to audiences abroad. But this narrative also influenced elites in China, many of whom were increasingly internalizing Western visions of a modern society. During this period, the continued reality of floods, drought, and famine on the North China Plain provided a stark contrast to the kind of society that a modern state and modern technology could advance. Thus, conditions on the North China Plain provided a narrative touchstone for political reformers and revolutionaries to criticize and attack contemporary institutional arrangements and practices.

Following the great 1855 flood, the Qing administration wavered in making a decision on how the Yellow River should drain to the sea. Should the new course that eventually drained into the Bohai Sea be the permanent course? Should the Yellow River be forced back to its old bed? Or a combination of the two? While the Qing were preoccupied by internal rebellions and by the challenge from the West, the river continued to roam over China's northeast. Although the Yellow River flowed into the channel of the Daqing River for its last 200 kilometers to the sea, the river was largely unregulated as it washed across the North China Plain. The Yellow River formed a shallow but expansive swamp in southern Zhili Province, giving an inkling of how the entire region would revert to a mosaic of lakes and marshes in the absence of human engineering. In other words, the North China Plain had returned full circle to the ecological conditions confronted by Yü the Great.[4]

Reports by a variety of European and American missionaries and travelers provide a montage of the North China Plain during this period. Alexander Wylie of the London Missionary Society was the first foreigner to visit the Yellow River's old (pre-1855) bed. Wylie was unaware of the river's shift in 1855, and, upon "ascending the bank [of the old Yellow River] which is 30 feet above the out-

lying area, instead of finding the rapid and turbid current he expected, a sandy plain stretched before him for a mile or more in
width."[5] In 1880 a similar story was reported by a certain Captain
Gill at a meeting of the Royal Geographic Society in London at
which Gill humorously recounted how Lord Elgin of the Royal
Navy, who had also been unaware of the river's course change, arrived with his fleet at the mouth of the Yellow River several years
after the 1855 course change to "prevent [Chinese] vessels coming
out of what was then a dry bed."[6] In another example, on his "journeys to the interior of China," James Morrison described the confluence of the old river bed and the Grand Canal: "When the Yellow River flowed in its old bed, the water-level of the canal was
some 20 feet higher than at present, but now this part of the canal is
nothing but an interesting relic of former days. Every here and
there one sees large boats which had entered the canal during a
period of high water, and are now lying rotting, having been de-
tained for years for want of water to get out."[7] Encountering the
swampy region to the north where the Yellow River now "flowed,"
British explorer Ney Elias reported, "The river at this point has no
defined bed, but flows over a belt of country some 10–12 miles in
width, having merely the appearance of a flat, level district in a
state of inundation. . . . For dreariness and desolation no scene can
exceed that which the Yellow River here presents; everything, natural
and artificial, is at the mercy of the muddy dun-colored waters as
they sweep on their course to the sea."[8] Rounding out his peregrina-
tions, Elias visited the region of the dike breach near Tongwaxiang.
He observed:

> [The area to the north and east of the breach was] that through
> which this section of the river [now] flows, and where we find
> many entire villages half-buried in deposits and deserted by
> the greater portion of the inhabitants; those who remain being
> in a poor and miserable condition. The houses are frequently
> silted up to the eaves, and have generally been abandoned. . . .
> As an example of this, I may mention a joss-house [local tem-
> ple] . . . where the level of the deposit was some 10 feet above
> that of the water . . . for some time the inhabitants had attempted

to accommodate themselves to the constantly diminishing height of the building, though since the last year or two apparently they had been compelled to abandon it. . . . The deposit on the inside was at precisely the same level as that on the outside, and was said by the villages in the neighbourhood to be 12 Chinese feet in depth . . . and to have been the work of 15 years, or 15 successive floods of the Yellow River.[9]

Because of hydraulic breakdown on the North China Plain, floods became endemic. Historically, however, it was droughts, not floods, that generated the greatest famines. And the frequency and severity of famine, either directly from floods or drought, was by all measures unprecedented in the late imperial period. The most deadly famine in Chinese imperial history was the 1876–1879 North China Famine, when a three-year drought followed serious floods. The "Great Famine" spread through Shandong, Zhili, Shanxi, Shaanxi, and Henan Provinces, claiming an estimated 9.5 million to 13 million lives.[10] The fragility of the ecological setting on the North China Plain was graphically chronicled by scores of Westerners, such as Timothy Richards, a member of the Baptist Missionary Society, whose work inspired the founding of the China Famine Relief Committee. Upon visiting Shanxi in 1877, Richards recorded in his diary:

> *January 30th. 290 li* [one *li* = ca. 1/3 mile] *south.*
> Saw fourteen dead on the roadside. One had only a stocking on. His corpse was being dragged by a dog, so light it was. Two of the dead were women. They had had a burial, but it had consisted in turning the faces to the ground. The passers-by had dealt more kindly with one, for they had left her clothes. A third corpse was a feast to a score of screaming crows and magpies. . . . One old man beside whom I climbed a hill said most pathetically: "Our mules and donkeys are all eaten up. Our labourers are dead. What crime have we committed that God should punish us thus."
>
> *February 1st. 450 li south.*
> Saw six dead bodies in half a day, and four of them were women: one in an open shed naked but for a string round her waist;

another in a stream; one in the water half exposed above the ice at the mercy of wild dogs; another half clad in rags in one of the open caves at the roadside; another half eaten, torn by birds and beasts of prey. Met two youths of about eighteen years of age, tottering on their feet, and leaning on sticks as if ninety years of age. Met another young man carrying his mother on his shoulders as her strength had failed.

*February 4th. 630 li south.*
Small wonder that I began to doubt my senses or my sanity, amid such scenes of horror. Was I among the living or among the tormented dead?[11]

Other accounts from the Great Famine of 1876–1879 described peasants who resorted to eating roots, bark, ground up stones, or adopted survival strategies that involved the sale or consumption of human bodies.[12] Beset with myriad internal and external challenges, the imperial government of the late nineteenth century had lost its capacity to manage ecological risk on the North China Plain. This was a stunning reversal from the High Qing era of the eighteenth century, when the state was able to manage subsistence crises with reasonable success. Whether such crises were generated by flood or drought, the Qing administrative bureaucracy was able to redirect food resources from the national granary reserve system and tribute grain from the south to alleviate price spikes that often followed inundation or drought. Granaries maintained by prefectures and by counties served to mitigate subsistence crises by releasing grain into the market at reduced prices, lending grain to farmers, and granting allotments of grain to individual households. Between 1753 and 1762, China's state bureaucracy orchestrated the distribution of nearly 6 million *shi* of grain (1 *shi* = 175–195 pounds).[13] However, redirected national priorities and the constrained financial resources meant that the Qing government in the late nineteenth century had little capacity to mitigate privation on the North China Plain.

By the first decades of the twentieth century, the North China Plain seemed to be in a perpetual subsistence crisis. Population growth, hydraulic breakdown, and withdrawal of state sponsorship all resulted in increased ecological risk. "Floods and droughts

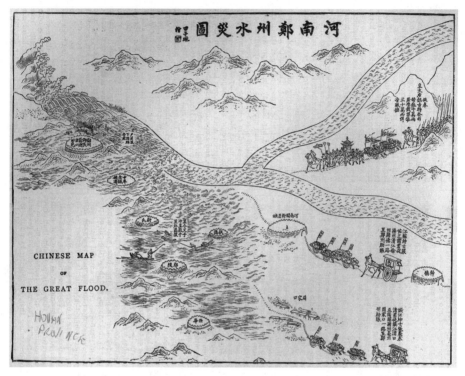

Engraving of the 1887 flood. From *The Leisure Hour: An Illustrated Magazine for Home Reading* (1888): 358. (Private Collection/The Bridgeman Art Library)

succeeded each other in rapid succession, blending seamlessly into what seemed like an endless catastrophe."[14] The conditions that ensued began to cement the image of North China as a "land of famine" in the minds of foreigner and Chinese alike.

In the southern portions of the region, mainly in the Huai River valley, conditions were particularly harsh. Localized cells of bandits, as well as more organized networks, ranged the desolate countryside of Jiangsu, Anhui, and Shandong Provinces in search of food. The most famous instance of social unrest on the North China Plain was the Boxer Rebellion (1898–1900). Myth, legend, and historical scholarship all contain truths and mistruths about the rise of the Boxers, but comparatively little attention has been paid to environmental factors.[15] It was certainly no coincidence that resentment of foreigners arose in a region of profound ecological upset. The point here is not to argue the specific causes of the Boxer

Rebellion but to show that ruptures of ecological order in the late nineteenth century held the potential for profound social and political unrest. The North China Plain had a celebrated tradition of nurturing rebel heroes, such as the founder of the Han dynasty, Liu Bang (256–195 BCE). Such historical narratives were object lessons for China's political elites who wished to promote a stable agricultural society. However, this statecraft was bankrupt, figuratively and literally, by the time the imperial system fell in 1911. Nevertheless, the state orthodoxy of maintaining rural stability through the promotion of hydraulic order remained a central concern throughout the twentieth century. Put slightly differently, every effort to reconstitute a central state in China during the twentieth century included an acute awareness of the political mandate to reorder the waters on the North China Plain. The major challenge, of course, was how to successfully implement this goal.

In the first half of the twentieth century, the collapse of central political authority, the rise of "warlordism," the incursion of Western economic interests, debates over traditional Chinese culture versus Western alternatives, and pervasive violence (culminating in the Japanese invasion in 1937) all contributed to a sense that China had reached the nadir of its existential struggle in the maelstrom of the "modern" world. To many observers, the landscape of the North China Plain was where this struggle was being "written." In 1921, for example, one of China's best-known twentieth-century poets, Guo Moruo (1892–1978), wrote "The Age of Flood" ("Hongshui shi-dai"), in which he describes contemporary China, being inundated with imperialism and warlordism, as experiencing the age of a second great flood. Amid the disorder, the narrator recounts the heroic determination of Yü the Great to reimpose order during that great flood in prehistoric times. In the poem, Yü and his companions cut trees, chisel rock, and "attempt to drive the furious flood on the earth to the sea/By using human strength."

In the familiar story in which Yü passes his home without entering, he selflessly responds to entreaties to visit his wife and home by stating:

> The universe is my home.
> Why should I have my own family?

I put my hand on my heart,
I put my feet on the earth,
How can I face my people
If I cannot control the flood?[16]

After 1911, China experienced a seemingly interminable cycle of flood and drought that generated a series of acute subsistence crises. Floods in 1917, droughts in 1920 and 1928–1930, the massive flood of 1931, the 1935 Yellow River Flood, and the breaching of the Yellow River dike in 1938—all these ecological disasters deepened the sense, domestically and internationally, that China was the "land of famine." In this popular perception, however, "land of famine" represented a more universal sense that China was a country deprived of all the qualities of a modern civilization. But at the same time, the North China Plain and the Yellow River valley were landscapes that were inscribed with the values of China's modern quest for survival and its rebirth as a great civilization. One particularly important element of this transposition of values upon the landscape was China's twentieth-century quest for a national identity—that is, the quest to transform China from a diffuse cultural concept to a discrete political entity based on a definable sense of the common attributes of a people and to draw on these attributes to form a nation. Here, too, the Yellow River valley played a distinct role, as it symbolized the historical origin of the Chinese nation. Beginning in this period, the Yellow River as the "mother of Chinese civilization" and as "China's sorrow" were both metaphors that existed side by side in contemporary Chinese historical consciousness. These constructions of landscape would be selectively employed to legitimize the reordering of the waters in this and later periods.

During the Republican period, the North China Plain was a stage where innovation met enduring patterns. This was an era when China was profoundly influenced by new forces of internationalization—for example, by modern science and technology as well as by new administrative methods, concepts of national identity, and assumptions of modernity. These modern forces negotiated with the imperative of ordering the waters of the North China Plain. Agents of modern hydraulic engineering, including a cadre of professionally trained engineers from Europe, North America, and China, introduced new technical approaches that held the promise of reestablishing hydraulic equilibrium with the tools of modern engineering, including concrete, turbine generators, and an array of pumping and other machines run by steam, petroleum, or electricity. These artifacts of modern science and technology held the potential for reasserting hydraulic order and increasing food production. At the same time, political and technical elites increasingly succumbed to the allure of multipurpose water management, premised on modernist notions of resource exploitation managed by the state for economic development. Although there were few accomplishments in "ordering the waters" during the Republican period because of limited central political control, administrative fractionalization, and Japanese invasion, these forces were further structured by the CCP after 1949 and ultimately shaped the management of water in the Maoist (1949–1976) and post-Maoist periods (1976–).

## Modern Science and Technology

During the late nineteenth and early twentieth centuries, the introduction of modern hydraulic science and technologies to China was an adjunct of foreign efforts to further penetrate China's economy as well as a deepening commitment to social welfare in China. Nineteenth-century European sojourners in China represented a range of constituencies that mutually reinforced the imperial project, including missionary organizations, scientific societies, and commercial concerns. Uniting the differing objectives of these constituencies was an abiding faith in the exclusive capacity of modern (Western) science and engineering to alleviate China's woes. Ney Elias (1844–1897), who, as we noted, reported on the new course of the Yellow River, arrived in China in 1866 in the employ of a British commercial firm. Elias took particular note of the potential for developing China's waterways. Before embarking for China, he had received instruction in geography at the Royal Geographical Society (suggesting the mutually reinforcing relationship between the geographical sciences and British colonialism). Marking perhaps the first instance of a foreigner wading into the great debates about the Yellow River after the 1855 course change, Elias argued for returning the entire flow to the bed of the old Yellow River. Ney went further: "That English engineers be employed and the whole works be placed under the superintendence of Englishmen for organization would seem to be an absolute necessity if the result is to be successful."[17] In light of the presumed weakness of the Qing administration and its lack of experience with modern science, the singular capacity of Europeans and of European science to address the predicament on the North China Plain was a theme sounded again and again. At an 1880 meeting of the Royal Geographical Society, Sir Rutherford Alcock, the president of the Society, articulated the contributions that could be made by European science and engineering acumen to controlling the Yellow River:

> [The Yellow River] not only menaced with sudden destruction the present inhabitants, but it affected the existence of whole generations, because when it left one part of the country and found its way to another, the district which it left generally became a desert of sand, and the portion which it overflowed became a shallow lake, in which nothing could grow, and over which nothing could sail . . . it was not beyond the reach of European engineering science to save the whole of that mighty plain from these disastrous overflows and changes of the river bed. . . . With its present knowledge of hydraulics and mechanical science in general, China was incapable of facing the difficulties of such a problem; but it was to be hoped that in their own interest, and in the interests of humanity, they would be induced sooner or later to call in the aid of European science, and then the gigantic task might be [assigned to those] . . . who could command all the resources of modern skill and science.[18]

In 1889, the first European technical delegation to visit China was dispatched by the Society for the Promotion of Dutch Engineering Works Abroad. The Society had as its "chief object the employment of Dutch capital, engineering and contracting skill on foreign works."[19] The team, composed of two civil engineers and a representative of a Dutch contracting firm, submitted a proposal to Li Hongzhang, viceroy of Zhili Province, that emphasized how the "hydraulic experience of Chinese experts might go hand in hand with the knowledge of Dutch engineers" and how it would "be astonishing if such a confederacy should not succeed in metamorphosing *China's Sorrow* into a blessing to the region through which it flows."[20] Although partly a sales pitch, the proposal also clearly suggests the importance to modern hydraulic engineering of the gathering of extensive hydrological data (e.g., velocity, discharge, chemical composition, and sediment load), using a variety of instrumentation. Premised on the success of river-training projects on the Rhine and Danube Rivers in Europe, the Dutch engineers reaffirmed the primacy of dikes. The engineers also argued for the selective use of reservoirs adjacent to the lower extremity of the main river channel for the settling of silt. On the question of a reopening a southern route using the old riverbed, the delegation did not preclude the future potential for splitting the Yellow River into two courses.[21]

The Qing government did not hire the Society for the Promotion of Dutch Engineering Works Abroad to carry out its scheme, but it is worth noting that the initial works suggested by the Dutch engineers did not depart dramatically from the orthodoxy established by Pan Jixun and further articulated by Jin Fu and others during the imperial period. What was different was the degree of refinement made available by the precision and range of data available from modern scientific instruments and survey methods.

Subsistence crises engendered by floods and droughts increased in frequency and intensity during the first decades of the twentieth century. Particularly severe famines struck North China in 1920 and again from 1928 to 1930. Severe floods occurred near Tianjin in 1917, 1924, and 1939 and in the Yellow River valley in 1925, 1933, 1935, and again in 1938. A particularly severe famine issuing from drought, the worst since the crisis of 1878–1880, struck the North China Plain in 1920. The void created by the lack of a functioning central government during the warlord era (1916–1926) was partially filled by relief initiatives emanating from domestic and foreign philanthropic organizations. Building upon the precedent of Timothy

Richards and expanding upon Chinese traditions of private welfare, foreign and domestic relief agencies proliferated in the early twentieth century, in response to the cascade of subsistence challenges on the North China Plain. Along with the American Red Cross and the Chinese Red Cross Society, the China International Famine Relief Commission (CIFRC) was organized out of a collection of regional famine relief agencies that had responded to the 1920 famine.[22] In addition to famine relief, the CIFRC launched water management projects in many locations on the North China Plain. Although these projects were modest in the context of the totality of need, they served as an entrée for a significant transfer of modern hydraulic engineering knowledge to China. Leading engineers from Europe and North America traveled to China to participate in water projects on the North China Plain. Among the most eminent of these practitioners were John Freeman and Olivier Julian (O. J.) Todd.

By the time he was hired as a consultant to the Chinese government in 1917, John Freeman (1855–1932) had developed a distinguished career as a civil engineer. After graduating from the Massachusetts Institute of Technology (MIT) in 1876, he participated in engineering projects including the Panama Canal, Hetch Hechy water supply system for San Francisco, Boston Harbor Improvement, and the Charles River Basin projects. Freeman was elected to the American Society of Civil Engineers in 1882 and subsequently served as director (1896–1898) and vice-president (1902–1903). He became president of the Society in 1922. Freeman's career was that of business executive, consultant, and public servant. All these roles came together as he traveled to China in 1917 to assume the position of consulting engineer to the China Grand Canal Improvement Board, which had been established by the government in Beijing to rehabilitate the Grand Canal. Freeman's appointment was an adjunct to his position as consulting engineer for the American International Corporation (AIC). Formed in 1915 to spearhead American investment abroad, AIC was the institutional expression of the interests of many of America's most powerful financial and industrial players, who saw a golden opportunity for the spread of American investment in China as European powers pummeled one another during World War I.[23]

While working on the Grand Canal project, Freeman indulged his growing fascination with the Yellow River. After conducting surveys of the river in 1917 and 1919, he wrote "Flood Problems in China," which was published in 1922 by the American Society of Civil Engineers and quickly became a seminal study of Yellow River hydrology. Freeman was stunned by his surveys of the Yellow River. He wrote that there seemed "to be hardly a 5-mile stretch of dike that has not been breached within the last few centuries."[24] Freeman concluded that the "fact of the readiness of the Yellow River to scour a deeper channel when properly confined between dikes, taken in connection with the practically unlimited depth of the bed of silt in which it can dig, together with the hydraulic law of increase of velocity with increase of depth, makes it feasible to maintain a *narrow channel* at all stages from drought to flood."[25] Freeman went on to argue that dikes should be constructed as close to the river channel as possible and that they should rarely exceed half a mile in width.

Pan Jixun would have been pleased. Here was modern science in service to a management orthodoxy established by Pan. Indeed, Freeman recognized as much when he stated, "All things considered, the Chinese have done wonderfully well in their river training by the methods they have followed for hundreds, and perhaps thousands of years; but it should be possible to make great improvements *by means of precise scientific observations* and by slow *painstaking study* in laboratory and field with modern instruments and methods."[26] But the spirit of rationalism, optimism, and positivism reflected in his cultural commitment to the progressive qualities of science and technology would lead to more profound interventions in the hydraulic landscape. This transition to the transformative tools and materials of modern technology were already leaving an indelible mark on the landscape of North America and Europe. They would soon do the same in China.

The second American to play a significant role in introducing hydraulic and hydrologic sciences to China was O. J. Todd (1899–1973). Todd grew up in Michigan and received his B.S. degree in civil engineering in 1908 from the University of Michigan. Todd became familiar with John Freeman through the latter's work on San Francisco's water supply. Freeman recruited Todd to work as assistant engineer on the Grand Canal Improvement Board. Todd accompanied Freeman on his trip to China in 1919, just after completing a two-year stint with the U.S. Army Corps of Engineers. China became a professional and personal passion for Todd over the next several decades. He served in a host of formal and informal positions in China, included chief engineer for the CIFRC (1923–1935), consulting engineer to the Yellow River Conservancy Commission (1935–1938), and, immediately following the end of the Pacific War in 1945, as advisor to the Yellow River Project of the United Nations Relief and Rehabilitation Administration in China (1945–1947).

By the time that Freeman and Todd traveled to China in 1919, there was a small but growing number of Chinese hydraulic technicians and engineers working on river conservancy issues. The growth of civil and hydraulic engineering in China was an adjunct of the modern industrial sector that was gathering momentum in coastal regions like Shanghai and Tianjin as well as of the famine mitigation efforts in the interior. In 1914, the Hehai Engineering Institute (Hehai gongcheng zhuanmen xuexiao) was founded in Nanjing. It later developed into one of the principal engineering institutes in China.[27] One of the first instructors to be hired at the Engineering Institute was Li Yizhi (1882–1938), who eventually became known as the "father of modern Chinese hydraulic engineering." Li was born in Pucheng County in Shaanxi Province near the confluence of the Wei and Yellow Rivers. According to one of his biographers, Li and his family were profoundly affected by drought conditions that hit the Guanzhong region in the late imperial period.[28] Following his graduation from the Beijing Teachers College (Jing shi daxue tang) in 1909, Li departed for Germany, where he studied engineering at the Berlin Royal Institute of Technology (Technische Universität Berlin). After a brief period back in Shaanxi in 1911, Li again embarked for Europe, where he visited engineering projects in Russia, Germany, France, and Holland. In Germany, he became familiar with Hubert Engels— considered the father of modern engineering in Europe. Engels also expressed a great deal of interest in the Yellow River. In 1913, at the age of thirty-two, Li entered Danzig Technical University

(Königliche Technische Hochschule zu Danzig). After completing his degree in civil engineering, Li returned to China in 1915 and immediately joined the industrialist Zhang Jian in establishing the Hehai Engineering Institute. In his eight years as professor and director of the fledgling institute, Li organized the curriculum and wrote or translated most of the texts and other curricular materials. Li was largely responsible for formulating a Chinese vocabulary for modern hydraulic science and engineering. In addition, Li introduced experiential learning techniques and fieldwork training that emphasized surveying and other data-gathering skills—a sort of specialized training that was a sharp contrast to the ideal of the amateur "scientist" in the imperial era. In short, Li is credited with institutionalizing modern hydraulic engineering training in China.[29]

A closer look at Todd and Freeman's work in China exemplifies the transitions taking place as modern hydraulic engineering was adopted in China during the first half of the twentieth century. Joining Freeman and Todd as they disembarked at Shanghai in 1919 for their Grand Canal survey work was Chen Tan, a graduate in civil engineering from MIT. Four years later, when Todd was named chief engineer for the CIFRC, he named C. P. Xue as his chief assistant. Like Chen, Xue had also graduated with a degree in civil engineering from MIT. Although the two Chinese engineers were from different regions of China, both found themselves on a similar trajectory when they were awarded scholarships to attend Qinghua College (later Qinghua University) in Beijing. Qinghua was established in 1911 as a preparatory college with funds from a portion of the Boxer Indemnity owed by China to the United States. Focusing on science and technical education, the school was initially staffed by faculty recruited by the American YMCA. As part of the funding arrangement, students were eligible to attend Qinghua by virtue of their success on nationally administered exams. Upon completion of their respective programs at Qinghua, students were could seek admission to U.S. institutions of higher education to complete their undergraduate degrees. Chen and Xue passed the entrance exam for the Qinghua program and shared first honors. Subsequently both received scholarships from the American Boxer

Fund Committee to attend MIT. As it happened, John Freeman was then on the MIT Board of Trustees and personally recruited both students (both of whom finished as the top two students in their college graduating class). After Freeman recruited Chen and Xue to work for him at his office in Providence, Rhode Island, both returned to China to work as hydraulic engineers. Chen continued to work with Freeman in various capacities over the next few decades, while Xue, after a stint with the CIFRC, assumed the important post of director of the YRCC under the Nationalist government (a post that had been held by the Englishmen Herbert Chatley for some years). Like the majority of Chinese engineers who had been trained abroad, both engineers stayed in China after 1949, convinced that "new China" needed their technical acumen.[30]

Many of China's newly minted engineers, either trained abroad or in the growing number of domestic training institutions, were hired by river basin organizations established by the Nationalist government. Perhaps the most prominent advocate of professionalizing water management was Li Yizhi, who actively recruited technical professionals to join the major river basin organizations in China. He called for harmonizing traditional and modern management approaches and early on encouraged technical training, centralized water administration, and basin-wide management approaches. Since his early death in 1938, Li has been placed alongside Pan Jixun in the popular pantheon of China's water heroes.

## Centralization: The Yellow River Conservancy Commission

From the perspective of China's recent historical memory, the 1910s and 1920s were perhaps the nadir of China's "century of humiliation." The humbling nature of the imperialist presence in China and the country's political disintegration into warlord polities suggested to China's educated elite that dramatic reform of China's traditional values and institutions was necessary for the country to survive in the modern world. China would be devoured if it continued to resist science, technology, and progress. Thus, for many of the so-called May Fourth generation who took to the streets in 1919 to protest China's ills, the values and animating spirit of science and democracy were the ultimate wellspring of China's reawakening. There could be no turning back. The imperial system was dead and buried. But for the moment, only social, political, and cultural disorder prevailed. In the maelstrom of imperialism and domestic political disintegration, however, long-standing cultural traditions of water management were never abandoned. On the contrary, there was a reaffirmation of the importance of managing the waters. From the swirling eddies of political, social, and economic disintegration emerged a revised national commitment to managing water. Indeed, the mechanisms for such renewed commitment were the techniques and tools of modern hydraulic engineering. The other expression of this revised commitment was a re-articulation of China's institutional arrangements for governing water. Again, drawing upon international models, while conditioned to valorize political and cultural unity, educated Chinese committed themselves to re-establishing central administration that could restore the nation to health.

The Northern Expedition of 1926 fueled hopes for a new era of political unity. Launched by a coalition of the Nationalist Party and the CCP, the Northern Expedition was a military campaign led by General Chiang Kai-shek designed to defeat or absorb warlords to unify the country. The Northern Expedition largely united southern and eastern China when it established a national government in Nanjing in 1927. However, facing myriad financial and other problems, Chiang turned what was a potentially revolutionary movement sharply to the right and purged the movement of his communist partners. This bloody crackdown in Shanghai and other urban areas decimated the ranks of the CCP. The split with the CCP would ultimately prove detrimental to Chiang and the Nationalist Party, but from the perspective of 1928 it appeared that the Nationalist Party was well on its way to unifying the country politically and was ready to begin a successful recentralization of administrative functions.

The first step in reasserting control over the waters of the North China Plain occurred in 1929, when the Nationalist government established the Huai River Conservancy Commission. The government was well aware of the traditional Chinese notion that political legitimacy could be enhanced by a state's success in "ordering the waters." The hydraulic and social problems of the lower Huai River valley were complex, but the work of regulating the lower Huai River appeared to be manageable, as opposed to the much larger task of ordering the flow of the Yellow River. The functional heart of the Huai River Conservancy Commission was the Engineering Office led by Li Yizhi and a cadre of foreign-trained engineers. Although international collaboration would continue to be a hallmark of Nationalist-era state building and infrastructure initiatives, the Nationalist government was keen on promoting native talent to lead its reconstruction efforts. The creation of the Huai River Conservancy Commission was a template that served to guide the subsequent creation of valley-wide administrative organs such as the Yellow River Conservancy Commission. Before these further administrative innovations could be implemented, however, an extraordinary flood hit the lower Yangtze and Huai River valleys in 1931.[31]

The 1931 flood remains one of the most devastating floods in Chinese history. Runoff from heavy early-summer rains was augmented by heavy snowmelt from the Himalayan highlands, generated massive upstream flows on the Yangtze and Huai Rivers. In July, seven cyclones hit the region (the summer average was two cyclones), and half the mean annual precipitation fell in that month alone. By August 1931, the downstream portions of the Huai and Yangtze Rivers were under enormous pressure as their dikes began to burst. During the evening of August 25, the eastern dike of the Grand Canal near Hongze Lake burst, sending water pouring through the lowlands of Jiangsu Province. Two hundred thousand people drowned in their sleep.[32] Government surveys conducted in the aftermath of the flood cataloged the depth and extent of the disaster: 4.2 million farm families out of a total of 7.9 million families in the region were affected by the flood—the equivalent of the entire farm population of the United States in 1931 (25 million individuals).[33] Other data were also suggestive: 40 percent of all survivors were forced to flee; houses were flooded for an average of 51 days, and the average maximum water depth of flooded areas was nine feet. Local reports recounted the search for human bones for fuel, the sale of women and children, tales of "parents eating flesh from the bodies of their children," and other reports of cannibalism.[34]

The 1931 flood was a highly politicized event. The enormous destruction and human misery were widely reported both at home and abroad, placing significant pressure on the Nationalist government to administer relief and to prevent such catastrophes in the future. As a corollary, there was the potential that such privation and suffering could exacerbate the social and economic instability of the North China Plain. The flood also hastened the impulse to direct the water resources of China from a central authority. In the immediate aftermath of the flood, the Nationalist government established basin-wide administrative organizations modeled on the Tennessee Valley Authority (TVA) in the United States.

The Yellow River Conservancy Commission (Huanghe shuili weiyuanhui, YRCC) was established in 1933. The creation of the YRCC, the first state-organized river management agency since the Yellow River Administration of the late Qing, signaled the return of a central administrative presence to the hinterlands of the North China Plain. This time, however, the return of the state was premised on goals quite different from those of the imperial state (which sought hydrological equilibrium as a precondition for stable rural society). To be sure, there was a great deal of rhetoric expended on how the state had a moral responsibility to end the horrific cycle of floods and famines on the North China Plain that had begun in the late nineteenth century. Just as importantly, Nationalist leaders were heavily influenced by state models in Europe and North America that embedded the project of "ordering the waters" within a modernist vision of development. As a representative of this point of view, Li Yizhi was appointed director and chief engineer at the commission's inception. Li strenuously advocated for a basin-wide organization that not only would create a pool of the best engineering and technical talent in China but would also have the administrative authority to transcend the provincial disagreements that would inevitably arise in basin-wide projects.

A second articulation of central planning during the Nationalist period was the creation of the National Economic Council (NEC) in 1935. The NEC was a suprabureaucratic organization, designed to centrally promote and manage key economic sectors by efficient and rational allocation of human and material resources. The NEC was largely construed as a way to lead rural rehabilitation or "reconstruction" of the North China Plain. This increased reliance on statism was also reflected by numerous polities around the world as governments of a variety of political stripes sought to mitigate the debilitating consequences of a global depression. The impulse to move toward greater central control of the economy came from a variety of experiments in state power, which ranged from the five-year plans in the Soviet Union and the New Deal programs in the United States, to the controlled central economies in Germany and Japan. Engineers naturally fit into this sort of developmental approach. Not only were they equipped with special domains of applied technical knowledge, but they were also imbued with a strong sense of "systems." Engineers were directors, possessing a conscious sensibility and pride about how component elements could be put together, or orchestrated, to comprise large, efficient, and rational systems like industrial plants, road systems, and hydraulic networks. In other words, the engineering of an industrial economy was the engineering of an industrial plant writ large. The sort of technocratic developmental approach that melded the respective talents of the engineer and the bureaucrat was attractive to technical specialists like Li Yizhi.[35]

On paper, the creation of the YRCC and its subsequent incorporation into the NEC represented the realization of Li's goals. The degree of centralized control that the Nationalist government imposed on the waterways of the North China Plain was unprecedented in history. In theory, all levels of water management projects—county, provincial, and central—were subject to technical approval by the NEC. In practice, however, actual administration over the Yellow and Huai River basins continued to be fractured and reflected the nature of Nationalist rule. The idea of basin-wide administration and centralized management remained only a blueprint, which frustrated Li and his colleagues.[36] At the same time, after the YRCC was formally folded into the NEC, technical professionals increasingly complained about an overbureaucratization of Yellow River management.

This level of central state involvement in water projects challenged technicians like Li Yizhi in their dual capacities as servants of the Nationalist government and as engineers guided by specialized knowledge. On the one hand, for educated elites, administrative service to the state was embedded in Confucian notions of good governance. On the other hand, there was little question that the modernizing goals of the Nationalist government resulted in the valorization of technical knowledge. Engineers in China, who had imbibed a cultural tradition of modern science and engineering, manifested a profound faith in the value of their unique contributions to the creation of "wealth and power" for a modern China. In many ways, then, the Nationalist period was an era when the state and what it deemed a group of critically important clients (namely, technically trained specialists) were engaged in negotiations: discussions of an identity of interest that could translate into a reasonable institutional structure that would achieve political legitimacy and modernity. In sharp contrast to the situation that would pertain after 1949 in China, or to the political arrangements in the Soviet Union, the Nationalist state was simply too weak to coerce complete compliance from the scientific and engineering communities in support of the state's modernizing dreams. At the same time, the increasingly tenuous social base of support of the Nationalist government impelled authorities to create a solidarity of interests with technical elites. Unfortunately for the Nationalist government, increased threats from internal and external enemies compelled choices and created realities that obviated a successful articulation of institutional arrangements that would have satisfied many of the technical personnel involved in water management.

## Multipurpose Water Management

Shortly after the Nationalist government was established in Nanjing, Chinese engineers began to call for water management approaches that transcended the single concern of flood prevention. Despite his early death in 1938, Li Yizhi still managed to publish over 200 articles and books that illustrated his growing concern for basin-wide technical approaches that would combine traditional flood control strategies with the principles of multipurpose river management that would consider benefits to shipping, irrigation, and hydroelectric generation.

Multipurpose river management was a framework developed in the West by the mid-twentieth century. Globally, the best-known model of river management that combined flood control, irrigation, power generation, and improved navigation was the TVA created by the U.S. Congress in 1933. The premise behind the TVA's management principle was that the full productive powers of the river should be exploited to promote economic growth. The tools of exploitation were modern technology. Thus, modern science and technology would be employed to control floods, generate electricity, irrigate fields, and transport goods. Management of such a complex undertaking, and the allocation of physical and human resources to plan, implement, and manage component projects, was invested in a valley-wide organization empowered by the central government and efficiently and rationally guided by scientific management principles. From a development perspective, multipurpose river management held the promise of rehabilitating poor regions, an even more attractive prospect amid the global depression of the 1930s. That such a set of assumptions and goals attracted the interest of modernizing technical and political elites in China is not surprising.[37] During the 1930s, Western and Chinese engineers were exploring water projects consistent with the spirit of multipurpose river development. With hydrological data and new construction material and techniques, Li Yizhi and O. J. Todd were imbued with new hope for maximizing the exploitation of river resources in accordance with the new multipurpose river development perspective.[38]

Chinese and foreign engineers began to discuss the potential for constructing dams and reservoirs in midstream sections of the Yellow River. The impulse to consider midstream reservoirs as a way of impounding floodwater led to plans for creating reservoirs to power hydroelectric generation facilities. Ideas about damming the Yangtze, or its tributaries, to promote regional development through irrigation and hydropower dated to nationalist leader Sun Yat-sen's (1866–1925) earlier vision of a reconstructed China. But damming the Yellow River, with its high silt content and its highly variable intra-annual flows, did not capture the imagination of engineers or nationalist actors until the Nationalist era, when the high-modernist sensibility coalesced in the minds of state and technical elites into a powerful vision of ordering and exploiting water resources in service to state and society.[39]

The earliest and best-articulated plan for hydroelectric generation on the Yellow River was prepared by Todd and his colleagues at the CIFRC. The proposal called for an overflow dam spanning the Yellow River channel with a maximum height of 45 feet at midchannel with a total length of 1,300 feet. Located 600 feet above the famous Hukou Falls in southern Shaanxi Province, the dam would generate 3.3 million kilowatt-hours of electricity. Premising the importance of seeking the "highest use" of available resources that would otherwise be "wasted," Todd mused that "to manufacture cotton into yarn, to grind grain into flour, to light cities and otherwise modernize this part of Shansi [Shaanxi] will be part of the benefit that these Yellow River Falls may confer on the nearby country."[40] It was perhaps the first time that the belief in the transformative potential of modern hydraulic engineering was on full display. Enthusiastically championed by Chinese technical and political elites, these visions would transcend the Nationalist and Communist periods and effectively mark a new era for water management on the North China Plain. Interestingly, Todd makes little mention of how a modern hydropower facility would coexist with silt. Such dreams of modernity would be troubled by this centuries-old conundrum for decades following Todd's initial proposal. Li was less enthusiastic about the prospect of constructing large structures across the Yellow River, citing the enormous cost and the lack of understanding of how such structures could manage silt. At the same time, Li clearly was convinced that, with further study, damming the Yellow River for irrigation, power, and navigation was imperative for regional and national development.[41]

Multipurpose water management implied access to a particular technology complex that included not only technological tools but also a management regime framed by certain assumptions and goals in which tools and institutional forms were adapted. From the broadest perspective, the developmental regime adopted in many polities by the mid-twentieth century was a "high-modernist" vision of economic development and state capacity that sought to efficiently and rationally exploit resources to power an industrial economy. Centralized mobilization of resources—financial, human, and technological—was a major tenet supported by Nationalist planners and engineers like Li. Significantly, however, there was no single technical approach to river management, or above all, to Yellow River management, upon which foreign and Chinese engineers agreed. There was, in other words, not a single technical prescription arrived at by the scientific objectivities of survey, measuring, or other tools of modern hydraulic engineering. In the 1920s and early 1930s, when the lion's share of attention was devoted to flood prevention in the downstream, engineering opinion coalesced around essentially two management options. Interestingly enough, these options largely evoked the basic choices that had guided river management for centuries in China: dispersing the water versus concentrating the flow. During the Republican period, there were variations on these persistent themes, to be sure; and in support of each approach, hydraulic engineers marshaled the tools of modern science. Yellow River management provided an arena in which differing technical approaches competed with one another: in this instance, the competitors included an international cast of characters (Americans, Europeans, and Chinese)—each with his own set of assumptions and priorities.

## Globalizing the Debate over Yellow River Management

Early- and mid-twentieth century debates about how to control the Yellow River exemplify the enduring challenges of river management that have transcended the perspectives of modern technologies or multipurpose development. During these years, some of the world's leading civil engineers became fascinated with the problem of the Yellow River. Most of the foreign and Chinese specialists who ventured opinions recognized the unique challenges that the

Yellow River presented to the field. None had encountered a river with that much silt. Yet most expressed confidence in the capacity of modern engineering to solve the problem. Engineers like John Freeman clearly expressed the confidence of a profession at the beginning of its golden era. Charles K. Edmunds, missionary educator and one-time president of Canton Christian College, recounts snippets of a talk by Freeman and provides a personal description of the engineer:

> "I want to be the Emperor of China." [Freeman said]
> He was an American of middle age, compactly and powerfully built, with friendly eyes and quiet voice. He was conversing in a Washington club and his hearers perceptibly leaned forward. It was evident that he had a real idea.
> "Emperor of China for five years," he continued. "And I want the job in order to do just one thing—to tame the Yellow River."[42]

Freeman's study, "River Problems in China," was the first compendium of data on velocity, flow, depth, sediment load, and so on. These data would become a baseline for subsequent studies, and Freeman's plan to restrict the Yellow River between narrow dikes was the foundation for subsequent discussions and disagreements about "taming the Yellow River."

Although Freeman distanced himself from later disagreements over Yellow River management, a remarkable network of foreign and Chinese engineers was engaged in spirited debate over Yellow River control. Hubert Engels, the famed German engineer, was one such participant. Freeman visited Engels's lab in 1913 and encouraged him to study the river.[43] For an engineer of Engels's status, the opportunity to study the Yellow River was alluring. By the early 1920s, Yellow River control had become the ultimate challenge among the international network of hydraulic engineers. Engels's initial results drove him to disagree with Freeman. He argued that Freeman's plan was wrong on several fronts. First, a straight river channel was "unnatural": in order for the bed to be stabilized, the river would need to adopt a winding course. Second, a narrow bed would not be able to contain a large flood surge. Third, Engels feared the potential for a catastrophic flood event in the midst of long and expensive engineering that would be required by Freeman's plan.[44]

Contributing to Engels's work in Dresden were two Chinese students, who, along with Li Yizhi, soon became the best-known engineers in China: Shen Yi and Zheng Zhaoying. After arriving in Dresden, Shen and Zheng quickly placed themselves under the tutelage of Engels. With Dresden as a center for Yellow River research, Shen and Zhang translated Engels's research into Chinese. Subsequently, the work of Engels ignited a heated discussion among Chinese engineers who had been considerably influenced by Freeman's perspectives on Yellow River management.[45]

In the meantime, one of Engels's former students, Otto Franzius, conducted his own studies of the Yellow River. Franzius, a professor of engineering at the Hanover Technical University, had recently accepted a new student from China, Li Fudu, a nephew of Li Yizhi. Franzius and Engels engaged in hostile debate over the management of the Yellow River. In several articles, Franzius rejected Engels's approach. Although he agreed with the latter about the need for a winding course, Franzius concurred with Freeman about the need for narrow dikes; and in some instances he recommended dikes even narrower than those prescribed by Freeman.

In 1932 Engels was commissioned by the Nationalist government to utilize a new outdoor hydraulic modeling facility in Obernach (Bavaria). Funded by a grant from Hebei, Henan, and Shandong Provinces, tests were conducted in 1932 and 1934 with the assistance, curiously, of Li Fudu. Why Li Fudu, a disciple of Franzius, assisted Engels is simply not known. According to the report submitted in 1935, Engels argued that the best method of Yellow River management was to maintain a winding course, with dikes positioned far away from the channel.[46] The river could concentrate its flow to flush silt while high water would allow sediment to settle near the dikes to strengthen these barriers.[47] The results of Engels's tests at Obernach threw a wrench into the YRCC's deliberations and delayed "the Chinese government's decision on how the river was to be regulated by levees."[48]

The Obernach Yellow River Model. (Courtesy of Technische Universität München)

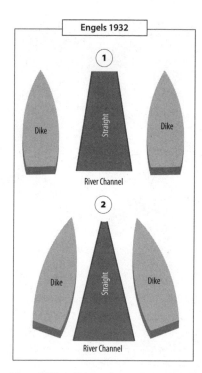
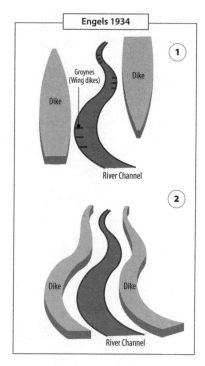

Four Yellow River channel configurations at Obernach. *Data source:*
Quan'guo jingji weiyuanhui shuili chu, *Engesi zhidao Huanghe shiyan baogao
huibian* (Nanjing: Quan'guo jingji weiyuanhui, 1935).

Technical collaboration between the League of Nations and the
Nationalist government in the early 1930s also exemplified how the
unique and enduring challenge of water management on the North
China Plain generated diverging technical prescriptions. Without
question, modern implements and methods for capturing hydrologic data, and their use in the application of modern technologies
(such as concrete, dredging, and earthmoving machines), suggested an extension of traditional river-based activities like irrigation, navigation, and flood control. At the same time, these modern
techniques and technologies introduced new management possibilities, such as the installation of large hydroelectric dams. The
challenge that the Yellow River presented to hydraulic engineering and to its practitioners in the early and mid-twentieth century
was this: the river did not yield to the universalizing principles of

modern science that so many technical experts took as articles of faith. League of Nations experts recognized this challenge when they stated:

> Theory and technique of river hydraulics in the West are based on a whole series of observations, experiments, studies, and experience systematically and uninterruptedly acquired in the course of several generations on a very large number of rivers and streams throughout the world, with the most diverse characteristics. There have thus been established a number of rules for the improvement of rivers and streams with characteristics that do not vary widely from the extremes of experience on which the general theories and rules of river hydraulics are based. The Hwang Ho [Huanghe/Yellow River] and other rivers of North China present hydraulic phenomena, which, in intensity and size, go beyond—frequently far beyond—anything of the same kind observed on Western rivers. Further, and more particularly, Chinese rivers present special phenomena, not observed on Western rivers, which may make the current rules of river hydraulics inapplicable to the Hwang Ho and other Chinese rivers.[49]

The problem, of course, was silt. European engineers had little experience with silted rivers. American engineers had the "big muddy" Mississippi and the Missouri Rivers, but the siltation problem of the Yellow River was on a different scale altogether. By the 1930s, the disputes over the fundamental principles of Yellow River management (which were often bitter) reflected the struggle of modern hydraulic theories to accommodate the challenges of the Yellow River.

## The Yellow River and the North China Plain in the Construction of the Nation

The midcentury transitions that served to reconceptualize the waters of the North China Plain included the construction of a national landscape. The Yellow River as the "mother of Chinese civilization" *(Zhongguo wenhua de muqin)* and North China as the "cradle of the Chinese people" *(Zhonghua minzu de yaolan)* were elements of a national mythos constructed by political and intellectual elites to craft a modern nation from the multiethnic and multicultural empire of the Qing dynasty.[50] This identity between North China and the nation would be challenged in the post-Mao era, but the symbolic value of North China in general, and the Yellow River in particular, continued to fuel a strong national mythos down to the contemporary period. For example, in a historical overview of the Yellow River published in 2006, the text begins by defining the Yellow River poetically:

> A river that is the eternal homeland, binding together the
> souls of the descendants of Yun Di and Huangdi [the
> Yellow Emperor, legendary king and founder of
> Chinese civilization];
> A river that bears the weight of 5,000 years of the
> Chinese people;
> A river that flows in the blood and life of every Chinese,
> spanning all places and times.[51]

There was, however, a disconnect between an idealized North China landscape and reality in the early twentieth century. In truth, the North China Plain was an ecological mess. Yet the Yellow River as "China's sorrow" went hand in hand with the river's role as the "mother of Chinese civilization." The river, absent effective regulation in the mid-twentieth century, was a symbol of the disorder and humiliation that China had experienced since the Opium Wars (1839–1842; 1856–1860). Ordering the river meant ordering China. The value of controlling the waters had a long tradition in imperial statecraft, but the identification of the river with the nation was entirely new. Defining how the North China landscape should look and determining what kinds of transformations were necessary to serve the nation in the modern world included dialogue between modernity and tradition, between national and international interests, and between political ideologies. All of these reinforced the transformations of the physical and cultural landscape of the North China Plain.

The emergence of the Yellow River as a cultural symbol of the nation was generated by a convergence of modern science and nationalist impulses that were fueled by warlordism and imperialism. Since the late nineteenth century, increasing numbers of China's educated elite believed that the survival of Chinese civilization depended on fostering a strong national identity—a notion that was easily expressed but whose realization proved enormously costly over the next century.

Roughly coinciding with the May Fourth movement of 1919, a series of protests against imperialist aggression that historians have considered a seminal event in the formation of a modern national consciousness in China, important new archaeological discoveries were interpreted to mean that North China lay at the center of *Chinese* ethnic identity and culture. Between 1923 and 1927, excavations at Zhoukoudian (50 km southwest of Beijing) led by the Swedish geologist J. G. Andersson unearthed fossil remains that became known as Peking Man *(Sinanthropus pekinensis)*. This finding built on an 1899 discovery of oracle-bone inscriptions near Anyang, Henan, on the North China Plain, which had substantiated the existence of an early Chinese state. During much of Chinese history, the Yellow River valley held special significance in the collective historical memory of Chinese civilization. These regions were the sites of the capital cities of China's earliest states. Many of the earliest extant historical records were also generated in this region. The eminent archaeologist Chang Kwang-Chih argued that the strength of this traditional notion of a "center" of Chinese civilization heavily influenced the interpretation of archaeological discoveries in the early twentieth century. These interpretations stressed the diffusion of Chinese cultural characteristics from one central area—the Yellow River valley. This view, however, was in sharp distinction to the view of Andersson, who argued that pottery excavated at Yangshao village in western Henan was evidence of cultural diffusion from western Asia. At one level, the archaeological work of Andersson and Davidson Black at Zhoukoudian, and subsequent interpretations emphasizing the diffusion of culture to China from regions outside China, ran headlong into the anti-imperialist sentiment that suffused the consciousness of many Chinese intellectuals and political elites. At another level, though very much related, was the long tradition of China's historical record. Thus, "Chinese archaeologists rejected foreign influence; for them diffusion outward from the Yellow River valley has continued to be a natural assumption because it gives to this region the same civilizing role that traditional historiography ascribes to it."[52] With some exceptions, the cultural-diffusion view of Chinese civilization largely held sway in China until the early 1970s.[53]

This admixture of archaeological interpretation and the search for national identity during the Republican period also engendered considerable debate over Chinese racial identity.[54] To many political elites of the May Fourth generation, the survival of China as a nation rested on forging and nurturing a sense of racial identity and unity. The problem for nationalist actors, however, was their commitment to maintaining the culturally and racially complex empire that had been bequeathed by the Qing. How could they create a single Chinese nation/race (Zhonghua minzu) from such a multiethnic, multicultural empire? Seeking to construct narratives that connected the present with the past, Chinese scholars of the Republican period sought evidence for national unity in ancient records and artifacts. Reflecting the ongoing debate over competing interpretations for the archaeological finds at Anyang and elsewhere in North China, ideas of Chinese unity were premised on creating "modern myths of national belonging."[55] During the 1930s, efforts to define the meaning of Chinese national unity was further impelled by the threat from Japan, especially after the Manchurian (Mukden) Incident of 1931 led to the effective Japanese occupation of Northeast China. Threats to China's sovereignty from Japan drove Chinese intellectuals (at least those with faith in the Nationalist Party) to recast Sun Yat-sen's call for arousing China's lost national consciousness (minzu yizhi) to enable China to survive in the modern world of the nation-state.[56] Faced with the loss of a strategic region (North China) that was populated with non-Han peoples, intellectuals and Nationalist Party activists such as Dai Jitao sought to integrate non-Han groups into a common "Chinese" ancestry. Dai and other racial nationalists argued that all ethnic groups, Han, Manchu, Mongol, Hui (Muslim), Tibetan, and others, were direct descendants of the Yellow Emperor. Other scholars challenged this facile idea of the racial construction of the nation.[57] For many Nationalist Party intellectuals, however, these objections ran counter to the imperatives of the time. To them, what was needed in a time of acute national crisis was a concise, simple rendering of national identity. The appearance of Peking Man in 1929 provided the evidence necessary to construct an argument for the origin of the Chinese people. For intellectuals such as Xiong Shili, Peking Man was irrefutable proof that all the peoples of China were "from a single progenitor who lived in the Yellow River valley some 500,000 to one million years ago and [whose descendants] then spread out in all four directions, populating China and possibly the rest of the world."[58]

The debate among China's intellectuals over national identity and racial unity was, in a sense, rendered moot by the full-scale Japanese invasion of China that began in 1937. Geopolitical realities helped forge a new Nationalist Party orthodoxy that was adopted by the CCP after 1949. As reflected in the 1943 publication *China's Destiny* (ghostwritten for Chiang Kai-shek), the official Nationalist Party line emphasized the consanguinity of all the peoples of China. All Chinese "stocks" had descended through a common ancestor or had been blended through intermarriage. Chiang explained in more detail: "The Chunghua [*Zhonghua*, Chinese] nation, as may be seen from its history has grown by gradual amalgamation of various stocks into a harmonious and organic whole. These various stocks, *originally of one race and lineage* . . . had developed different cultures, and this in turn accounted for their different characteristics. However, during the last five thousand years a continuous process of amalgamation has been going on through frequent contacts and constant migrations so that they have now become integral parts of one nation. . . . This was how the Chinese nation came into being in ancient times."[59]

Chiang's linking the identity of Chinese people with North China would be a myth sustained by the CCP in its revolutionary and post-revolutionary periods. The perpetuation of the national myth fed the nation- and state-building aspirations of both Nationalist and Communist elites. The historian Edward Friedman

writes, "Chinese, before the 1930s, did not see themselves as a nation happily born in the north who peacefully, through a superior
culture, amalgamated the Manchus and succeeded as a people because they were protected by a Great Wall. That was a twentieth-
century construction, a particular reading of history in the service
of anti-imperialist nationalism."[60] By identifying North China with
the origins of Chinese civilization, and by linking the North China
region with the peasantry and agriculture, Chinese Communist
leaders added additional moral layers to the mythic landscape of
North China that, after 1949, were employed in support of hydraulic transformations.

## Inundations and the Tragedy of Huayuankou

As the Nationalist government struggled to establish a unified national administrative structure, as engineering and technical elites
sifted through competing plans for managing the Yellow River, as
multipurpose water development planning spawned images of prosperity, and as nationalist intellectuals and political actors struggled
to create a sense of national identity—through all of this—the Yellow, Hai, and Huai Rivers continued to flood. The Yellow River
experienced acute floods in 1933 and 1935. After multiple dike
breaches in the lower valley, the government officially reported
12,700 dead, 2.73 million "affected people," and 600 square miles
inundated by the 1933 flood. The 1935 Yellow River flooding resulted in even greater losses: 6,000 square miles of flooded farmland
affecting 4 million to 5 million people.[61]

The even more devastating flood that fanned across the North
China Plain in 1938, however, not only was unprecedented in the
scale of its destruction but was unique because it was deliberately
instigated. Two years after the 1935 flood, Japanese forces quickly
swallowed up large chunks of North, East, and Central China. By
the end of 1937, Japanese armies had captured Shanghai and Nanjing. The offices of the Nationalist government relocated to Wuhan
in southwest China and struggled to mount an effective defense.
Eventually settling on a "space for time" strategy premised on Nationalist leaders' conviction that there would be an eventual Ameri-

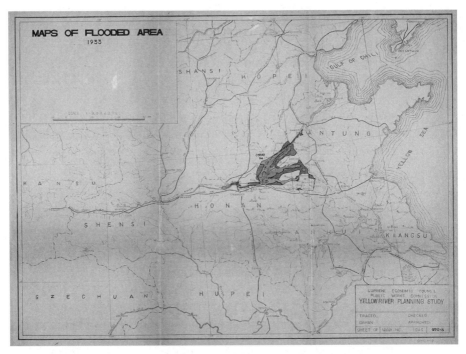

Map of the 1933 flood. (Percy Othus Collection on Chinese Hydrology, Courtesy of Historical Collections and Labor Archives, Special Collections Library, The Pennsylvania State University Special Collections)

can intervention, Chiang Kai-shek endeavored to stall the advance of Japanese troops with the time-honored tactic of breaching the dikes of the Yellow River, a decision borne of desperation.

During the first five months of 1938, Nationalist forces and regional armies loyal to the central government engaged in a bitterly contested series of battles known as the Xuzhou campaign. Japanese troops totaled two hundred thousand, with large numbers of reinforcements arriving later. Japanese forces encircled the city of Xuzhou in northern Jiangsu Province and were poised to advance on the strategically important cities of Kaifeng and Zhengzhou. Chiang Kai-shek had few military assets north of the Yangtze to hurl at Japanese forces. As the historian Diana Lary comments, "Only her [China's] geography and her vastness could save the country, and then only parts of it. To protect the central government at Wuhan (and possibly to enable it to move even further west, behind the great gorges of the Yangzi [Yangtze], into Sichuan province), time was desperately needed. The River was used to buy time."[62]

Beginning on June 12, 1938, the Nationalists' Central News Service issued dispatches that the Yellow River dikes had been destroyed by Japanese aerial bombing. Japanese military and civil officials quickly denied it. According to Chinese sources published years later, not until Japan surrendered in 1945 did credible evidence emerge that Chiang Kai-shek had ordered the destruction of the dikes.[63] The possibility of utilizing a Yellow River flood against Japanese troops had been discussed by Chiang and military officials in the early months of 1938. The decision to blow up the dikes was finalized by early June, when Japanese forces had advanced to within 25 miles of Zhengzhou. Two abortive attempts to breach the southern dikes near Zhaokou (Henan Province) were made near the location of the dike break in 1887 that had generated one of the most deadly floods in the lower Yellow River valley. There was a certain irony in the Nationalist inability to generate a flood along a river that so frequently burst its dikes. The first attempt failed because the explosives created too narrow a breach; the second failed because the saboteurs chose a section of the south dike that was one of the most fortified in the region (precisely because this had been a region of multiple breaks in the past). The third attempt, near Huayuankou, succeeded. With Japanese troops gathering on the north dike, Nationalist troops eschewed explosives, instead using the labor of the Fifty-third and Thirty-ninth Corps to dig through the dike. At first, water eased out of the narrow opening, but by June 9 it rushed out in torrents. Three-quarters of the Yellow River flow at Huayuankou poured southward, once again, toward the Huai River in Anhui and Jiangsu Provinces.[64]

The Nationalist government hastily staged a filmed production replete with explosions (purportedly bombardment from Japanese planes). The production also highlighted troops and locals gallantly striving to close the breach. Foreign journalists were brought to the scene to witness yet one more example of Japanese atrocities. As the filmmakers edited the scenes, the extent of the flood came into sharper focus. The small rivers that carried floodwaters to the Huai

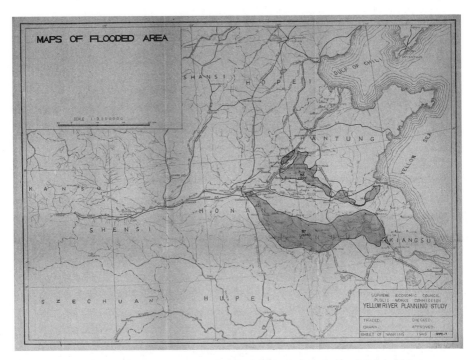

Map of the 1935 and 1938 Huayuankou floods. (Percy Othus Collection on Chinese Hydrology, Courtesy of Historical Collections and Labor Archives, Special Collections Library, The Pennsylvania State University Special Collections)

River quickly overflowed. The water carried by the Huai overwhelmed hydraulic defensive systems (some of which had been recently reinforced by work in the 1930s), inundating much of northern Jiangsu Province. Wartime conditions from 1938 to 1949 prevented any accurate description of the flood's impact, but subsequent counts by Nationalist and Communist governments reported between 500,000 and 1 million dead, 12 million people "affected," and 3 million to 4 million refugees. Wartime conditions also meant that repairing the breach and restoring the river to its "old" (pre-1938) bed had to wait until the war's end. Thus, for the better part of a decade, vast swaths of Henan, Anhui, and Jiangsu were a depopulated no-man's-land, with land and villages buried in silt. Large numbers of refugees fled to Shaanxi, where many experienced profound famine conditions.[65]

The Huayuankou breach. Huanghe shuili weiyuanhui, *Shiji Huanghe* (Zhengzhou: Huanghe shuili chubanshe, 2001).

In the end, this "use of water to substitute for soldiers" *(yishui daibing)* had marginal military value. The Japanese advance on Zhengzhou was stalled, but the primary Japanese objective, the capture of Wuhan, was accomplished in October 1938. However, Huayuankou was used by Japan, the Nationalist government, and the CCP in portrayals of enemy callousness to human suffering. Japan flew foreign journalists over the inundated region to portray the brutishness of the Nationalist government. The Nationalist government valorized the heroic efforts to stem the flood and the sacrifices that the Chinese people were making in the face of Japanese aggression.[66] Perhaps the greatest beneficiary of the flood, however, was the CCP. The social and economic dislocation generated by the inundation provided fertile ground for the recruiting efforts of the Party as it organized guerrilla warfare behind enemy lines on the North China Plain. After it became widely known that Chiang had ordered the dike destruction, CCP propaganda relentlessly cited stories, images, and data from the flood to delegitimize the "bandit Chiang Kai-shek" and the Nationalist Party. After 1949, "Huayuankou" became a potent image of Nationalist illegitimacy (and CCP legitimacy) and was widely employed to mobilize the region and the nation to sacrifice to the cause of ordering the waters.

## Communists and Japanese on the North China Plain

Shortly before the events at Huayuankou unfolded, the CCP established a base camp in the loess region of the Yellow River valley. Mao arrived in Yan'an in 1935 with several thousand CCP members and Red Army members who had survived the Long March. Covering 8,000 difficult miles, the Long March began with the exodus of over a hundred thousand CCP and Red Army members from their base in South-Central China after Nationalist forces had encircled it in what appeared to be a final "extermination campaign" against the CCP. It almost succeeded. Judging by the number and condition of the survivors who arrived in Yan'an, the odds that this ragtag group would lead a successful revolution in less than two decades seemed slim indeed. But history unfolded otherwise, and the Yan'an period proved critical to the success of this revolutionary insurgency.

During the Yan'an years, CCP leaders established important new patterns for making and remaking revolution. With a membership that grew rapidly after the Japanese invasion of 1937, Mao established mechanisms to maintain Party leadership and discipline. In Yan'an, the CCP fashioned organizational structures that ensured the primacy of the Party over the military, conceptualized and instituted "rectification campaigns" to ensure Party discipline, mandated that all artistic production serve revolutionary goals, and experimented with rural policies designed to create an identity of interests between the Party and the peasantry. Equally important was the careful crafting of a revolutionary ethos, an interpretation of Party history that could advance the momentous task of building communism. One such historical reading was of the Long March. The Party survived against all odds. The lesson of the Long March was that humans can, if properly motivated, accomplish anything. To Mao, this animating spirit could energize a Communist transformation of China despite daunting obstacles.

The formative experiences of the CCP in Yan'an fostered the construction of a "revolutionary North" centered on the Yellow River valley. As we have seen, Chinese civilization and early Chinese states arose in this region. And so strong were the attractions of China's civilization that all challengers from across the Great Wall were largely assimilated into Chinese civilization. Over the centuries, the North China landscape, visited by regular cycles of flood- and drought-induced famine, generated a resilient rural society that

regularly suffered (*chi ku*, "ate bitterness"), but nevertheless perse-
vered and survived. The Chinese Communists extended the cultural symbolism of landscape that had begun during the 1930s and
inscribed the Yellow River valley and the North China Plain with
Chinese, peasant, and revolutionary values. The Party embraced
the sorts of cultural production that would deepen this imagery.
For example, Lu Xun was a member of the May Fourth generation
celebrated by Mao during the Yan'an period. Lu Xun's short story
"My Old Home" ("Gu xiang"), written in 1927, reflects the tensions
experienced by some May Fourth intellectuals. On the one hand,
cultural iconoclasts argued that traditional Chinese values impeded
the country's quest to join the modern community of nations; on
the other hand, some intellectuals committed to this notion wor-
ried about the consequences of a wholesale negation of Chinese culture. "My Old Home" reflects this tension as the protagonist returns to his remote village under the "yellow sky" to accompany
family members as they are about to relocate to his urban home. By
recalling the same yellow sky described by Cressey during his visit
to the North China Plain in 1921, the narrator indicates that his
hometown is indeed in North China. The narrator meets a childhood friend, but the two men, one urban and the other rural, traditional and modern, have become estranged. The city is seen as the
future, and the countryside as the past. But the narrator laments the
loss of rural values as China modernizes—the values of innocence,
hard work, and honesty. The preference for the values of rural society, the values in this instance of North China, coincided with the
self-identity of the CCP and found resonance in the Party's emerging social base of revolutionary support, the peasantry.[67]

Identifying the Yellow River valley and the North China Plain
with a nationalist and revolutionary script was reflected more directly in Xian Xinghai's (1905–1945) "Yellow River Cantata"
("Huanghe dahechang") composed in Yan'an in 1939. Xian studied
music at Peking University and later traveled to Paris, where he was
the first Chinese student accepted at the Paris Conservatory to
study composition. After returning to China in 1935, Xian composed a number of patriotic works inspired by Japan's occupation of
Manchuria. Intrigued by the rapid regeneration of the Communist

movement after the Long March, Xian made his way to Yan'an. In 1938 he became dean of the Music School at the Lu Xun Institute of Arts. One year later, he composed the "Yellow River Cantata."

Music as a medium to communicate national identity began in China during the early twentieth century. After the emergence of a public school system during the first decades of the century, patriotic Chinese elites promoted the inclusion of song in the school curriculum. Simple didactic tunes such as "Freedom" ("Ziyou"), "Man of China" ("Zhongguo nan'er"), "Song of the Ancestral Homeland" ("Zuguoge"), "Love One's Country" ("Aiguo"), and "The Yellow River" ("Huanghe") were sung by children to inculcate the meaning of "nation"—its territory, people, and principles.[68] As the threat from Japan intensified, the Nationalist government encouraged patriotic songwriting and supported the public performance of song to buoy national spirits and to increase national resolve. Musicians and composers organized the Singing for Resistance against Japan and National Salvation Movement (Kang-Ri jiuwang geyong yundong). Artists in Shanghai organized the People's Singing Society (Minzhong geyonghui) and the Chinese Nationwide Singing to Resist the Enemy Society (Zhonghua quanguo geyongjie kangdi xiehui). Organizers rounded up participants to sing at factories, schools, and antiwar rallies. After the full-scale invasion by Japan in 1937, singing societies and troops went into the hinterlands to rouse villages. Xian Xinghai was a member of the Second Drama Traveling Troupe and ventured to the CCP Base Camp in Yan'an during a tour of the interior.[69]

There was a blend of influences in the repertoire of the Singing for Resistance against Japan and National Salvation Movement. Many patriotic composers recast traditional folk songs using new patriotic verses. Others, classically trained like Xian, melded Western classical forms with traditional motifs and instrumentation. Although such a synthesis of artistic influences would raise objections in hyperradicalized moments later in the Party's history, during the Yan'an period Mao was satisfied that the "Yellow River Cantata" met the mandate laid down in Mao's "Talks at the Yan'an Forum on Literature and Art" (1943) that all artistic production should advance the cause of revolution.[70]

The emergence of song in China as social imaginary was consistent with patterns of patriotic expression in other cultural contexts. Simple, catchy, and emotionally charged, the music was meant to be sung and to be sung by as large an assemblage as possible. The performances by mass choruses were meant as "symbolic expressions of order and musical performances as active means of organizing people, drawing upon widespread beliefs that music can stir as well as depict emotions, can create as well as represent community. Going beyond the image or text, music adds a performative dimension, an active means by which to experience the nation, by which to feel and act national."[71]

In 1939, the CCP Second Anti-Enemy Performing Arts Troupe was selected to perform the premiere of the "Yellow River Cantata." In six movements (later revisions brought it to nine), the lyrics were adapted from a patriotic poem of the period. Because of a lack of instruments, particularly violins, the orchestral accompaniment was played mostly on traditional Chinese instruments. Mao was in the audience for the second performance at Yan'an. The piece begins with a narrator asking:

> Friends! Have you ever seen the Yellow River?
> Have you crossed the Yellow River?
> Do you remember the scene as the boatmen fought a
>     life-and-death battle against
>            The swift currents and high waves?
> If you have forgotten, listen to this.[72]

From the very outset, the Yellow River is posited as a challenge. The crossing of the turbulent river can be read as a metaphor for the revolutionary struggle: as a journey from feudalism to modernity, from imperialism to national autonomy, from Nationalist to Communist governance. Here, crossing the river is a passage to Yan'an and to the revolutionary struggle. The central protagonists are the boatmen, representing the laboring, the downtrodden, and the exploited, who lead the struggle against the "swift currents and the high waves" in their efforts to reach the other bank. The pace and force of the music complement the lyrics that chronicle the heroic efforts of the boatmen to cross the river. Celebrating collective action, the rowers urge one another on: "Pull hard, pull together. Pull hard, pull together." But conditions deteriorate as "lowering storm clouds veil the sky" with "waves as high as mountains"; "biting cold wind beats our faces," and "dashing foam against the boat ribs." "Steersman, hold firm at the helm" to navigate the "Perilous waters, surging, billowing, perilous waters." In case the audience and chorus members did not understand the extended metaphor, the chorus sings, "Rowing is like fighting at the front." As the first movement reaches a resolution, the boatmen have arrived at the opposite bank: "Turn your heads? See the raging waters! Ours the victory!"

After the boatmen defeat a Yellow River poetically imbued with reactionary, feudal, and imperialist qualities, the second movement evokes an image of the river restored to its historic role in the formation of Chinese civilization. Xian employs the historical and cultural role of the great river to serve the revolutionary goals of the present. With the musical notation *con affetuoso* (affectionately), the verses continue: "The majestic course [is] like a sinuous dragon/ . . . You're the cradle of our people,/ Five thousand years of ancient culture draw their food from you . . . How many heroic deeds have had their scene along your austere banks." Just as the rise of Chinese civilization occurred along the Yellow River, Yan'an, which lies near this same river, shall once again be the cradle of a reborn China, this time a Communist China. The second movement ends with a return of the narrator's voice:

> O Yellow River!
> You rage a hundred miles an hour, unstoppable,
> Throwing great arms out [to] either side.
> You modeled our great nation's spirit.
> Daughters and sons of our Fatherland will
> Learn from you to be as great and strong,
> As great and equally strong.[73]

Throughout the next four movements, the river is the provider of sustenance, an ancestral home, and a sacred altar of the motherland. The landscape is also a symbol of the outrage of the oppressed, of

Chinese civilization itself. The myth of the Yellow River and North China as the cradle of Chinese civilization and the womb of its rebirth could hardly be more simply and powerfully communicated ("North China, defend our whole great land!"). How this imagery would be manipulated in support of massive human intervention in the Yellow River valley was a critical component of the post-1949 narrative.[74]

## Japanese Wartime Planning on the North China Plain

As the Communists fought behind enemy lines and sang in Yan'an, Japanese planners and engineers pondered how to exploit the water resources of the North China Plain as they attempted to administer occupied territories in China. In 1939 the Japanese East Asian Research Institute organized the Yellow River Investigation Commission (Diaocha Huanghe weiyuanhui). Over the next five years, the institute compiled historical material on Yellow River control and collated all available data. Its final report, "General Report on Investigations of the Plan to Regulate the Yellow River" ("Huanghe zhiliguihua de zhonghe diaocha baogaoshu"), was the first to fully reflect the goals of multipurpose river management.

Proposals to regulate the river focused on flood management and sediment control. The Institute first advocated the restoration of the Yellow River to its pre-Huayuankou channel (pre-1938). In terms of channel control and flood mitigation, the report called for the construction of flood retention basins on the main channel of the Yellow River at Sanmenxia (Three Gates Gorge) (Henan), Balihutong (Shanxi), or Xiaolangdi (Henan), behind dams of varying heights. The plan also called for a widening of the Yellow River dikes east of the Beijing–Hankou railroad by 60 meters (to serve as a flood retention basin), as well as utilizing downstream rivers in the vicinity of the main stem as flood diversion channels. The report also stressed the need of reforestation in the loess highlands to control sediment runoff to the river and its tributaries.

The very heart of the proposal, however, was its focus on developing the productive powers of the river for irrigation, transport, and power. For the first time ever, the Japanese researchers proposed extensive irrigation development in the downstream area of the North China Plain. Their plans called for establishing five irrigation districts that would water 37.55 million *mu* (6.2 million acres). In terms of navigation, emphasis was placed on the need for the efficient transport of minerals throughout North China. The Yellow River would serve as the central artery of a water transport system that included a rejuvenated Grand Canal. The greatest stress in the report, however, was placed on hydroelectric capacity. These plans were premised on the needs of a developing network of chemical and mineral industries, focused on fertilizers, synthetic petroleum, carbide, aluminum, and steel. The proposal outlined two possible plans. The first envisioned a series of fourteen "staircase" dams and power plants between the Qingshui River and Yumenkou. The second prescribed a series of eleven staircase dams from Qingshui to Xiaolandi. Discussion of plans for a large dam at Sanmenxia and reservoir centered on the sedimentation problems of the reservoir and the height of the dam. Initial discussion focused on a dam of 86 meters, with a reservoir capacity of 40 trillion $m^3$. Because the annual sediment transport was 15 million square meters, the effective life of the reservoir would be 38 years. Ultimately, with concerns about the "backup" effect that such a large reservoir would have on the Wei River and the Tongguan region, the proposal settled on a two-stage construction of a dam at Sanmenxia with an initial height of 61 meters followed by an additional 15 meters to be added later. The proposal was an influential first step toward articulating the full exploitation of the water resources of the Yellow River and the North China Plain that would have profound consequences for long-term resource supply in the region.[75]

## Postwar Reconstruction: Communists, Nationalists, and Americans

Following Japan's surrender in 1945, American and Chinese postwar planners quickly turned their attention to addressing economic and social dislocation in China that could engender civil war between the Nationalist and Communist forces. The United Nations Relief and Rehabilitation Administration (UNRRA) was the institutional expression of this goal. UNRRA identified the North China Plain as a critical region for economic rehabilitation. And the lynchpin of this endeavor was closing the Huayuankou breach that was the source of massive immiseration in the region. In order to better understand the complexity of closing Huayuankou, it is necessary to explore the struggle between the Nationalists and Communist forces to lay claim to the North China Plain during and after the war.

In the early twentieth century, economic and social instability on the North China Plain was due to a combination of environmental deterioration and struggles between local warlords. Agricultural populations reacted to precarious ecological conditions by forming social networks that engaged in banditry as a form of self-preservation.[76] Elsewhere, local self-defense organizations such as the Red Spear Society (Hongqiang hui) sprang up in Henan, Hebei, and Shandong Provinces to protect villages from rapacious bandit and warlord activity. The formation of Red Spear societies accelerated during the 1920s as warlord strife intensified. The struggle to protect scarce resources from appropriation by warlord armies united local elites, landowners, and tenants. During the First United Front (1922–1927) of the Nationalist and Communist Parties, Communist organizers were sent to rural areas on the North China Plain to politicize many of these local self-defense groups and to encourage support for the joint Communist–Nationalist Northern Expedition, a military operation intended to reunify China. After Chiang Kai-shek's violent purge of the CCP in April 1927, referred to in CCP historiography as the "Shanghai massacre," the leadership of the Red Spear movement largely identified with the Nationalist government. Nevertheless, the organizational experiences of the CCP on the North China Plain would ultimately benefit the party in its efforts to politicize the peasantry during the Japanese occupation.[77]

Following the April 1927 purge, the CCP, at least what remained of it, increasingly focused on developing a peasant-based revolutionary movement. Surviving CCP members established base camps in several locations in China, typically located in border areas beyond central state control. These areas were also ecologically unstable regions, had a fragmented social structure, and were the-

aters of contestation between regional power holders. CCP organizers attempted to work with existing power structures to restructure local society through radical reform policies, such as land
redistribution. But CCP members were generally frustrated by the
conservative nature of such rural societies. The CCP cast about
with little success to find approaches to restructure rural society.

The war with Japan completely changed the terms of such challenges. For example, a quick glance at CCP membership in Henan
suggests a dramatic change in the Party's fortunes in the late 1930s.
At the time of full-out Japanese invasion of China in 1937, the CCP
counted 97 members. Two years later, membership had surged to
16,000.[78] The war provided an opportunity for the CCP to alter its
revolutionary strategy. The Party established bases behind enemy
lines, where it replaced radical policies like land redistribution with
moderate reforms such as rent reduction and tax reform. During
the war, the North China Plain experienced a power vacuum. The
region had no central authority, and Japanese occupiers were spread
thin. Social disorder created spaces where CCP cells could form
partnerships with militarized local structures. In Henan Province,
rural discontent grew over Japanese economic extraction, labor
conscription, and a system of rural cooperatives that returned only
a subsistence-level quantity of their own agricultural production to
the peasantry. This was the North China area that had experienced
nine years of flooding following the rupturing of the dikes in 1938
by Nationalist forces. It provided fertile ground for the organizational
efforts of the CCP. The flooded areas of Henan-Anhui-Jiangsu is
where the anti-Japanese Yewansu Base Camp (Kang-Ri yewansu
genjudi) emerged that organized guerrilla activity against Japanese
occupiers. To the east, in northern Jiangsu, CCP organizers led water
conservancy efforts after the Communists gained a foothold in that
region. Conscious of the legitimizing power of "controlling the waters," CCP cadres organized "progressive gentry" and "moderate elements" from several counties to join with peasants and Party members to execute dredging and irrigation projects in the Hongze Lake
area. The relative success in these endeavors generated substantial
goodwill for the Party.[79]

At the end of the Sino-Japanese War in 1945, Yellow River water
continued to stream from the breach at Huayuankou southward to

the Huai River while the CCP was successfully establishing a rural presence in much of this region. Chiang Kai-shek was fully aware of the connection between the two issues. By 1943, Chiang was already focused on postwar realities, and for him those realities centered on the challenge of the CCP. Indeed, a member of the Nationalist government joined representatives from forty-three other countries at the White House in November 1943 to discuss the economic, social, and political challenges of the postwar world. Out of that meeting came an agreement to organize UNRRA. Predating the Marshall Plan by several years, UNNRA's primary mission was to provide the populations of formerly occupied territories (and returned refugee populations) with immediate relief through the supply of food, fuel, clothing, shelter, and medical necessities. Rehabilitation would be achieved by providing material and expertise for rebuilding communication, agriculture, and industrial infrastructure to catalyze economic recovery. A fundamental tenet was that material and technical aid be extended to populations irrespective of "race, creed, or political belief" and that at no time should "relief and rehabilitation supplies be used as a political weapon."[80] The pronouncements of Nationalist government officials seemed to confirm that it had everything to gain by upholding the integrity and efficient operation of UNRRA in China.[81]

In November 1943, the first members of UNRRA's China team departed Washington, DC, to establish an office in the Nationalist government's wartime capital in Chongqing. Chinese officials requested the dispatch of foreign technical specialists in a wide array of sectors, including mining, railroads, agriculture, and water conservancy. In all, some 2,000 foreign experts were sent to China. UNRRA also sent $17.8 million worth of supplies.[82] Although impressive in real terms, this assistance obscured the complexity of launching an international aid effort in a country with such fiercely competing political constituencies and also masked the difficulties of addressing the enormous needs of China's people (particularly on the North China Plain).

The UNRRA-sponsored project with the greatest visibility, both within and beyond China, was the closing of the mile-wide Huayuankou breach. The Huayuankou project was promoted vigorously by officials of the Nationalist government because the Communists held *de facto* control of large chunks of the North China Plain. Chiang Kai-shek hoped that reimposing hydraulic stability would engender rural stability and ultimately would generate goodwill for the Nationalist government in this CCP-dominated region. Closing the breach and rehabilitating some 2 million to 3 million acres of farmland was also consistent with one of UNRRA's primary objectives—namely, to assist China in becoming self-sustaining in agricultural production, a goal that took on additional importance given worldwide grain shortages in the postwar period. The Communists were initially cool toward the project, which they perceived as an exclusive U.S.-Nationalist partnership. UNNRA representatives in China, however, insisted on CCP participation. At a series of conferences involving the three parties throughout 1946–1947, Nationalist representatives were tepid about CCP involvement, but the fact that CCP forces held sway in the lower Yellow River valley gave the Nationalists little ground to resist CCP participation. The Communist representatives demanded that dike work below Huayuankou be organized by local CCP representatives with UNNRA food supplies and material. Another complicating factor was that downstream dikes had deteriorated significantly during the war. Any return of the Yellow River to its pre-1938 channel would mandate extensive dike rehabilitation downstream. According to Chinese sources, Zhou Enlai, China's long-serving premier after 1949, demanded that appropriations of UNNRA money and material support the relocation of half a million peasants who farmed and lived in the former riverbed (now dry).[83]

The Huayuankou project began in January 1946, and plans called for the repair of the breach before the summer high-water season. The American hydrologist O. J. Todd was invited back to China by UNNRA to be chief advisor. The project quickly fell behind schedule after an unexpected flood hit the area in June. Work resumed in early fall with more than 10,000 laborers sustained by UNRRA provisions. UNRRA also allocated all heavy equipment and supplies, including 1,000 long wooden piles from Oregon, 800,000 board-feet of lumber, 2,243,000 sandbags, and 23,000 rolls of wire mesh, along with bulldozers, trucks, barges, and other heavy equipment. In March 1947, the breach was sealed, and the entire flow of the Yellow River returned to its old course.[84]

There was initial uncertainty as to whether the downstream dikes would hold. Dike-strengthening work in Communist-controlled areas was dogged by a variety of problems in 1946 and 1947 that were expressions of a deepening civil war. Both the Nationalists and the Communists engaged in aggressive actions that delayed progress. Several years later, reflecting on his experiences as the director of UNRRA's China Office, J. Franklin Ray wrote:

> In the spring [of 1947] an offensive by Government forces had initiated bitter fighting along the politico-military border. UNRRA shipments had therefore to be dispatched under the flag of truce. . . . Military passes were required for these relief convoys departing from Government-held bases and their issuance was long delayed by Government field commanders. Even after their issuance in response to appeals to top military authorities of the Chinese Government, these passes were often cancelled without explanation. Upon their final reinstatement the ultimate sanction was imposed by still another branch of the Government to prevent the relief of famine in an "enemy" area. The Chinese air force repeatedly bombed and strafed UNRRA-CNRRA convoys bound for communist destinations.[85]

As acrimony increased, Nationalist officials delayed or withheld shipments of UNNRA flour and grain to Communist-controlled areas on the North China Plain. The official institutional history of UNRRA, written in 1950, recapped the challenges:

> The most formidable issue in implementing UNRRA's policies in China arose in connection with the attempt to achieve non-discriminatory distribution to the people living in Communist-controlled areas. This presented an exceptionally difficult and perplexing problem since the work entailed was subject to forces beyond the control of UNRRA. . . . The course actually followed by UNRRA . . . was [to] exert continuous pressure

upon the [Nationalist] Government to distribute equitable quotas of UNRRA supplies to the people in Communist-controlled areas, and to furnish practical assistance . . . in carrying out such distribution. UNRRA's effort to induce the Chinese Government, engaged in civil war, to adhere to the principle of nondiscriminatory distribution, as applied to the people in "enemy" territory, was, however, successful only to a slight degree.[86]

Through it all, the dikes held. Two hundred thousand workers moved a total of 22 million m³ of earth to shore up the dikes. Six million refugees returned to the formerly flooded area.[87] The only question that seemed to remain by mid-1947 was which political faction could most effectively lay claim to the legacy of Huayu-ankou. Which party could appropriate the Huayuankou disaster to inform its narrative of reconstituting the Chinese state and nation? In the CCP narrative composed during the Chinese Civil War, the Nationalists' destruction of the Huayuankou dikes and the CCP's central role in rehabilitating the lower Yellow River were touchstones, central to the Party's history of creating a revolutionary identity with peasants of the North China Plain. During the Civil War, the North China Plain region remained controlled by the CCP. After the breach was closed, millions of returning refugees found a landscape transformed. They found "people gone, the livestock drowned, crops destroyed, houses, roads, bridges, public buildings all ruined. Much of the land was uncultivated . . . the irrigation channels had been destroyed and the soil was covered with silt."[88] The CCP embedded dike rehabilitation in a militarized revolutionary struggle. Slogans among dike workers in Communist-controlled areas of the Yellow River valley reflected this ideal: "With one arm we carry a rifle, [with] the other we carry a shovel"; "With blood, sweat, and tears the workers battle the Yellow River and Chiang Kai-shek." Such slogans equated the river with the "bandit" Chiang and also signaled that it would be the workers (many of them peasants) who would achieve the victories over the waters and over him. This sort of revolutionary discourse would be reproduced to advance the transformation of the Yellow River and the North China Plain after 1949.[89]

## Postwar U.S. Engineering Advice

On January 5, 1947, a delegation of American engineers visited the Huayuankou worksite. The Americans were in China on a related but quite different project. They traveled to China as proponents of multipurpose river development that was one component of U.S. postwar commitment to global development programs. The delegation's visit was devised by T. V. Soong, president of China's Executive Yuan (Cabinet), and Ralph Tudor, chief engineer of the Morrison-Knudsen Consulting Group in China. Morrison-Knudsen was best known as a principal in the construction of the Hoover Dam. Several years later, in 1954, a *Time* magazine cover story entitled "Construction: The Earth Mover" described Harry Morrison, co-founder of Morrison-Knudsen, as "the man who has done more than anyone else to change the face of the earth."[90] Earlier, the American-trained engineer Shen Yi, serving as a senior Nationalist official, argued that China should use postwar access to U.S. aid to develop the Yellow River on the TVA model.[91] In the charge given to the American engineers Shen described the Yellow River as "China's greatest and most important development problem.... [Any plan] should be considered on a broad basin-wide basis, taking into account developments and benefits in every respect, including flood control, irrigation, navigation, power development and also industrialization, economics, and the complex sociological aspect."[92] The complex "sociological aspect" was a clear reference to the CCP presence on the North China Plain. Elsewhere Shen wrote that multipurpose river development was the only way to maximize benefits to mankind. He pointed to numerous examples in the United States, including the Central Valley Project (California), the Big Thompson–Colorado River project, and the Boulder Canyon Project (Hoover Dam). Shen also noted that other countries, such as the Soviet Union and India, had already implemented this type of development scheme.[93]

The American engineers, constituting the Yellow River Consulting Board (Huanghe guwen tuan), were committed and experienced apostles of multipurpose water development. Eugene Reybold had served in the U.S. Army Corps of Engineers since 1926 and had been appointed chief of engineers for the U.S. Army during World War II. Reybold gained renown for his work on the Mississippi River and was decorated for a stint with the U.S. Army Corps of Engineers. James Growdon was a captain of the U.S. Army Corps of Engineers during World War I, when he earned a Distinguished Service Cross for completing a bridge over the River Vesle under German machinegun fire. By 1946, Growdon was the chief hydraulic engineer for the Aluminum Corporation of America (Alcoa). However, of the three engineers, John Savage was perhaps the best known. Savage spent most of his career at the U.S. Bureau of Reclamation, where he designed the Hoover, Shasta, and Grand Coulee Dams. John "Dam" Savage had been on the forefront of disseminating the gospel of multipurpose river development around the world. In addition to projects in India, Afghanistan, Australia, and elsewhere, Savage had developed plans for a massive "Yangtze River Gorge Project."[94]

On January 17, 1947, the "ambassadors with bulldozers" submitted their report. The Consulting Board's "Preliminary Report on the Yellow River" emphasized the construction of reservoirs in the middle stream, along with a fixed riverbed in the downstream, which would be straightened as much as possible.[95] The consultants also suggested a series of dams and reservoirs in several gorges on the Wei and the Yellow Rivers. These projects would help control flood crests and downstream sedimentation, as well as encourage hydroelectric generation and irrigation development. In recommending project sites, the Board specifically rejected the Sanmenxia site (most rigorously put forth in the earlier Japanese plan), arguing that a project at this site would not be helpful in managing silt behind the dam. The Americans similarly rejected the entire series of dam construction projects proposed by the Japanese, by arguing that their lifespan would be rendered unacceptably short by reservoir silting.

The American emphasis on large-scale construction of dams and reservoirs was a fundamental tenet of multipurpose engineering. The focus on midstream engineering also represented one pole of the debate over Yellow River management as it developed in the postwar period. Perhaps the extreme expression of this perspective

Ambassador with a bulldozer: Harry Morrison. *Time* Magazine, May 3, 1954. (© 1954 Time Inc. Used under license.)

was the plan formulated by the Japanese East Asian Research Institute that focused on the development of hydroelectric capacity to fuel resource extraction (mining) and industrial development. The other side of the debate focused on the flood- and sediment-regulating capacities of dikes and stabilized beds. The report disagreed with the notion of upstream reservoirs, arguing that silt deposition in the downstream bed would inevitably occur as multipurpose use reduced downstream flows.

An American engineer, John Cotton, was invited by Shen Yi to comment on the delegation's report. Cotton disagreed with his

American colleagues about flushing silt to the sea. Cotton believed this would be a waste of a valuable resource. Cotton's emphasis was on soil conservation measures that would stabilize silt deposits on the loess highlands. He also proposed constructing a series of silt-detention reservoirs on Yellow River tributaries. Instead of flushing the silt down the Yellow River, the detention reservoirs would eventually fill up with silt that could then be cultivated. Cotton supported the idea of the eventual construction of a large hydroelectric facility at Sanmenxia, but not until the silt load of the Yellow River had been adequately reduced.[96] All these engineers believed in the gospel of multipurpose water development and were fully committed to its realization. But within this broad agenda, there was no consensus on how to reach this goal.

As we have seen from our review of differing Japanese, American, and Chinese water management plans, what mattered was the definition of ultimate goals: economic, social, and political. In other words, the precise terms of resource exploitation would be constructed on the basis of larger social, economic, and political aspirations. In the immediate context of postwar China, the resolution of these issues, or at least the parameters of a continuing debate about water development, would be powerfully conditioned by the defeat of the Nationalist government by Mao Zedong and Chinese Communist forces. After Mao's formal declaration of the founding of the People's Republic of China on October 1, 1949, the values and goals of the new regime would shape the management of water resources for the Yellow River valley and the North China Plain.

The century between the course shift of the Yellow River in 1855 and the revolutionary victory of the CCP in 1949 witnessed profound, interdependent ecological and political transformations. From the broadest perspective, the breakdown of Yellow River control resulted in increasing economic immiseration for villages on the North China Plain. By the mid-twentieth century, the economic privation caused by floods, droughts, and famine was aggravated by war that sparked wide-scale migrations of local populations from the North China Plain. For those who stayed, it was a bitter struggle to survive.

This was the century of "China's humiliation." Following its defeat in the Opium Wars in the mid-nineteenth century and an enormously destructive conflict with the Taiping rebels, the imperial state struggled to formulate a new statecraft that could incorporate China into the international order of the late nineteenth century and that could cope with the increasingly heterodox social and cultural forces introduced by this process. After the 1911 revolution, this struggle only intensified, manifesting itself in the country's political disintegration during the warlord period, Japanese invasion, and Civil War. The landscape of the North China Plain and the political, cultural, and social landscapes of China were all intertwined. Each acted on the other in a dialectic that, in the end, generated more misery and deprivation on the North China Plain. Pressures on the hydraulic system were already building by the 1855 disaster. And by the early nineteenth century, Qing expenditures on regulating the Yellow River had skyrocketed to well over 10 percent of all imperial expenditures, suggesting a desperate though inevitably insufficient attempt to maintain equilibrium in the river system.[97] Over time, the imperial government abandoned its traditional policy of redistributing internal resources to support ecologically vulnerable regions of the empire, such as the North China Plain. At the same time, the weakened imperial government withdrew from its traditional role as guarantor of the ecological stability of the North China Plain. The emergence of the Yellow River as "China's sorrow" and of North China as the "land of famine" were the consequences of a deliberate change in statecraft priorities, the incapacity of central and local institutions to assume managerial responsibilities, and the inexorable forces of water and silt that continued to debilitate river control structures.

However, it was precisely during this period of hydraulic breakdown and national humiliation that the germs of a reconstituted system, aqueous and political, emerged. The twentieth-century internationalization of China introduced a set of ideas, practices, and institutional forms that were at once contradictory, legitimizing, and adaptable to tradition. The notion of river training (which had been widely employed in Western ideas about river management beginning in the late nineteenth century) had a long tradition in a cultural orthodoxy in China that valorized the conscious shaping or training of character, both of the individual and of the natural world. Even more, this tradition legitimized human intervention in the natural world to create a sense of harmony with, or complementarity to, the agricultural pursuits of China's farmers—the economic and moral backbone of the realm. Therefore, during the late imperial and Republican periods, the introduction of Western practices of modern science and technology was perhaps less revolutionary than transitional—in the sense that it extended imperial-era commitments and practices in new directions.

Western science and technology has engaged in a dialogue with native traditions in a variety of cultural contexts to generate a synthetic outcome.[98] But such a recognition should not obscure the transformational potential that the incorporation of modern science and technology into enduring frameworks could have on the landscape of the North China Plain. The physical transformations fueled by this synthesis would be on full display after 1949, but many of the component elements leading to this resolution first appeared during the Republican period. For example, the tools and methods of measurement supplied data on which to base the design of structures. New developments in steel, concrete, fossil-fuel-powered earthmoving equipment, and turbines all suggested a dramatic expansion in the range of engineering solutions available to hydraulic engineers. These methods and tools were introduced to China by foreign technical specialists, but they were quickly adopted and embedded in Chinese practice by a growing cadre of civil and hydraulic engineers who had been trained in Europe and North America and by a growing number of engineering and technical training institutions established in China during the Republican period. In terms of governance, engineers such as Li Yizhi melded the traditional ideal of centralized sponsorship of Yellow River control with the emergence of valley-wide administrative structures (based on U.S. models like the TVA) to argue for the formation of a centralized Yellow River management agency. This kind of statism envisioned by Nationalist leaders, and ultimately accomplished after 1949, drew on the theories and techniques of modern hydraulic engineering and once again focused attention on developing the interior spaces of China—a goal that had been largely abandoned in the mid-nineteenth century.

The second transition in water management that occurred during the Republican period that shaped the transformation of the North China waterscape was the cultural redefinition of the landscape. The centering of the Yellow River and North China in the conception of the nation during the Republican period complemented the myth of Yü the Great and his role in creating the basis of Chinese civilization by ordering the rivers of the North China Plain. Archaeological discoveries (such as Peking Man) in North China were seized upon by many intellectuals to affirm the notion of a singular Chinese race and culture, which had been nurtured in and which had spread out from the Yellow River valley. The seeds of a state orthodoxy identifying North China with Chinese civilization would reach full maturation in the late 1940s. North China was where the CCP "ate bitterness" in difficult ecological conditions. This was where the Party staged guerrilla warfare against the forces of imperialism and where the party launched its decisive battle against the "reactionary" forces of Chiang Kai-shek. The new China that arose from the disorder of the century of humiliation was a China where the rivers ran unregulated. The power of the history of this landscape and its identity with the Chinese nation would be unceasingly employed by the CCP to exploit the productive potential of water in rebuilding the wealth and power of "new China" after 1949.

Last, the full exploitation of the productive potential of waters was a notion adopted by hydraulic engineers and economic planners in China during the Republican era. Trained in China or abroad, Chinese engineers were part of an international network of engineering expertise that was increasingly influenced by the perspectives of multipurpose water projects. Transport and irrigation had been developed along the Yellow River during the imperial period, but the new technologies available to twentieth-century technical experts and economic planners dramatically expanded these possibilities. There was no single engineering recipe for all regions. Engineers like John Freeman, Hubert Engels, Li Yizhi, and Shen Yi became aware of the unique challenges of the Yellow River and realized that these challenges necessitated variations on what might otherwise be considered immutable laws of hydrology and hydraulic engineering. Such engineering plans were firmly embedded in the particular economic, social, and political goals held by their sponsors. In other words, engineering was socially constructed. The broader economic, social, and political goals of the CCP would influence how water was managed on the North China Plain. But did a new synthesis of tradition, science, technology, and institutions initiated during the Republican period lead to a new hydraulic and social equilibrium on the North China Plain after 1949?

# 4  *Making the Water Run Clear*

IMMEDIATELY AFTER 1949, the CCP embarked upon a program of consolidating a national identity and building a state dedicated to achieving communist modernity. It was a difficult project. Under the leadership of Mao and a small cadre of extraordinary revolutionary leaders, the Party had already managed what seemed impossible a decade earlier. It was under this leadership that China freed itself from the shackles of imperialism and "feudalism." Engineering the economic and social bases for a breakthrough to communism required continuous struggle and revolution. This would be Mao's life work. When we look at the events of October 1949, when Mao and senior leaders stood high upon a rostrum outside the gates of the old imperial palaces of the Forbidden City announcing the birth of the People's Republic of China, we can better understand how Mao and his fellow revolutionaries believed that such a profound transformation could be possible. The privations of the Long March, the heroism of guerrilla warfare, and the victory in civil war inculcated a seemingly unbounded optimism. If the Chinese people believed the promise of egalitarian abundance, if they were disciplined in the arduous task of struggle, and if they retained fealty to Party leadership, all obstacles, whether material or cultural, could be overcome to recast China. The hard part for Mao and the Party was *how* to get there: What would be the means to these ends? How to inspire? How to enforce discipline? Where to invest? How to develop resources? Different actors had different answers to these questions, and the struggle between the participants in these debates periodically threw the country into paroxysms of wide-scale violence.

The struggles to build communism in China transformed the landscape. All Chinese who were sufficiently concerned with China's existential struggle argued that enriching the country by increasing economic output was an imperative. Most argued that primacy be given to developing a modern industrial sector premised on robust agricultural output. Growth in both sectors required aggressive mobilization of resources. The broadly articulated goal of generating greater national product necessarily meant an increased call on the water resources of the North China Plain. Did the particular social and political goals of the CCP, namely, building a modern communist polity, lead to indelible changes in the environment? Adding a further layer of complexity, did alternative development paradigms championed by different constituencies within the Party generate unique consequences for the ecology of the North China Plain?

In exploring how the water resources of the North China Plain were managed after 1949, two related questions arise: How did state- and nation-building goals shape the adminstration of water on the North China Plain; and what were the longer-term consequences for the region? The first question has a material and moral dimension. Management schemes were a direct outcome of particular economic goals. The goals drove technology choice and organizational form. At the same time, the moral dimension focused on how the quest for these material outcomes furthered the creation of communist man and society. In other words, how did the shared effort and sacrifice of mass-based digging of irrigation ditches create a moral and social consciousness necessary to build communism? The division of the question into material and moral dimensions is indeed forced, and the differentiation is perhaps more implicit than explicit. But to the degree that we can come to an understanding of the relationship between state- and nation-building and the management of water resources, we will be able to broach broader questions about continuity and change of water management in China after 1949. For example, were water resources managed in a unique fashion? Was there a certain content and form of communist hydraulic engineering that was distinct from traditional approaches? Was there a recognizable Chinese/communist approach to managing water versus a Western/capitalist pattern? Perhaps more tangibly, what were the comparative dynamics of constructing a big dam on the Yellow River and one on the Colorado River? Were there unique means to unique ends? And, ultimately, were there unique environmental consequences from these means and ends?

The answer to the various iterations of the continuity/discontinuity question is a definitive "yes and no." There is little doubt that China's modern institutional patterns, technology choices, developmental assumptions, and cultural values were adapted from tradition and contemporary global contexts. But a complicating factor lies in the dramatically different development trajectories that resource management followed during different periods in China after 1949. We can identify certain dualities, or alternative approaches, to the management of water that reflected different paths to building communism. At different moments, we see the pendulum swinging back and forth between capital-intensive and labor-intensive construction, international cooperation and self-reliance, technical expertise and ideological commitment, technical knowledge and indigenous or local practices, and central and local management. The different technology complexes (i.e., technology choice, administrative organization, use of cultural symbols and myths) were driven by values that were, in effect, engaged in a technological dialogue. At certain times, Mao consciously formulated the desirability of a synthesis, or coexistence of both complexes, when he called for development to "walk on two legs."

At the same time, the landscape of the North China Plain was a canvass upon which to employ symbols and myths to create an idealized landscape of production. In epic discourse, with formative roots in Yan'an, the North China Plain was the landscape of a "second creation"—a recasting of the original epic narrative of the genesis of Chinese civilization that centered on the myth of Yü the Great. Second creation stories, or narratives of conquest, often celebrated "regeneration through violence" and through stories that define the self against an "alien other."[1] Borne of the violent crucible of Yan'an, the Party came to power at the barrel of a gun and continued to make revolution after 1949 in a series of violent struggles. The "alien others" of this struggle were Japanese imperialists, Chiang Kai-shek, American provocateurs in Korea, and the uncontrolled and underutilized waters of the North China Plain. The Yellow River was a totem, a cultural touchstone for collective identity. An uncontrolled Yellow River and an impoverished North China Plain had to be rectified, a process of transformation through a violent re-creation of the landscape to serve the nation and the building of communism, an altered landscape that, both in its means and ends, symbolized the productive and egalitarian possibilities of "New China."

We can find symbols, myths, and values, transcribed upon a landscape in a variety of cultural contexts that serve the aspirations of modern states and nations. But we must also recognize the uniqueness of the Chinese context. Along with the Nile, Amazon, and Volga Rivers, few other rivers have become so identified with national consciousness as the Yellow River. Few other comparable agricultural regions have existed in such an ecologically vulnerable state as the North China Plain, and no other landscape in the world, in both its spatial and temporal profiles, has been as transformed by humans as the North China Plain. Finally, no other region has required as much investment to retain some semblance of ecological equilibrium as the North China Plain. This commitment to ordering the waters, one that has been sustained for centuries, must necessarily continue. A complete breakdown of the artificially managed hydraulic system of the North China Plain would create the sorts of disorder that polities are not likely to survive. Thus, this commitment to water control is a distinctive Chinese reality, representing a long continuity.

## The Huai River Project

The political transition of 1949 did not stop the floods. In late 1949 and in 1950, both the Huai and Yellow Rivers flooded. The immediacy of the inundations motivated the new government to move quickly to stem the flood menace on the North China Plain. Party leaders were acutely aware of the social and political instability of the Yellow and Huai River valleys. After all, they had capitalized on this sense of immiseration to establish a revolutionary base in the region. But CCP leaders did not want to tempt fate. They sought to stabilize the region as quickly as possible.

Several meetings in late 1949 and early 1950 set the immediate agenda for the country's management of water. At the Consolidated Meeting of Water Conservancy in the Liberated Areas, the minister of water conservancy stated that the fundamental direction for water conservancy was flood protection. Also in late 1949, Zhou Enlai, addressing the Sixth Meeting of the Government Administrative Council of the Central People's Government, implored water conservancy cadres to focus on preventing floods as a precondition to the productive potential of multipurpose development. In that address and at a talk before representatives from the natural sciences in 1950, Zhou retold the legend of Yü the Great, who had passed the door of his own house three times without stopping as he labored to harness the great rivers of the North China Plain. Zhou roused the comrades by declaring, "Our achievements will be nothing less than those of Yü the Great."[2]

Like the Nationalist government two decades earlier, the new government had control of the Huai River as its first priority of water management. Control structures on the Huai were perhaps the worst in the region and had not been reinforced since the 1855 Yellow River course change. In 1950, after a large flood, Mao reportedly initiated water control efforts on the Huai with the pithy statement: "We must control the Huai" (women yiding ba Huaihe xiu hao). This was the new communist state's first attempt to organize a mass labor effort. Although in later years, the mass mobilization of human and material resources on projects on the North China Plain would demonstrate an unprecedented capacity to manipulate local society, the 1950 Control the Huai Movement (zhi Huai yundong) suggested a state and party struggling with the challenges of bureaucratic organization, labor recruitment, and management. The immense task of implementing a management plan that included large-scale reservoir and dike building was complicated by the sociological dynamics of the North China Plain, particularly in the Huai River valley. After decades of environmental breakdown and social chaos, local institutions had collapsed. The CCP had indeed been able to capitalize on the breakdown of social, economic, and political order to organize guerrilla warfare against the Japanese

Model workers from the Huai River Engineering Project.
(Courtesy of the International Institute of Social History,
Amsterdam)

and to defeat Chiang Kai-shek. But after 1949, this instability presented a challenge to Party cadres as they carried out the more pro-
saic tasks of building water control structures. Archival sources
reflect both enthusiasm for work accomplished and an admission
of problems. Although originally planned for two years, the entire
project lasted five years and focused on constructing upstream reservoirs and reinforcing downstream dikes.[3]

The focus on upstream storage and downstream dikes was premised on preventing famine. Upstream reservoirs would provide
water for irrigation during dry periods, as well as retention of upstream flood crests. The challenge that the new government faced
in restoring agricultural and political stability was suggested in a
variety of reports. A major problem with the Huai project centered
on worker mobilization. Initially, relief was extended to all volun-
teers willing to exchange labor for wages and sustenance. However,
it quickly became apparent that work efficiency was low. According

to a 1953 report, "cadres lacked experience, time was too short, cadres and workers too rushed, tasks too many, workers ill-prepared for work, and workers not motivated." The same report summarized worker attitudes: "seven days thinking, eight days staring, nine days waiting, and ten days absent." Further, "the workers didn't want to be sent too far away, they couldn't return home for the Chinese New Year, they didn't want to work hard, and they didn't want to work too long."[4] At some locations, tools were lacking, fights broke out between peasants from different villages, cadres lacked sufficient technical knowledge, they were capricious in work assignments and pay, and wages were simply too meager.

Party workers sent to villages and work sites engaged in a variety of "educational efforts" and used other motivational tools to increase efficiency. These efforts included the identification of model workers, posters, shortwave radio broadcasts, slogans and songs, speeches, newspaper readings, stories, and political instruction based on the CCP general line (zonglu xian) and the draft national constitution. The embedding of Huai water management into a discourse of historical and contemporary enemies of the Chinese peasantry permeated these motivational efforts. The destructiveness of the Huai River was a result of China's feudal, semifeudal, and semicolonial past. "Those landowners under feudalism squeezed people and seized their lands. The peasants had to find places to survive, so they planted on mountains but were not aware of water and soil balance. Moreover, some local tyrants and despotic gentry occupied mountains. They cut trees and filled lakes to create farmland; they fought to control water. . . . The Huai system was destroyed because of feudalism and outside invaders."[5]

Constructing external threats that equated water conservancy with saving the nation infused newspaper reports and the exhortations of high officials. The Huai flood of 1950 coincided with the U.S. entry into the Korean conflict, and Chinese official rhetoric linked Huai water control efforts to fighting American imperialists:

the American empire was a threat to our nation, and Chairman Mao's desire to control the Huai . . . [was a means] to resist the [American] empire. Only a country that knew the needs of its people could be strong, and the people would do everything for their country against their enemies. . . . His people knew he only thought of them. Those who fled the Huai River valley [during the Anti-Japanese War] returned. . . . People from places not affected by Huai disasters voluntarily helped on this Huai project. They all gathered and worked hard, not only for this Huai project but also to resist the American empire. Working on the Huai project and resisting America were for the same reason—to give people a better life. Working on the Huai sprang from the same spirit as resisting the American empire.[6]

The drive to complete the Huai River project quickly was also conditioned by a lack of resources. The implications of this scarcity were most evident in the construction of large upstream dam and reservoir projects. Sixteen dams with a total storage capacity of 32 billion m$^3$ were envisioned for the long term, with initial construction of three dams/reservoirs in the first stage (1950–1952). Shimantan was the first project completed, followed by Fuzeling. From the recollections of one participant, Zhang Chongyi, we get a sense of how these early projects were pieced together. Party organizers expended considerable effort in creating an identity of interests between peasant workers and the goals of the state. Zhang recalled mandatory study of Mao's "Theory of New Democracy," "Discourse on Coalition Government," and "On People's Democracy" "that safeguarded completion of the dam [Fuzeling] and of maintaining the quality and quantity of work." On the more mundane matter of construction, Zhang stated that there was inadequate knowledge of how to build this first arch dam in China: "The only thing we had was an arch dam design from America. However, we wanted to follow the idea that Chairman Mao had of following the mass line [*qunzhong luxian*] . . . we could not find any information about conservancy from the books in the country." As a result of the rather ad hoc nature of the work, "frustrations and accidents occurred" during the first phase. "The central government summoned us, I was one of those people . . . and we were told that we should start over." Lacking sufficient technical expertise, the project turned into a training site. Attempting to address the shortage of expertise, a group of more than fifty students from Shanghai Jiaotong University, and a professor were sent to the site in 1951. They originally came for one year of practical training, but the students were so helpful and [were] needed for construction. Therefore, they [project leaders] formulated a way to keep the students. They devised a Fuzeling University. The construction site was like a university. The students were able to learn skills [ranging] from hydrology to planning and design. Students would work and study at the same time. All the people, intellectuals and workers, were teaching and learning among themselves. In order to learn technical skills, the technicians studied Russian and tried to gain some knowledge from Russian books.[7]

Themes of self-reliance and economy pervaded Zhang's recollection of the Huai construction. He explained that most of the project material came from China, and added, "We wanted to make sure we saved expenses on both materials for the dam and on the cadres. We also tried not to use the 'public' money [state funds]. The only time we spent 'public money' was on a Chinese opera from Shanghai to play a show for the workers." Economy was practiced at all levels, for example, "the workers picked up nails that dropped on the floor after pulling farm shelves down [from relocated households]. After they gave the nails to the team leader, they were ordered to pick up more nails. . . . Another thing was that lots of wood floated away because of a large flood in 1953. The command post sent a group of people searching for the wood."[8] The results of these efforts were mixed. The dam at Fuzeling reservoir required reinforcements over the next several decades. Two decades later, the Shimantan Dam, along with a series of Huai valley dams, failed disastrously in 1975.

The sense of "learning on the fly" was reinforced by the administrative and organizational adjustments that were made throughout the Huai Project. Perhaps the greatest adjustment that affected worker mobilization was a series of wage increases that ameliorated problems of recruitment and work discipline. Wage increases resulted in worker efficiency that was so much greater that the project needed fewer workers. Recruitment of workers declined to 4 percent of affected agricultural populations from 10 percent.[9] Lower recruitment targets allowed Party organizers to sign up better workers and served to increase the presence of the Party (CCP members constituted between 5 and 12 percent of the workers). Increased efficiency and smaller numbers of workers also allowed planners to eliminate midlevel supervisors who had come under increasing criticism for their lack of technical knowledge and experience.

The Huai River project set a precedent that would largely be replicated, with important variations, in the decades to come. Most generally, conducted in the context of financial stringency and imperialist pressures, it suggested the spirit of the Long March. Slogans at the worksites emphasized speed, size, and self-reliance. The mass mobilization of roughly 2 million laborers every season of the project suggested the coercive reach of the state and party bureaucracies, and their capacities to organize large-scale projects in a relatively short period. A strong sense of self-reliance was reflected in verbiage that conflated the war on the Huai with the war on American imperialists in Korea. Indeed, conquering nature was the moral equivalent of Chinese troops battling the forces of imperialism. For a Party conditioned by violence, for a Party that grasped victory through violence, the projection of imagery and metaphors of violence upon the landscape was an extension of this formative culture. Participating in such "battles" on the home front and abroad was to contribute directly to the creation of a New China under the leadership of Chairman Mao and the Party.

## The Yellow River Conservancy Commission

After the revolution, the Yellow River Conservancy Commission (YRCC) maintained the institutional structure of its Nationalist-era predecessor. However, with regard to personnel, there was both continuity and change. Li Yizhi, the father of modern hydraulic engineering in China, was dead. Some YRCC officials, like Shen Yi, had fled to Taiwan with Chiang Kai-shek and the Nationalist government. Others, like Zhang Hanying and Li Fudu, remained on the mainland to serve the new government. But a new cadre of technical experts filled increasingly important leadership roles in the YRCC. Two of them, Wang Huayun and Qian Zhengying, managed to have long careers through the political turbulence of the Mao era. Numerous reports, published during his career and afterward, anointed Wang (1908–1992) as the latest in the venerated line of water heroes, a worthy successor to Li Yizhi. Wang became the first director of the YRCC, after having served as the director of the Ji-Lu-Yu (Hebei-Shandong-Henan) Liberated Area YRCC since 1946. Much of the work of Wang and his colleagues focused on strengthening dikes along the pre-1938 Yellow River channel, in anticipation of the plugging of the Huayuankou dike breach.[10]

The other expert, Qian Zhengying (1923–), was born in Zhejiang Province. Her father received a doctorate in engineering in the United States and returned to China to work for the Nationalist government. Qian studied civil engineering at Datong University in Shanghai. Following a brief stint working for the Nationalist government, she joined the CCP in 1941 and immediately served in various technical capacities with the Red Army in the northern Anhui/northern Jiangsu region. Her dogged determination on the Huai River during the Anti-Japanese War and her efforts along the Yellow River in the postwar period propelled a rapid rise in the ranks of the pre-1949 Communist bureaucracy as she was appointed to lead various river conservancy units. In 1950, Qian was dispatched to participate in the Huai River Project, in which she was a leading designer of large dam and reservoir installations, such as Shimantan. Only thirty years old, and one of the few women to reach such a position, Qian was appointed vice-minister of the Ministry of Water Resources (later the Ministry of Water Resources and Electric Power) and later minister. Her fingerprints can be seen on all the major water projects on the North China Plain.[11]

In the years immediately following the 1949 Communist victory, the YRCC turned most of its attention to controlling downstream floods. The years between 1950 and 1957 witnessed the first concerted dike construction under the new government (two additional efforts were staged, in 1962–1965 and 1978–1985). As in the Huai River project, the principal concern was to mitigate the risk of famine. Organization and labor recruitment benefited from the lessons of the Huai project, but the particular challenge on these Yellow River projects was a technical issue: how high to configure the dikes. The guiding principle for downstream drainage was "broaden the channel, strengthen the dikes" *(kuanhe, gu ti)*. More similar to the perspectives of Hubert Engels and Li Yizhi than to those of Pan Jixun and Freeman, this notion of broadening the channels also included the idea of creating downstream flood diversion reservoirs to relieve peak flows.[12]

In the early years of the People's Republic of China (PRC), annual dike reconstruction projects on the Yellow River were organized during the postplanting (March–June) and postharvest (October–December) periods. Each county assigned rural cadres to organize labor teams drawn from riparian villages. Local Party officials also conducted propaganda, requisitioned equipment, and organized food and rest facilities for the workers. Cadres and subcounty-level leaders were also required to engage in political study and on-site training. At the worksite, YRCC personnel were responsible for monitoring the quality of work and for making adjustments to compensation. The evolution of the worker remuneration schemes suggested flexibility and willingness to offer material incentives proportional to group and individual labor output. Drawing from a traditional system of levies placed on local villages and a sense of voluntarism inculcated during the anti-Japanese War, a new system was ratified in 1951 that paid labor battalions according to the amount of work done. In turn, the battalion distributed pay to individual members, again, according to the output of labor. Under the revised system, the general rule of thumb was that every 16 wheelbarrows of dirt carted equaled 1 square meter (m²). For every 1 m² of dirt workers were paid 1.25 *gongjin* of millet (1 *gongjin* = 1 kilogram). Healthy workers could haul 5 to 7 m² per day, thus earning 7 to 8 *gongjin* of millet each day.

Workers who regularly exceeded quotas were idealized. Model workers and heroes were praised for their sacrifice to the nation. Wu Chonghua was a village leader in Pingyuan who, in the winter of 1950, led 170 peasants from his village to participate in dike rebuilding. The team worked 46 straight days and hauled an average

$8.21 \text{ m}^2$ of earth per day. Wu set a record of $25 \text{ m}^2$ in a single day. That spring, Wu's village suffered serious flood damage, but the 37,000 kg of millet that his villagers earned on Yellow River dike work sustained them through the disaster. Recognized as a model worker, Wu was rewarded for his sacrifice with a trip to Beijing and an audience with Mao Zedong, who was reported to have said, "If we accomplish the fixing of the dikes, only then can we prevent the Yellow River from flooding; if we can stoutly defend the Yellow River, only then can we safeguard our family and defend our nation."[13]

A 1952 story in *Renmin Ribao* (People's Daily) expressed the relationship between the control of nature and Party leadership in its report of the heroism of Xiao Zhenxing. As the secretary of a local chapter of the Communist Youth League, Xiao and several other comrades were walking to a Youth League meeting when they observed rising waters that threatened to burst a dike. Xiao volunteered to stay at the scene while he dispatched his companions to mobilize local villagers. As he walked atop the dike, a 5-foot break suddenly opened. Xiao flung his body into the breach. Holding back the great force of water, he staunched the flow until villagers, armed with tools, arrived to plug the gap. A villager was quoted as saying: "In the past our crops were drowned every year, then the people's government constructed levees and my crops have been safe. This year, thanks to Xiao, our crops can survive."[14] After declining offers of food and rest at the home of a local farmer, Xiao and his comrades hustled off to convene the Youth League meeting.

In 1950 Pingyuan Province organized a meeting to recognize 107 heroes and model workers (*ying mo*) for their role in China's liberation from the "bandit" Chiang Kai-shek, imperialist America, and the destructive Yellow River. The honorees were archetypes of the ideal peasant and worker under the guidance of the CCP whose qualities included selflessness, tenacity, and "class consciousness," as well as the capacity to organize, lead, and mobilize others. Reporting on the gathering, *Xin Huanghe* (New Yellow River) provided a representative profile of heroes and model workers. Demonstrating superior qualities of worker organization and mobilization, as well as a willingness to make personal sacrifices, Ji

Chengxing was the leader of a team that "worked well, worked much, started early, and ended late."[15] In May 1949, Ji experienced a serious arm injury. After refusing his comrades' pleas to rest, Ji proclaimed, "Even if my arm is in a sling, I can still work!" He devised a rig that allowed him to participate in tamping work with one arm. He and his team were awarded red banners, and Ji was selected as the Number One Hero of 1949 dike work. These brief narrative accounts of heroes and model workers are imbued with the spirit of Yan'an. The heroes overcame seemingly insurmountable odds and endured enormous personal sacrifices. The success of their efforts (through organization and mobilization) was celebrated time and again in a language and tone that evoked the epic struggle of revolution.

The celebration of heroes and model workers was accompanied by an exhibition of Yellow River flood control work in Xinxiang City (northern Henan) in 1950. The exhibition highlighted the goal of "transforming the Yellow River from a harmful influence to a productive power to benefit the masses."[16] Banners at the exhibit enunciated the purpose and proposed outcome of the show: "[People should] emulate the deeds of heroes and model workers, work hard to create new experiences, raise political awareness, raise technical ability, increase work efficiency, embody revolutionary heroism, gather the wisdom of the heroes and model workers, and further develop an activist spirit."[17] The exhibit embedded Yellow River control in a revolutionary narrative of a constructed past and an idealized future. The first room focused on the history of the Yellow River. Visitors viewed photographs that depicted "[the] bandit Chiang Kai-shek and the American imperialists plugging the breach at Huayuankou and allowing the uncontrolled release of water down the Yellow River channel into CCP liberated areas, the bombing of dikes by Nationalist warplanes in CCP liberated areas, and the pitiful sight of the people. You can see the righteous indignation of bitter hatred [for Chiang and the Americans] . . . the people's courage is aroused even more to march into the future." An article in *Xin Huanghe* quoted visitor reactions from the exhibition comment box. One visitor wrote, "I have never seen the Yellow River; today I have seen the grievous cruelty that the masses in the

Yellow River received from bandit Chiang. I am totally filled with hatred. After today I will certainly work hard to eliminate these cruel elements [of society]."[18]

The second and third rooms of the exhibition focused on the present and future. Tables and maps showed how flood control would result in large gains in agricultural production. After seeing maps on wind direction, rainfall amount, and sediment, one contributor to the comment box stated: "I couldn't believe controlling the river was so complicated—all these considerations!" According to the *Xin Huanghe* article, what particularly captured the attention and support of the people were maps of the People's Victory Canal (Renmin shengli qu), China's first large-scale water diversion and irrigation canal on the North China Plain, suggesting how the water of the Yellow River could be utilized for irrigation. Another comment noted: "I give the technician my respect! I give those dike workers my respect! I give the future workers my respect! I also want to become a worker!"[19]

Yellow River dike work was also used by the Party as a proxy for the creation of a new social and economic order. Party leaders encouraged the writing and performance of songs at the worksite to energize labor. Songs such as "The Great Race to Resist America and to Restore Dikes Song" ("Kang-Mei fu ti da jingsai ge"), "Building Up Dike Defense Song" ("Zhuqi tifang ge"), and "Tamping Earth to Control the Yellow River Song" ("Zhi Huang mingong dawo ge") reinforced, in simple and repetitive stanzas, the celebration of labor and its contribution to the realization of ideals of mutual cooperation and communal effort.

At the various flood control project sites, labor was socially transformative and created a sense of identity through shared sacrifice of the peasantry and the Party. Yellow River dike work was perhaps the largest single undertaking in the early 1950s to incorporate women's labor. Newspapers and reports made explicit comments on the contributions of women in controlling the Yellow River. Men and women were encouraged to invest the labor necessary to increase agricultural production and thus to strengthen the nation. State emphasis was placed on projects that benefited the most. The basic principle was that "responsibility was large for those projects

Women's Yellow River Dike Construction Brigade. From Shuili bu Huanghe shuili weiyuanhui, ed., *Renmin zhili Huanghe liushi nian* (Zhengzhou: Huanghe shuili chubanshe, 2006), 102.

that benefitted more" *(duo shou yi duo fudan)*. The ideal was a blurring of distinctions in authority by politicizing the act of labor. Emphasis was placed on the importance of peasant labor in creating the basis for robust agricultural production in service to the nation. Within the matrix of people's democracy, a unified country, a stable social order, land reform, and mutual cooperation, the Party had created the conditions for leaders and peasants to form a single whole. Employing a metaphor of the body, Minister of Water Conservancy Fu Zuoyi equated the integrated masses as a spine *(gugan)* held together by the vertebrate-like links *(gang)* of organization and technical expertise of Party cadres—and these links, individually and collectively, support the single body that digs, carts, and tamps earth to expand the agricultural basis of the nation.[20]

Newspaper reports spared no effort in reporting achievements in dike rehabilitation. In simple graphs and charts, *Renmin Ribao* summarized the accomplishments of China from 1949 to 1954: 2.9 billion $m^3$ of earthworks, 19 million $m^3$ of stonework, and 8.4 million levee and dike worksites.[21] Citing data from 1947 to 1955, Wang Huayun claimed 1,822 kilometers of restored and augmented dikes, which had entailed the application of 1.3 billion $m^3$ of earth and 2.6 million $m^3$ of stone, in service to guaranteeing the livelihood of 18 million people in the lower valley.[22]

In 1952 Mao toured the North China Plain. In fact, this was his first domestic excursion since the founding of the People's Republic on October 1, 1949. Mao and his entourage of senior Party and government officials departed Beijing in October 1952. After stopping in Shandong to observe newly completed dike work, Mao paused in northern Jiangsu to view the old joint bed of the Yellow and Huai Rivers. The next stop was Tongwaxiang, the site of the infamous 1855 rupture that forced the Yellow River to its northerly course. Perhaps the most frequently reproduced image from the trip is one of Mao sitting atop a summit of Mang Mountain overlooking the Yellow River near Zhengzhou. The Beijing–Hankou railroad bridge crosses the river in the background while Mao gazes upstream, seemingly toward Yan'an. Mang Mountain is also the site of imperial tombs from the Han, Sui, and Tang periods. The photo captures the aspirations of creating a modern nation, melding the recent experience of revolutionary struggle with the memories of an ancient race and civilization all centered on the North China Plain. Mao also visited the People's Victory Canal in northern Henan. After viewing the project, Mao reportedly stated that every county in China ought to have such facilities. He also instructed Wang Huayun, director of the YRCC, to "think big" about detention reservoirs, instructing Wang to build 90,000 small reservoirs instead of the 9,000 that Wang had suggested.[23] Reflecting an early recognition that the water resources of the Yellow River and North China Plain would be insufficient to sustain long-term development, Wang suggested to Mao that water could be diverted from the Yangtze to the Yellow River. According to Wang, Mao concurred: "The waters in the north are scarce, but water in the south is plentiful, so it's okay to borrow a little from the Yangtze."[24] Mao would subsequently return to the Yellow River valley in 1953, 1954, and 1955, to encourage Wang and his colleagues to re-engineer the waterscape.

## The People's Victory Canal

In addition to the Huai River and the Yellow River Dike Reconstruction projects, the early 1950s also witnessed the creation of an epic narrative emanating from the construction of the People's Victory Canal, which Mao also visited in 1952. Formally called the

Mao at the Yellow River, 1952. From Shuili bu Huanghe shuili weiyuanhui, ed., *Renmin zhili Huanghe liushi nian* (Zhengzhou: Huanghe shuili chubanshe, 2006), 102.

Yellow River Irrigation and Diversion project (Yin Huang guan'gai jiwei), the plan called for construction of a 200-kilometer canal from the Yellow River (west of the Beijing–Wuhan railroad bridge) to the Wei River (near Xinxiang in northern Henan). Roughly half the 50,000 m³ of water withdrawn from the Yellow River would be used to irrigate some 100,000 *mu*² of land. The other half would be released into the Wei River, eventually flowing to the Grand Canal. The project was the first expression of the overall post-1949 development thrust of transforming river waters from a "threat into a benefit." It was the success of this first large-scale diversion and irrigation project on the North China Plain that suggested to Mao and other Party activists the potential for similar engineering projects in the future.

The notion of diverting large amounts of water for irrigation on the North China Plain was not a new concept after 1949. During the Nationalist period, engineers familiar with pumping technologies led design efforts for a number of projects that siphoned Yellow River water for irrigation. In 1928, pumps were installed near Kaifeng that could irrigate 10,000 *mu*. In 1929, west of Huayuankou, similar irrigation pumps were installed. In 1933 and 1934, there were also pumps siphoning water to fields in Shandong.[25] But it was largely as a result of the Japanese occupation that plans were originally sketched for what would become the People's Victory Canal. In the early 1940s, the Japanese East Asian Research Institute compiled plans to develop the mineral and agricultural resources in China's interior provinces to feed industrial plants in the northeast. In addition to drafting preliminary plans for a large hydroelectric dam at Sanmenxia, Japanese engineers designed blueprints for a project that would divert Yellow River water for irrigation and transport.[26]

The Japanese proposal spurred vigorous debate among Chinese engineers after 1949. Those urging caution argued that the limited funds appropriated for water conservancy should be focused on the most urgent need: flood prevention. Others noted the risk of opening up sluice gates in the protective dikes of the "hanging river." Such an audacious project had never been attempted before and could compromise the integrity of the dike at times of high water. Still others argued that silt would clog intake gates at all points in the system. Supporters countered that such a complex project could be accomplished with modern engineering practices. Other supporters reasoned that humans could never completely eliminate floods, so it would be better to begin efforts to realize the productive benefits of the river.[27] Although documents shed little light on how the disagreements were ultimately settled, one has a sense that the project was promoted by Party and state actors operating under pressure to claim legitimacy for the Party by organizing the human, material, and technological resources that could chart a path toward agricultural and industrial modernization.

The rationale for the People's Victory Canal project was premised on a future of material abundance. In the Xinxiang region, for example, the "past" was a series of yearly flood and drought disasters, particularly during the rule of the "reactionary Nationalist

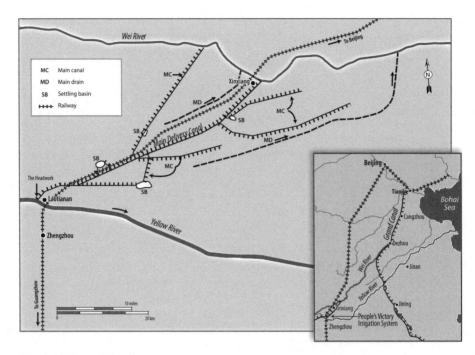

People's Victory Canal

clique." These conditions, plus the exploitation of the "Nationalist bandits," left a population destitute and homeless. In Dongzhuang village successive years of drought drove 106 households (out of a total of 147) to flee; 70 people starved, 16 households sold daughters, and 80 houses and 400 *mu* of farmland were sold. During all this, "landlords and the capitalist class took the opportunity to plunder and rob, concentrating large amounts of land into their hands."[28] At the same time, constructing the canal would afford the opportunity to tackle all forms of technical challenges that would inform future projects, including the management of silt. The political impulse took a familiar form: "Even more important [than the technical considerations], the project is the start of the desire of our great leader of our great Party, Mao Zedong, to transform the pernicious Yellow River into a beneficial river. Accomplishing this pioneering engineering feat will manifest the unarguable superiority of the basis of Marxist-Leninist theory as [the] guiding thought of the socialist system."[29]

In more practical terms for state and party, the potential contributions of the project to agricultural and industrial growth were critical. The central focus was cotton. Economic planners largely adopted Japanese development plans for the North China Plain—that is, increase agricultural output to feed industrial development. The Wei River region already produced significant quantities of cotton, but the goal was greater output that could be shipped to industrial centers like Tianjin via the Wei River and the inner Grand Canal. The need for the Yellow River Irrigation and Diversion project was dramatically underscored in July 1951, when an urgent directive from the Central Committee of the CCP called for an increase in cotton production because it had continually failed to meet demand from the yarn and textile industries. In response, the first phase of the Yellow River project was fast-tracked for completion by spring 1952, a year ahead of the original schedule. Instructions from the Central government read, "Water should be released in April 1952. From the perspective of the economy the purpose is to expand cotton growing fields in order to expand cotton production; from a more general perspective it is a method to overcome the problem of industrial resources; from the political perspective it is to combat the economic embargo of imperialist countries and to strengthen the alliance between industry and agriculture; and it is an important undertaking for the people's democratic dictatorship."[30] In contrast to the Huai project, there is little evidence to judge the level of coercion employed to complete the massive project in such a short period of time. The numbers at the worksite for the first phase (ending in April 1952), however, suggest the importance of the project to the state: It involved more than 500 CCP cadres, 10,000 laborers, 5,241 water channels, 4,099 kilometers of channels, 775,000 m$^3$ of earth, 80,000 horse carts, and 26,000 oxcarts.[31]

There was extensive coverage of the project in the state media. A recurring trope was the power of the Great Leader, who commanded the moral force to instruct, guide, and reform individuals, often using the idiom of Confucian familial roles of authority. For example, at the ceremony to celebrate the opening of the canal in April 1952, a local village representative is quoted by *Renmin Ribao:* "When digging began on the channel the masses said: 'The water of the Yellow River has no consciousness [*mei you liang*]; if you don't plug a breach well, the water will flow out.' Now after we see the Yellow River waters evenly flow past, the masses say: 'Under the leadership of the Chairman CCP Mao Zedong, the river is well-behaved. Just as we take care of our parents, we must certainly take care of the irrigation channel.'"[32] A similar report on the opening ceremony described how "people saw the even and regular way the water flowed through the channel, and they sang and shouted: "In the new China under the leadership of Mao Zedong, not only can he reform the loafer [*er liu zi*] into something good, but he can reform the Yellow River into something good."[33] The same article employed a simple anecdote to convey the moral and material connection between Party, Mao, and peasant: "There were peasants that ran to the canal and quietly ladled out some water, then they returned to their home to carefully taste the water to see if it was bitter or sweet, and to see if the precipitate was mud or sand. Once they tasted the water [they realized] it was sweet and the sediment was not sand but mud. They then ran back out to the street jumping and leaping."[34]

The Party had liberated China from foreign and domestic enemies and had liberated peasants through land reform. Now the Party was leading the construction of a socialist economy and society by defeating the repressive forces of nature on the North China Plain. Although agricultural output had improved since 1949, farming was still dependent upon the heavens for rain. "The peasants still were not able to conquer nature." However, because of the success of the Yellow River Irrigation and Diversion project, the area was now liberated from the constraints of nature, and the Yellow River Irrigation District "was leading the peasants down the path of guaranteed happiness and prosperity. . . . The success of the Yellow River Irrigation and Diversion project eliminated the traditional fear of the river, and allowed them [the peasants] to believe in the power of the masses under the leadership of the Party, to use scientific methods, and to grant them the strength to conquer nature and develop agricultural production."[35] In 1954, after two additional main feeder lines were added to the project, the "Communist Canal" and the "Red Flag Canal," Zhang Hanying reported phenomenal increases in cotton and millet production in the Yellow River Irrigation District. Zhang also stated that this was just the beginning, as he offered a simple vision of Yellow River water serving industrialization through the development of hydropower.[36] In addition, the presumed success of the irrigation project suggested greater potential for the reorganization of local society to advance efficiency in the agricultural sector. Indeed, the Party had begun promoting the formation of agricultural production mutual aid societies to encourage sharing of material and labor among households. The labor demands of irrigated agriculture were also cited by the Party in its promotion of mutual aid societies.[37] In a few short years, the presumed logic of expanding irrigated agriculture led to the communal organization of the countryside.

## The Soviet Union and the Creation of a Technology Complex

The first few years of PRC water management practices suggested a melding of continuity and innovation. At the same time, state policies reflected a certain caution predicated on the need to rebuild a war-torn economy while taking steps to forge a new social structure. Thus, land reform, occasionally violent but comprehensive in its outcomes, coexisted with tempered policies toward the private sector. A similar sense of restraint characterized state and Party approaches to ordering the waters on the North China Plain. To be sure, Mao and other Party elites were signally aware of the traditional legitimizing power of setting the rivers right. Other reflections of continuity included labor recruitment and organization and the continuance of an emphasis on downstream flood control within a system of dikes. At the same time, however, state planners and engineers championed the modern goals of multipurpose river development, with the ubiquitous call to "turn harm into benefit." Upstream and downstream flood control reservoirs were built on the Huai and the Yellow Rivers. Modern technology was utilized to build the People's Victory Canal, while state and Party bureaucracies penetrated local society in unprecedented ways to compel compliance with state goals. This admixture of continuity and change was not without tension. There were cases of local resistance to management efforts, as witnessed in the case of Huai flood control projects. There were also tensions centering on questions of technical knowledge and the politicization of professional engineers. And there were disagreements over the relative value of flood control versus multipurpose development. In many ways, these tensions were to be expected. They reflected foundational debates inherent in the efforts of the new state to define the structure and operation of technological systems that had been designed to support the creation of a new state system and new national identity. The long road to creating a new orthodoxy after 1949, plus a technology complex to buttress that orthodoxy, was a torturous journey.

One critical component of the technological complex of these early years (1949–1954) was Soviet technical advice. As suggested earlier, the notion of a technology complex connotes the entire range of goals, actors, material, institutions, and tools organized to support a single project or a set of projects.[38] The normative mid-twentieth-century multipurpose approach to water management in Europe and North America included components such as hydrologic science, material sciences, building technologies, structural design, mechanized building equipment, supply chains, transportation infrastructure, generators, sluice gates, rational organization, central administration, and careful management of information dissemination. The objective of this technology suite was to advance industrial development by exploiting the full productive powers of water. Considerable influence over China's technology complex during the 1950s emanated from the Soviet Union. Soviet technical advisors arrived in the early 1950s under the provisions of the Sino-Soviet Treaty of Friendship, Alliance, and Mutual Assistance of 1950.

A Chinese translation of the 1950 Russian text *Organization and Practice of Water Conservancy Engineering*, introduced the principles and ideals of water conservancy in the postwar Soviet Union. The text reflected the technical culture of the Soviet advisors dispatched to China, shaped by Joseph Stalin's aggressive push to construct large-scale installations as outlined in the first postwar five-year plan. The plan called for the reconstruction or construction of

thirty hydroelectric facilities. As laid down by various pronouncements (going back to the mid-1930s) on the execution of these large
engineering projects, the Soviet textbook articulated five guiding
principles of water management: plan-*ism*, industrial-*ism*, mechanized construction-*ism*, rapid construction-*ism*, and Stakhanov-
*ism*. Plan-*ism* called for all projects to be approved within the structure of the five-year planning processes. In the Soviet Union,
central planning allowed for the rational coordination of resources
within the overall developmental objectives of the national economy. Industrial-*ism* reduced costs and construction time. Instead of
relying on small-scale localized production facilities, construction
teams could adopt a contract system, in which the team entered
into agreements with units that had sufficient technical capacity to
manufacture and supply component parts and equipment on a
large-scale, standardized basis. This simplification of the supply
system reduced costs on any given project and provided a supply of
standardized, boiler-plate parts to accommodate multiple projects.
Mechanized construction techniques offered four advantages: increased labor efficiency, for equipment easily transcended the limitations of human strength; rapid construction; improved quality of
construction; and lowered construction costs. The 1946–1950 five-
year plan established standards for mechanization: 60 percent of all
earthwork, 95 percent of concrete preparation, 90 percent of mortar
preparation, 60 percent of concrete curing (with water), and 50 percent of painting. Workers should be organized on the principles of
mechanization—that is, labor organization should run with ma-
chinelike efficiency. Last, Stakhanov-*ism* referred to the rational-
ization of labor and technology to ensure that work was executed
with utmost speed and efficiency.[39]

Comparing the cultural context of large-scale Soviet technology
with that in the industrialized West, there is a similar inclination
toward a "fascination with technology as a panacea for social and
economic problems."[40] Although these desires are attributable to a
generic twentieth-century modernizing impulse, the social and political goals of the Soviet Union generated a unique "technology style"
by which Soviet political elites were obsessed with an "almost unbounded faith in science and technology. . . . The style of Soviet

"Today's USSR Is Our Tomorrow." (Courtesy of the International Institute of Social History, Amsterdam)

technologies was characterized by an aesthetics based on two concerns: an exaggerated interest in mass production owing both to
egalitarian ideological precepts and resource scarcities . . . and the
gigantomania that grew out of the fascination and commitment to a
technology of display."[41]

In the first years after 1949, there were signs that China was developing the first stages of a technology complex that drank at the
trough of Western and Soviet modernist notions of technology
while, at the same time, extending traditional practices. What
emerged in China was a hybrid technological complex informed by
modern science and technology, much of which was transmitted by
engineers who had been trained in China and abroad before 1949;
modern science and technology as conditioned by the Soviet experience; traditional Chinese technological approaches; modernist notions of the mutually supportive role of the state and of technology
in pursuit of political and socioeconomic goals; and traditional
statecraft practices in China. This technology complex can be gen-
eralized in the pithy Maoist characterization of "walking on two
legs." But the manic exercise of power between 1949 and 1978 complicated any linear progression of this synthetic approach. Conditioned by power struggles at the top of the political hierarchy, the
evolution of China's technology complex was characterized by acute
periods of disequilibrium in the country's approach to water management. For example, different periods after 1949 witnessed differing impulses in the relative balances of capital-intensive and labor-
intensive projects, self-reliance and international cooperation,
technical expertise and mass mobilization, and central planning
and local initiative. In other words, although the ends remained
consistent (i.e., building a modern communist society), the means
bounced between these dualities. These tensions were clearly evident in the differences of tone and emphasis offered by cadres in
the early 1950s on the issue of senior technical personnel. Writing
in 1952 on the importance of studying Soviet experiences, Zhang
Hanying asserted:

> Our water conservancy personnel, because most of them came
> from the old society, all bring with them thinking that is di-

vorced from politics and a technical perspective that is divorced from reality. The hydraulic engineering of capitalist countries is for the exploiting classes to extract profit. They [the engineers] cheat people by concealing their true nature as exploiters. . . . During the past period of rule by the Nationalist reactionary clique, there was very little water conservancy work, and there was very little for technical personnel to work on. The nature of this history allowed most of our old-style technical personnel not to have any connection in thought and feeling with the working people.[42]

Writing two years later in *Renmin Ribao*, Minister of Water Conservancy Fu Zuoyi struck a different tone: "I have run into many elderly engineers born in old China, and each one of them spent half a life in frustration. Some of them studied foreign hydraulic engineering, but until they became old there was never an opportunity to participate in any engineering [projects]. Today, however, there is engineering everywhere, and everywhere there is a need for technical talent. Old engineers everywhere receive respect. Young engineers have an even greater future. Those people who have studied technical specialties will affirm what I have said."[43] The differing attitudes toward professional engineers was a relatively early suggestion of the dual nature of the technological complex that would become manifest in the years to come.

On the ground, floods in 1952 and 1953 indicated continuing water problems on the North China Plain. Despite the endless reports enumerating the number of workers recruited and wheel-barrows of earth moved in dike and reservoir construction between 1949 and 1953, it became evident by the end of 1953 that the "regime's . . . dreams of China's age-old ambition: the regulation of the big rivers" remained unrealized.[44] The admission was made in the opening address by Fu Zuoyi at the 1953 All-China Conservancy Conference. The rather startling recognition of errors was made in the context of defining the role of water conservancy in China's first five-year plan (FYP), then under discussion.

The five-year planning process was an effort to replicate Soviet-style central planning. Like Soviet developmental ideals, the "general line" of China's first FYP emphasized socialist industrialization, national defense, and the development of heavy industry. Expanded agricultural output would generate capital for investment in the heavy industrial sector. Increased agricultural production would also supply raw material for industrial production and provide sustenance to urban industrial workers. And, last, heavy industry would be fueled by hydroelectricity. In the development of socialist agriculture, Fu argued that greater efficiency of irrigation through mutual cooperation would hasten the creation of agricultural cooperatives. Fu admitted that water control had earlier been planned and carried out in a rash manner, giving rise to problems in labor organization, output, and remuneration. These specific shortcomings, as well as a lack of sound engineering principles, led to construction outcomes that were sloppy and perhaps even dangerous. Echoing Soviet-inspired goals of efficiency and rationality, Fu demanded that water conservancy projects be based on a regularized bureaucratic process that emphasized data gathering, comprehensive planning, and efficient management. This Soviet technology complex stressed large capital-intensive projects with a decidedly multipurpose objective. The Yellow River Plan was heavily influenced by this planning approach.[45]

## The 1954–1955 Yellow River Plan

After several years of study and design by the YRCC, the legislature, the National People's Congress (NPC), approved the Multipurpose Plan for Permanently Controlling the Yellow River and Exploiting Its Water Resources (Guanyu genzhi Huanghe shuihai he kaifa Huanghe shuili de zong'er guihua de baogao) in July 1955. In his report to the Second Session of the First NPC, Deng Zihui, vice-premier of the State Council (Cabinet), stated, "In the Yellow River basin Chinese civilization began. It was the cradle of Chinese civilization."[46] Deng noted that 1,500 breaches and inundations over the previous three thousand years had wreaked havoc across the North China Plain. Particular emphasis was placed on the destruction caused by the series of floods in the 1930s under the rule of Chiang Kai-shek, including the Huayuankou breach that killed over 800,000 people. "In the past, the scourge of the Yellow River was inseparable from the crimes of the reactionary ruling class [*fandong tongzhi jieji*]."[47] The heroic deeds in the CCP liberated areas after the defeat of the Japanese initiated remarkable changes in the downstream area. Since that time, "the scandalous state of the old dikes—defective, broken, flimsy, shaky and hole-ridden—which symbolized the corruption, impotence, and disregard for human life of the reactionary ruling class has been fundamentally changed" by the trinity of Party, government, and the People's Liberation Army.[48] The Party's claim to legitimacy was also affirmed by China's very recent successes in flood prevention: "In Kuomintang [Guomindang, or Nationalist Party] days, in 1934, the heaviest flow of [Yellow River] water was only 8,500 cubic metres per second [near Zhengzhou]. In that year the Yellow River breached dikes in four places in Henan. . . . In 1935 when the heaviest flow was 13,000 cubic metres per second, the Yellow River carved a massive breach in Shantung [Shandong], and overran a dozen counties. However, after the founding of the People's Republic of China, in 1954, the heaviest flow reached 14,000 cubic metres per second, but no damage was wrought."[49]

Deng also carefully placed the transformational efforts of peasants and water heroes in the deep history of controlling the waters in China: "The heroic spirit that our ancestors displayed in flood fighting . . . [is reflected by] the teeming millions who live along the lower reaches of the Yellow river [who] have always worked intensively on flood-prevention . . . striving to restore agriculture and engaging in other productive activity."[50] Deng reconstructs the narrative of Yü the Great, emphasizing Yü's selfless dedication as he stayed away from his home for eight years.[51] Deng also quoted from a poem by the revolutionary writer Guo Moruo that celebrated Yü the Great: "If now I cannot overcome the floods,/How can I meet the people's expectations?"[52] Other water heroes were invoked. Deng drew particular historical sanction from Yü the Great and Pan Jixun by affirming the fundamental correctness of these water heroes. But Deng stressed that it was the historical mission of the Party to achieve the absolute taming of the river and to bend the will of the water in service to the construction of communism. He wrote: "The way of guiding water and silt to the ocean was ancient, an age-old idea as we read in the legends about Yu's dredging the Nine Rivers. The famous watchword of Pan Chi-sun [Pan Jixun] to 'Build dikes to hold water in check, and let the water carry away the sand' showed that this was the limit of what they [the ancients] sought to do."[53] In other words, given available technical and political constraints, Yü's and Pan's methods achieved all they could. According to Deng, no simple "policy of guiding water and silt can tackle the problem at its roots. And it would be utterly wrong for our People's government [renmin zhengfu] to stick slavishly to this policy when we have the scientific and technical advantages of today. What we have to do is to find a thorough and comprehensive solution to the problem, not just to eliminate floods but also to develop water resources."[54]

This careful reading of the past was intended to legitimize CCP claims to authority. Flood and drought management was critical for a region that totaled 40 percent of the nation's farmland and that was the source of 62 percent of China's wheat production. The North China Plain produced half of China's cereals (millet, oats), 57 percent of its cotton, and 67 percent of its tobacco. In addition, within the Yellow River basin were rich deposits of coal, petroleum, iron ore, aluminum, and other minerals that could help stoke industrial development plans in the interior. There was no hydroelectric generation before 1949, and river transportation was restricted to small wooden boats (10–75 tons) and inflated sheepskin rafts.[55] With the development of multipurpose river use, the future would be profoundly different. "Yellow River floods and the disaster they bring can be completely stopped."[56] All silt would be captured in large upper- and midstream reservoirs, thus rendering downstream water clear and riparian areas more stable. "Everybody living in the lower reaches will from then on be free from the present burden of flood-prevention."[57] Dams on the main stem of the river will generate 23 million kilowatts of electricity (average annual output of 110 million kilowatt-hours), or "ten times the nation's total electric power output in 1954."[58] Large reservoirs would increase irrigated farmland from the "present 16,500,000 mou [mu] to 116 million mou, a good 65 per cent of the total area in the Yellow river basin that needs irrigation." Since the Yellow River could not sustain irrigation on the remaining 35 percent, Deng warned, "We shall have to consider supplementing the water of the Yellow River by diverting water from the Han [a tributary of the Yangtze] or other rivers."[59] And, last, after all the dams were constructed a 500-ton ship would be able to navigate the river upstream to Lanzhou. Deng concluded the vision by declaring, "It will not be long before all the deputies here and the people of the whole country see the day when the age-old dream comes true—when the Yellow River runs clear."[60]

The Yellow River Plan of 1954–1955 was nothing if not audacious. Premised on the basic principle of water storage, large and small reservoirs would be created on the main stem of the Yellow River, on its tributaries, and in the ravines of the loess highlands to regulate water flow for flood control, irrigation, hydropower, and silt control. "Everywhere, section by section, from highland to gully, from tributary to the main river, water is to be stored and soil retained."[61] The plan called for forty-six dams on the main stem of the Yellow River system, including four large multipurpose complexes. This series of dams and reservoirs would form a staircase, providing the capacity to carefully coordinate flow rates for multipurpose use. "Everything possible is to be done to use the river water for industry, agriculture, and transport and to leave loess and rain water where it is needed—on the farmland."[62]

Because the Yellow River Plan was projected to take decades to complete, a first phase was delineated that would be implemented under three FYPs (i.e., from the mid-1950s until 1967). This initial phase focused on the construction of two multipurpose projects at Sanmenxia and at Liujiaxia; three dams on the river to irrigate a million *mu* in Qinghai, Gansu, Shanxi, Shaanxi, Inner Mongolia, Henan, Hebei, and Shandong; and a series of soil conservation initiatives including 638,000 check dams, tree planting on 21 million *mu*, and other farmland improvement efforts. The dam and reservoir at Sanmenxia was the showpiece, by virtue of both its size and its critical water and silt regulating functions for the downstream (i.e., North China Plain). According to Deng's report, "The Soviet Union has agreed to design the Sanmenxia project. Comrade Koroliev, leader of the group of Soviet experts who are helping us draw up the Yellow River plan, will be responsible for the design."[63]

Construction was scheduled to begin in 1957, with completion in 1961. With a maximum depth of 350 meters (1,150 feet), the reservoir would generate 1 million kilowatts, "sufficient to meet industrial and other needs in Shaanxi, Shanxi, and Henan [Provinces] for quite a long time."[64] In dry seasons, the reservoir would "step up the minimum flow in the lower reaches from 197 to 500 cubic meters per second, to provide enough water for irrigation and navigation in Henan, Hebei, and Shandong."[65] When filled, the reservoir would inundate over 2 million *mu* of farmland and would displace 600,000 people. But "these re-settlers will assuredly earn the deepest gratitude of the 80 million people living under constant menace from the Yellow River."[66] The reservoir would "reduce the heaviest imaginable flow of 37,000 to 8,000 cubic meters per second."[67] It would also serve as a massive silt detention pool that would render the Yellow River clear downstream. Finally, said Deng, "[The] Sanmen Gorge [Sanmenxia] reservoir will last fifty to seventy years, perhaps even longer, thanks to its large storage capacity, [to] the silt-detaining dams on the tributaries of the river, and above all, to the works on water and soil conservation in the middle reaches."[68]

The Yellow River Plan bore the imprint of a particularly aggressive Soviet-style technology complex of the postwar years. Chinese sources state that a team of seven Soviet experts arrived in early 1954 to help draw up the Yellow River Plan.[69] The plan was among 156 national projects to receive Soviet assistance. Sanmenxia was the result of several years of intensive surveys up and down the river, including a much-publicized 1952 YRCC survey mission that discovered the origin of the Yellow River. The survey team also made measurements of the Tongtian River and concluded that it could serve as a conduit to divert Yangtze River water to the Yellow River.[70] This was the genesis of Mao's comment to Wang Huayun concerning "borrowing a little water" from the Yangzte during Mao's North China tour of the same year. Although there was no formal south-to-north water diversion component of the Yellow River Plan, planning discussions for the massive project reflected the large-scale engineering culture of the Soviet advisors. As Stalin

The 1955 Great Plan to transform the Yellow River. (Courtesy of the International Institute of Social History, Amsterdam)

stated, "With the lapse of time and the development of human knowledge, when man had learned to build dams and hydro-power stations . . . man learned to curb the destructive forces of nature, to harness them . . . to convert the forces of water to the use of society and to utilize it for the irrigation of fields and the generation of power."[71] Stalin's words echo those of Deng, who argued, "We must conquer the Yellow River and change the natural conditions of its whole basin, in order to bring about a fundamental change . . . to meet the demand on the resources of the Yellow River made by China's national economy during the present period of building of socialism and the future building of communism."[72] Though such sentiments would have rung equally true for large-scale resource exploitation projects the world over in the postwar period, the Yellow River Plan found particular resonance with the "Great Volga Project" of the Soviet Union.

The "Great Volga" scheme was initiated shortly before World War II and was restarted with vigor after the war. The plan was to create a cascade of reservoirs behind a series of large multipurpose dams. The first dam was completed in 1953 (Gorky Dam) followed by the massive Kuybyshev Dam in 1958 and the equally large Stalingrad facility in 1960. At that time, one-third of Soviet hydroelectric capacity was generated by these three facilities. When the scheme was fully articulated, with additional dams on the Volga and its tributaries, it would generate electricity equal to the country's total 1957 production.[73] The fundamental "transformation of nature" suggested by the Great Volga Project was both a philosophical and cultural feature of Soviet economic management. As a basic element of Marxist-Leninist theory, such transformations were defined as "the purposeful change of the geographical environment consciously undertaken for the improvement of natural conditions in the interests of mankind."[74] Such sentiment led to florid descriptions: "Let the fragile green beast of Siberia be dressed in the cement armor of cities, armed with the stone muzzles of factory chimneys, and girded with iron belts of railroads. Let the taiga be burned and felled, let the steppes be trampled. Let this be, and so it will be inevitably. Only in cement and iron can the fraternal union of all peoples, the iron brotherhood of all mankind be forged."[75] Maxim

Gorky's exhortation to make "mad rivers sane" welled from the same philosophical-cultural spring.[76] In the immediate postwar period, the Soviet impulse for rapid transformation was heightened by escalating cold war pressures. The Volga plan and Stalin's "Great Plan for the Transformation of Nature" (a massive intervention that included extraordinary soil conservation and irrigation projects, as well as promethean river-diversion plans) reflected the rise of High Stalinism (1945–1953), a developmental paradigm that rearranged nature in unprecedented ways.[77]

Before Soviet advisors arrived in China en masse beginning in 1954, the vast bulk of technical exchange flowing from the Soviet Union to China took the form of translations of Soviet technical texts, including, as noted previously, texts on hydraulic engineering. These texts were permeated with High Stalinism. China's government and Party leaders simply had little organizational and managerial experience in running a national economy. For practical and ideological reasons, they were inclined to apply the Soviet model as quickly as possible. There were five features of the High Stalinist Soviet managerial model imbibed by the CCP, government, and technical personnel: Party control over management (and Party control over all economic decisions at every level); mass organization methods, such as production campaigns; militarization of management; worker education and re-education; and the equation of labor with patriotism. "[High Stalinism] was a time when the Soviet Communist Party was completely in control of all spheres of production, as well as all aspects of life outside the factory. The most important consequence of translating this literature into Chinese was that the CCP was able to learn and use the High Stalinist lessons of total Party control, and it was able on this basis to justify its own complete usurpation of power, in both the factory and society."[78]

Total Party control was the central message of Soviet material translated in China during the early 1950s. In an article published in *Shuili Fadian* (Water and Hydropower) in August 1955, a Chinese engineer suggested the ways in which the Yellow River plan was a manifestation of Soviet five-year planning projects. He argued that Chinese engineers lacked the confidence to tackle difficult engineering planning. The author stated that it was only through the rational and scientific planning approach of the Soviet advisors that the Yellow River Plan had come to fruition.[79] There were indeed many characteristics of the Yellow River Plan that suggested a Soviet model based on central direction, robust bureaucratic organization, and the critical role of technical personnel. These characteristics included, for example, adopting multiuse development; selecting sites based on analytical evaluation of benefits to power, irrigation, flood control, and transport; technical designs based on consultation with relevant functional ministries; and construction designs based on evaluation of meteorological, geologic, and hydrologic data.

The settlement of technical disputes also suggested a particularly Soviet approach. Early debates between Chinese and Soviet engineers centered on the location of the first dam and reservoir project. Some argued for Mang Mountain, and others for Sanmenxia. Soviet engineers prevailed. They argued that Sanmenxia was the only site suitable for a dam and reservoir of a sufficient size to accommodate multiple use. Two other technical debates related to the choice of site centered on population relocation and siltation. A large structure at Sanmenxia would force the relocation of hundreds of thousands of peasants. Soviet advisors argued that the costs of relocating peasants would pale in comparison with the benefits of the project. The silt issue was also easily resolved. Soviet engineers were reported to have defended the soil conservation measures prescribed in the plan—that is, check-dams and silt retention reservoirs on Yellow River tributaries in the loess highlands. According to the Soviet advisors, these installations would dramatically reduce sedimentation of the reservoir. Besides, they said, the reservoir was designed to accommodate some unavoidable sedimentation. Other features of the Soviet technology complex that informed the Yellow River Plan were the impulse toward gigantism and powerful central control that could induce bureaucratic compliance across government agencies.[80]

Immediately after the approval of the Yellow River Plan by the NPC, a public education campaign was launched to broadly communicate the goals of the project. A speech to the NPC by Song Qingling, the widow of Sun Yat-sen, referred to as the "mother of the nation," was widely cited by the state-run media. Song argued that the Yellow River Plan would "transform the whole face of China's backwardness." She asserted that to control the Yellow River would result in taking "a river that . . . meant endless calamity to our motherland, and changing it to be a well-spring of happiness for coming generations."[81] Li Jishen, vice-chair of the NPC, sustained the optimism in remarks that were also widely quoted: "A plan that promises to permanently control a river that for tens of thousands of years has been uncontrollable and has resulted in the most serious devastation gives me incomparable happiness. The magnificence of this plan is not only unprecedented in our history, but is also rarely seen in other countries in the world. I am profoundly affected. Through the leadership of the Party and Chairman Mao we can now conquer what have historically been unconquerable natural disasters."[82]

It would be difficult to overestimate the optimism of the moment. The promises of modern technology; the organizational capacity of the Party; the strength of Party and state to protect the nation from external enemies (e.g., in Korea); and the friendship and technical guidance of the Soviet Union, which had repulsed fascist invaders, exploded an atomic device, and re-engineered nature for industrial and agricultural purpose—all of these factors engendered robust optimism in China by 1955. This sense of strong purpose and identity, in turn, contributed to a feeling, as reflected in the Yellow River Plan, that China was on the brink of a new and productive relationship with environmental forces that had heretofore been a monumental struggle to control.

This "high tide" of optimism encompassed the mobilization of vast human and material resources to transform the landscape. Articles and editorials in the state-run media continued to hammer away at this commitment. An editorial in *Renmin Ribao* titled "A Great Plan to Conquer Nature" recounted how the state and Party had redressed thousands of years of debilitating floods and droughts, while proving false the notion that the river has the capacity only for harm. The editorial quotes one of China's most famous poets, Li Bai (701–762): "The waters of the Yellow River from Heaven come, / Hastening they flow to the sea never to return."[83] According to the editorial, from a modernizing perspective, water simply should not flow to the sea. It must provide irrigation, electricity, and transportation through regulated reservoirs. "[Under the Yellow River Plan] a large amount of water will not even reach the sea, as it will be diverted to irrigate both sides of the river."[84] The editorial noted that the Yellow River Plan reflected new socialist conditions and technical capacities that would give humankind the boundless strength to control and utilize nature. The current generation must win this battle. The battles would be many and great, but the task was glorious. The article concluded, "We have the resolute will to conquer all difficulties. Our slogan is to make the mountain bow down and make the waters yield."[85]

One day later, *Renmin Ribao* ran a lengthy polemic by Zhang Hanying titled "A New Milestone for Controlling the Yellow River" ("Zhili Huanghe de xin de lichengbei"). According to Zhang, during the Ming dynasty, the works of Pan Jixun were guided by two objectives: to improve grain transport and to protect imperial tombs. Feudal ruling classes had, in effect, sacrificed the livelihood of the peasantry. Although modern science and technology had accompanied the invasion of capitalism, "semifeudalism" and "semicolonialism" blocked any solution to the ecological wreck on the North China Plain. During the reactionary Nationalist period, Freeman, Engels, and Franzius had proposed plans that tended toward fantasy. They had little understanding of the Yellow River or of the processes of sedimentation. It was only the Chinese engineer Li Yizhi who understood the imperative of basin-wide management and the particular importance of controlling sediment runoff with soil conservation in the loess highlands. Zhang navigated a careful course. He preserved a nationalistic tone in his statement that Chinese were aware of the sources of siltation centuries ago, but that feudal interests and a lack of scientific knowledge had blocked the development of remedial measures. Zhang gave other examples of imperialist avarice, including the plans formulated by the Japanese East Asian Institute and the work of the American Yellow River Consulting Board that worked in support of the Chiang Kai-shek regime. Zhang explained that the Japanese had crafted their plan (focused on hydroelectric development) to support the extraction and processing of industrial material. And, Zhang said,

"Make the Mountains Bow and the Rivers Yield." (Courtesy of the
International Institute of Social History, Amsterdam)

the American proposal for placing a dam at Balihutong was not credible. The American advisors had simply flown over the site. In Zhang's words, the American advisors were "thorough dupes" (yun-nong ren).[86]

Chinese articles on the Yellow River that poured forth in late 1955 expressed the dual nature of the river in multiple ways. The river was a parent that nurtured generation after generation of Chinese. Generations were born (seeded), matured, and died (harvested) in the Yellow River valley. And one article noted, "[As the] famous song the 'Yellow River Cantata' relates, the Yellow River is the cradle of our nation."[87] After regulation by Yü the Great and the rise of Chinese civilization, however, the river lost its moral path. Raging, surging, and violently roaring down the valley, it caused thousands of years of human misery. Until "Liberation" (1949), the river rejected any attempt to rectify its behavior. In "A Thousand-Year Dream Realized" ("Qian nian mengxiang you shixian"), published in Renmin Ribao, Liu Hao concluded, "Our Yellow River, a vast and mighty river, nurtured us, yet was a river that gave us sadness and misery. With the determination of the great CCP, we shall transform it to become a source of well-being for our people. For thousands of years how many people have dreamed of this ideal, now this dream is within the grasp of our generation!"[88]

In "Sanmenxia," also published in Renmin Ribao, Hua Shan linked the transformation of the Sanmenxia Gorge directly to the efforts of Yü the Great, who, with great determination and an iron will, chiseled the opening of rivers through rock. The author also explored how Sanmenxia was a critical portal for goods and grains coming from Southeast China during the imperial period. But navigating the gorge was dangerous. Rocks and shoals caused losses of up to 80 percent of grain shipments. A saying developed that "to pass [through] Sanmenxia was to risk life" (sheming guo Sanmenxia). For centuries, people thought that it was impossible to chisel through the igneous hardness of the gorges. They had dreamed for centuries of "hoisting the sails from the jade ocean to the river's source." But now this rock was viewed as a precious stone—a precious stone upon which the largest dam in the world could be built: "When the Yellow River runs clear! This dream was the same as saying, 'The sun will rise in the west.' Since ancient times people used to say this when something was impossible. Now, not only do we dare to say it, but we have it in our grasp to realize."[89]

In 1954 the YRCC organized an exhibition called the "Controlling the Yellow River Exhibition" (zhili Huanghe zhanlanhui), first shown in Zhengzhou in April 1955, then in Beijing at the Zhongnanhai government complex adjacent to the Forbidden City. The exhibit opened in time for the Second Session of the First NPC, when the Yellow River Plan was approved. With electronic displays, photographs, charts, paintings, and texts, the exhibit celebrated the Soviet-style technology complex. It was imbued with the values of central planning, modern technology, scale, modern scientific data, and transformative technology. It explicitly represented the Yellow River valley as the cradle of Chinese civilization, but also suggested that the river flowing through multiple cultural regions symbolized a singular national identity that united all of China's peoples in its quest for modernity. The exhibit reinforced the master narrative of the rebirth of Chinese civilization by the CCP.

The original exhibit in Zhengzhou included three rooms, while the exhibit in Beijing and subsequent stops in 1955 and 1956 occupied two. One entered the exhibit to strains of the "Yellow River Cantata," as the characters bian Huanghe wei le (transforming the Yellow River to a beneficial river) were written on a large banner above the entrance door. The first room presented electronic models, photographs, paintings, charts, graphs, and ancient books. Upon viewing the first room, one journalist recounted, "Originally the Yellow River was our source of wealth, but because of the historical devastation of reactionary ruling classes, the vast grasslands and forests were destroyed, causing sediment runoff and the creation of a destructive river."[90] Another commentator noted, "[With a pointing stick directed at an electronic model of the valley] the exhibit guide told us that in the past three thousand years, the Yellow River has broken through its dikes 1,500 times; on 26 occasions the river changed course. . . . It is unknown how many lives and how much production was lost."[91] From the same electronic model, the journalist saw a fanlike array of red lines in the lower valley, representing the vast alluvial fan created by the multiple course

Mao at the 1955 Yellow River Plan Exhibition. From Shuili bu Huanghe shuili weiyuanhui, ed., *Renmin zhili Huanghe liushi nian* (Zhengzhou: Huanghe shuili chubanshe,
2006), 154.

changes of the lower Yellow River. He noted, "We [also] saw a
short green line cutting across the river, cutting the Yellow River
into two lengths. The guide told us: 'This is where the Sanmenxia
dam and reservoir will be built. The dam will end thousands of
years of flood disasters, and allow us to use excess water to generate electricity and irrigate fields.'"[92] Upstream of the green line
(Sanmenxia), the Yellow River was rendered in yellow, downstream
in green, indicating that the water below Sanmenxia would run
clear. The model also illustrated the monumental work of dike restoration and augmentation that the peasantry had accomplished
since 1947. It also showed the People's Victory Canal and gave related data expressing the length of irrigation canals in the system,
the number of irrigated *mu*, and data about harvests and new shipping lanes that had been created when water was diverted to the
Wei River. The electronic model indicated irrigation projects in

Shandong, as well as soil and water conservation efforts, hydrologic measuring stations, and data on economic and social losses from past floods.

Among the books displayed in the first exhibition room were those describing the labors of Yü the Great and Pan Jixun. As the journalist suggested in his article, to focus on past efforts of control was to imply continued failure. The two historical alternatives, dividing the water *(fen liu)* or concentrating the flow *(du liu)* to guide water and silt to the sea, were akin to treating the symptoms but not the disease *(tou teng yi tou)*. The journalist concluded, "With the power of modern science and technology we can now utilize the water and silt by storage and conservation."[93]

The first exhibition room also held two oil paintings that contrasted the past with the present. One was of a pre-Liberation dike, damaged and run down *(popo lanlan)*, drooping and weak *(you di you bao)*. On the inside of the dike was a badger hole and a fox hole. Everywhere on the dike were military trenches and pillboxes. Wild beasts were everywhere *(yeshou)*. Trees had been felled. The article continued, "This painting sufficiently explains that the crimes *[zui'e]* of Yellow River disasters and the reactionary political ruling class cannot be separated."[94]

In the second room of the exhibition, the guide pointed to an electronic model illustrating where dozens of dams would be built on the river: "After these are built the entire turbid river will run [clear as] jade."[95] The guide enumerated the benefits of dams and reservoirs: a tenfold increase in the country's electricity supply that would fuel industry, commerce, cities, and villages; irrigation of more than 100 million *mu* of farmland; production of 130 billion *jin* of grain and 1 billion *jin* of cotton; and the sailing of 3,000-ton ships between Zhengzhou and the sea. According to one journalist, "After soil and water conservation measures have been carried out in the Northwest, the land will be fertile, the waters will run clear, and people will forever enjoy secure and contended lives. The arid climate of the loess highlands will change. Everywhere will be clear with fragrant smells of flora; everywhere will be scenes of prosperity."[96]

Another journalist distilled the exhibition's central message: "The Yellow River valley is the origin of Chinese civilization and the economic and political center of China. But for centuries now the river has been a source of calamity. People could only wonder when a sage would appear to control the waters of the Yellow River. But now, after seeing the exhibition on Controlling the Yellow River, we can proudly feel that the dream of the ancients will finally be realized."[97] A second such summary concluded, "Visitors to the exhibition will understand how the project reflects the nation's determination, the large investment necessary, and the massive scale of the project."[98] The exhibition, which ran from April 1955 to May 1957, traveled to six provinces and was visited by over a million guests.[99]

## The "High Tide" of Water Conservancy

While the Yellow River Exhibition was celebrating the promise of Soviet-style engineering, a mass movement was launched for an irrigation campaign that promoted a contrasting technology complex. Soon after the Yellow River Plan was approved in 1955, the country embarked on a "high tide" of water conservancy (shuili gaochao). On the one hand, the large capital-intensive projects exemplified in the Soviet-influenced Yellow River Plan developed apace, with the primary focus on the Sanmenxia Dam. On the other hand, however, under the direct leadership of Chairman Mao, a mass-based irrigation campaign was launched, founded on the voluntarism of the Yan'an period. For several years, the slogan "walking on two legs" suggested a unique Chinese development path. On one side were the state planners, who inclined toward the "beliefs of Stalin and his successors [that] underlined state construction, social order, and Marxist laws of development."[100] On the other side of the planning arena was Mao, "[whose] thinking always returned to issues of individual will, human capabilities, and mass actions . . . [as he] sought to create social instruments in which the creative abilities of man were released, often prompted by anger at tradition or foreign oppressors."[101]

The High Tide irrigation campaign was launched in 1955. It was premised on the goal of rapidly increasing agricultural production in order to support China's industrial development. The new irrigation campaign was a corollary to the collectivization effort that sought to increase the efficiency of agriculture through collective labor and community sharing of farm implements and resources. The Chinese campaign was inspired by a Soviet model. During 1954–1955, the Soviet Union was also engaged in an aggressive push to raise agricultural production with its "Virgin Lands campaign." This campaign focused on the Central Asian region of the Soviet Union, where vast tracts of virgin land were plowed under to grow wheat.[102] The respective campaigns of the Soviet Union and China employed similar statist goals and exhortations, but the context of land and labor in China was different from that of the Soviet Union. First, the Soviet model was characterized by low population densities in the countryside coupled with nascent industrial production in the cities and by a level of agricultural mechanization on the steppes of Central Asia that made further mechanization relatively easy. Second, there were no large tracts of "virgin land" in China. To be sure, there were great hopes for expanding agricultural land through conservation measures in the loess highlands, but rapid gains in agricultural production could be gained only by intensifying inputs in areas where agriculture had traditionally been practiced. Because China had little access to foreign capital and insufficient capacity to produce large quantities of chemical fertilizers, Mao favored the development of irrigation through mass labor inputs to rapidly expand agricultural production.[103]

Mass-based development of irrigation networks did not, however, proceed independently of multipurpose waterworks that were dependent on large capital outlays, central management, and technical expertise. Irrigation networks relied on dams and reservoirs to control floods, to regulate water availability, and to power irrigation and drainage pumps. The aggressive development of irrigation districts in the Yellow River Plan were premised on the completion of a network of dams on the Yellow River and its tributaries. For example, the "Red Banner" irrigation scheme, designed to hydrate some 10.7 million *mu* along the Grand Canal, depended on technically sophisticated regulating works. All across the North China Plain, irrigation projects were planned that melded Soviet-style engineering with mass-based labor. The irrigation campaign was an effort premised not only on state initiative but also on traditional organizational practices. Peasants on the North China Plain had long been organized by the state (or by its local proxies) to manage local water projects. Lacking modern technological inputs to further raise agricultural production, and committed to maintaining hydraulic order, the imperial state provided material and moral support to local society for water projects that would sustain agricultural output in ecologically vulnerable zones like the North China Plain. In addition, Chinese traditions of agrarian cooperation, impelled by the labor demands of seasonal sowing and harvesting, as well as by the frequency of acute environmental events like floods and droughts, engendered a familiarity with collective effort in the Chinese countryside. Thus, technologically, organizationally, and culturally, the mass-based irrigation campaigns of the 1950s drew on a long tradition of state and village experiences.[104]

Although state and Party representatives articulated plans to aggressively expand irrigation immediately after 1949, it was not until the 1955 High Tide that irrigation development began in earnest. Up to that point, the major concerns of the state were flood control and drafting of plans for large multipurpose projects. In the pre-1955 period, several large projects were designed, in part, to expand irrigation on the North China Plain. Examples include the Fuzeling Reservoir on the Huai River, the Guanting Reservoir near Beijing, and the People's Victory Canal in the Yellow River valley. But because investment priorities were directed at flood control, and with the country's limited capacity to produce large amounts of steel and concrete, large-scale irrigation projects were few in number. "Reluctant to reject Soviet advice, at the same time economically unprepared to follow this advice which stressed big projects, and perhaps not yet capable of mobilizing the masses for irrigation projects requiring labor-intensive investment, the CCP did not make a concrete effort in this area until the fall of 1955."[105]

By 1955, several trends had converged that signaled China's potential to mobilize the peasantry on the North China Plain for local, small-scale projects that would collectively transform the agricultural landscape. First, between 1952 and 1954, famine conditions, in the wake of floods or droughts, hit areas of the North China Plain. In each of these years, Fu Zuoyi and other officials referenced the need to develop small-scale irrigation projects.[106] For example, after serious famine struck following the 1954 floods, Fu remarked, "Small irrigation works must be developed energetically. If the power and resources of the masses are well utilized, small irrigation works can be built in large numbers, far and wide. If the projects are undertaken by the people with government aid, the state does not have to spend much but the results will be great."[107] Second, the successful mobilization of peasants for flood relief in 1954 proved the Party's capacity for mobilizing and organizing peasant labor. Third, the critical role of agriculture to fuel the development of heavy industry was reinforced by the dismal harvest of 1954. One commentator stated that "during the initial stage of the first Five-Year Plan we still did not have a full understanding of the importance of agriculture. It was only as we put the Five-Year Plan into practice that we began to fully realize the serious significance of agricultural development to socialist industrialization."[108] Last, the potential for launching a mass-irrigation campaign coincided with the "high tide" of collectivization. Party leaders including Mao (who was an ardent advocate of collectivization) reasoned that mobilization for small-scale irrigation projects would be more efficient under the rubric of the collectives. And the speed of rural collectivization in China was indeed stunning. By 1957 over 90 percent of the rural population was enrolled in agricultural cooperatives, which produced a "certain optimism among the leadership concerning its ability to achieve quick results by relying on the masses."[109]

In January 1955 clear signals emerged from the center of an impending mass-based irrigation movement. In delineating the country's annual water management goals, the report of the 1955 All-China Water Conservancy Conference (Quan'guo shuili huiyi) stressed mass-based small-scale irrigation projects. The report stated that all levels of water conservancy governance should rely on the masses to overcome the tendency to valorize large projects over small ones. The report called for a strengthening of provincial and local water governance organizations to lead these efforts. Mass-based projects would rely on the "masses assisted by the state" *(min ban gong zhu)*. According to the report, "Mass-style [irrigation projects] . . . relied upon the knowledge, labor, and investment of the masses. Technique (or technology) would be transmitted to the masses." In this context "assistance from the state" meant organizational leadership, technical direction, and financial assistance.[110] Two months after this report was published, in March 1955, the minister of agriculture stated that the government did not have the financial or technical wherewithal to focus exclusively on large projects. Instead, he argued that increasing agricultural production must rely on the energy of the masses to carry out small-scale irrigation projects.[111]

The preference for mass-based irrigation campaigns was a reflection of a change in the developmental paradigm championed at the very top of the political hierarchy, specifically by Mao. The "Little Leap" of 1955–1956 was an early indication of a future pattern of Chinese politics that "can be seen as a series of experiments in decentralization: top leaders sensing opposition from their colleagues bypassing these impediments by appealing to an increasingly broad coalition of decentralizing actors."[112] Beginning in 1955, we see increasing debate over management approaches with their associated technology complexes. These debates became inextricably intertwined with political struggle. In the brief history of CCP rule until 1955, there was a duality of the statist model of the Soviet Union and the more populist and voluntarist ideals that could be conjured from the memory of Yan'an and peasant revolution.

The pendulum between the two developmental impulses (statist versus populist-voluntarist) was decidedly pushed to the left in 1955. Up to that point, consensus among Party leadership continued to advance Soviet-style development schemes that emphasized central planning and resource allocation, heavy industry, and modern science and technology in service to large-scale projects of resource extraction. This consensus continued despite forces that persistently challenged the Sino-Soviet relationship, including a rocky personal history between the leaders of the two countries (first, between Mao and Stalin and then, after 1953, between Mao and Nikita Khrushchev). Despite these tensions, until early 1955 Mao continued to write that the road to socialist construction was guided by the five-year planning process, that it was premised on borrowing from the Soviet Union, and that it would require decades to build an industrialized socialist country.[113]

In mid-1955, however, the substance of Mao's pronouncements about the road to socialist modernization changed significantly. He began to argue that a fundamental reorganization of rural society was necessary to rapidly increase agricultural output. The specific issue that elicited Maoist intervention was rural collectivization. The process of socializing agriculture began in 1949 with the creation of Mutual Aid Teams in North China. The goal was the pooling of labor. By 1952, roughly 40 percent of all peasant households were enrolled in MATs. As a corollary to land redistribution, largely completed by 1953, new socializing institutions called agricultural producer cooperatives (APCs) were initiated. APCs collectively farmed land, but peasants were rewarded with a dividend for household contributions of labor, land, farm implements, and animals. Party leaders explicitly sought to avoid the egregious excesses of the Soviet collectivization process of the 1930s that largely wiped out Russia's highly productive *kulak* class.[114] In contrast, the path toward full socialization of China's countryside was intended to erode the position of wealthier peasants more slowly, while exacting a relatively gentle appropriation of the peasants' agricultural surplus for investment in heavy industry. The process of collectivization in China, however, was uneven. Despite a period of rapid formation initially after 1953, by 1955 the pace of collectivization slowed and reports suggested that the vitality of the APCs was flagging. These developments prompted Mao in 1955 to urge greater alacrity in reorganizing rural society.[115]

Although China's leadership broadly agreed that adoption of socialized agriculture would best advance agricultural output, the pace for this transformation was keenly debated in early 1955. The minister of water conservancy, Deng Zihui, urged caution. In late July 1955, Mao gave a speech to an assembly of provincial and local Party secretaries stating that China should accelerate the collectivization drive. He wasted no time in attacking the opposition's credibility: "Throughout the Chinese countryside a new upsurge in the socialist mass movement is in sight. But some of our comrades are tottering along like a woman with bound feet, always complaining that others are going too fast. They imagine that by picking on trifles, grumbling unnecessarily, worrying continuously, and putting up countless taboos and commandments they are guiding the socialist mass movement in the rural areas on sound lines. No, this is not the right way at all; it is wrong."[116] While issuing several additional stinging rebukes to those who counseled caution, Mao stated later rather casually, "In the spring of 1955 the Central Committee of our Party decided that the number of agricultural producers' cooperatives should go up to a million. This means a little more than a 50 per cent increase adding 350,000 to the original 650,000. Now I feel this increase is a bit too small. Possibly the former figure of 650,000 should have been roughly doubled. The number of cooperatives ought to be increased to 1,300,000."[117]

Mao's intervention had a strong impact. At a pace that stunned even him, the formation of APCs rapidly spread throughout China. At the same time (and with Mao's strong support), the CCP leadership also launched a political campaign that sought to silence opposition to the new push to collectivization. Fearful of being branded as "rightists," local cadres pushed the collectivization effort with zeal. A stream of injunctions to the villages was issued from the center. In his preface to the *Socialist Upsurge in the Countryside*, published in December 1955, Mao urged vigilance against "rightist opposition" to the wave of collectivization sweeping the nation. This "upsurge" of rural economic and social restructuring also induced state and Party officials to compose a Draft Program for Agricultural Development in the People's Republic of China, 1956–1967, which laid out the tasks and responsibilities of APCs throughout the country. The APCs envisioned in the Draft Program were the "higher-stage APCs," which by late 1956 constituted the majority of collectives. Such APCs represented full collectivization of agricultural production. All property was collectively shared, and farmers were paid according to their labor, not according to dividends determined by their share of the property that was collectively managed. By the end of 1956, over 90 percent of all of China's households were enrolled in higher-stage APCs, two years earlier than Mao had thought possible—an unchallengeable endorsement of his vision for a "high tide of socialism."[118]

This massive collectivization of rural China was expected to generate increased agricultural output. For example, in North China, grain output was slated to increase from an average of 150 *jin* per *mu* in 1955 to 400 *jin* by 1967 (the last year of the Program for Agricultural Development).[119] Aside from certain measures reputed to have been successfully pioneered in the Soviet Union (e.g., deep plowing, close-row planting, land reclamation, soil and water conservation, and large inputs of fertilizer), the only reasonable option to expand agricultural output on the North China Plain was to aggressively exploit water resources to irrigate vast tracks of dry soils. For Fu Zuoyi, and other attendees at the 1956 All-China Water Conservancy Conference, a retreat from grandiose visions of multipurpose water facilities to a supportive role in the mass irrigation campaign was inescapable. With the focus clearly on the irrigation campaign, delegates were told that the specific goal for North China was to average more than one irrigated *mu* of land per person.[120] This retrenchment did not mean that plans for the Sanmenxia dam were slowed or halted, but a reorientation was made clear by Fu's admission that water conservancy work had to be rectified from an undue reliance on central planning and large projects. His report stated, "Up to 1954 we have not developed a plan for each province to develop small-scale agricultural projects. . . . But even after 1954 when we strengthened leadership . . . because we insufficiently regarded the activism and strength of the masses, our plans remained conservative and backward so that we could not press forward. We have not sufficiently utilized the latent power of the masses."[121]

Chinese sources in early and mid-1956 abound with data suggesting herculean successes in the High Tide irrigation campaign and the agricultural production drive. For example, reports stated that during the winter of 1955–1956, irrigated areas in China increased from 370 million *mu* to 480 million *mu*; 4.5 million wells were sunk; and 27,000 small reservoirs were constructed.[122] It was clear that Party cadres at the provincial and local levels pursued their High Tide mandates to expand irrigated agriculture with zeal. However, it was also clear that there were problems. The hyperaggressive

goals of the campaign generated disappointing results that were unacceptable to most Party leaders, including Mao. Already by mid-1956,
momentum was building within the leadership to temper the High
Tide. The reasons for this retrenchment were financial constraints
and the questionable outcomes of irrigation construction. During
the 1955–1956 campaign, capital expenditures in the agricultural
sector had sharply increased. Such outlays, which contributed to a
budget shortfall in 1956, could not be sustained. Without loans,
APCs simply could not install regulating equipment for irrigation
systems. In addition, reports from rural cadres painted uneven results in the utility and efficiency of many irrigation systems that
were built in haste with insufficient technical guidance. Indeed,
despite the data quoted above suggesting impressive gains in the
irrigation infrastructure, agricultural production increased by just
4.6 percent, well shy of national targets. Fu Zuoyi produced considerable justifying verbiage in his review of the results of the
1955–1956 irrigation campaign as he trumpeted the 150 percent
increase of irrigated farmland on the North China Plain. But he
also broached problems emanating from the mass campaign. He
noted that of the 1.04 million wells sunk in Henan Province,
47 percent malfunctioned. Likewise, he said, the quality of channels, reservoirs, levees, and other irrigation infrastructure suffered
from the enthusiasm of the masses absent sufficient technical
inputs.[123]

The promotion of High Tide policies and the subsequent retreat
signaled an emerging division within the leadership over developmental priorities. The retreat was a victory of sorts for those who
favored the more incremental and centralized approach of Soviet-
style planning. It is notable that the opening paragraph of Fu's report to the 1957 All-China Water Conservancy Conference focused on planning for Sanmenxia. Still, the High Tide farmland
irrigation campaign of 1955–1956 clearly suggested the potency of
Maoist entreaties to mobilize the Party apparatus and the peasantry and to engage in unprecedented exploitation of water resources. The reported "success" of this approach would provide the
model upon which to mount ever-larger Maoist campaigns in relatively short order.[124]

## The Struggle over Sanmenxia

Shortly after Mao's mid-1955 speech urging haste in the collectivization effort, the YRCC entrusted the Lenin Design Institute in Moscow with the design of the Sanmenxia Dam and electricity generating station. The year and a half between August 1955 and the start of dam construction in April 1957 were spent in back-and-forth discussions, intended to resolve disagreements between Soviet and Chinese engineers over dam height and reservoir size. The initial framework set forth in the Yellow River plan was for a reservoir depth of 350 meters with a capacity of 360 million m³. Less than a year later, in April 1956, the Lenin Institute released its "Report on the Preliminary Design of the Sanmenxia Key Project." The report recommended a minimum reservoir depth of 360 meters within eight years of completion (i.e., in 1968), which would reduce downstream sedimentation by 20 percent, and which, fifty years after completion, would reduce downstream sedimentation by 50 percent.[125] Even though construction began in April 1957, a host of conferences were convened during the following year to discuss the size of the dam and reservoir. According to Chinese sources, the debate hinged on three questions: the number of people and amount of land to be displaced and the capacity of the reservoir to accommodate silt. A change of reservoir level by even a few meters would have important consequences. For example, if the reservoir maintained an average depth of 350 meters, over 2 million *mu* of farmland would be inundated, with dislocation of 600,000 people. With a depth of 330 meters, less than half this farmland (940,000 *mu*) would be lost; and only 272,000 people would be displaced, less than half of those affected by the alternative proposal. By far, the economic impact of a larger reservoir would hit Shanxi Province the hardest. The reservoir could potentially back up the Wei River at its confluence with the Yellow River near Xian. Upstream-downstream political issues dominated the debate.[126]

Several engineers offered opinions that had long-term significance for their own careers and for the long-term perception of China's management of water. Huang Wanli was a Qinghua University hydraulic engineer with an MS in Hydrology from Cornell and PhD in Engineering from the University of Illinois (1931). Before returning to China to work for the Nationalist government, Huang was employed for several years as an engineer by the TVA. After assuming his position at Qinghua in 1949, he served as consultant for a number of projects in China. In May 1956, Huang drafted a paper for the YRCC in which he concluded that none of the arguments over dam height and reservoir size were based on an exhaustive cost-benefit analysis. Huang maintained that any reservoir with a depth of more than 360 meters would not generate economic returns sufficient to offset the heavy losses of farmland and the outlays required to relocate hundreds of thousands of people. Huang also argued that sedimentation would build up much more quickly than the design could accommodate. Ultimately, the dam would be forced to flush sediment downstream. Huang directly challenged the mantra that after the Sanmenxia Dam was built, the "Yellow River would run clear."

In later discussions of large-scale dams during the late 1990s, when the Three Gorges Dam was constructed on the Yangtze River, domestic and international observers lauded Huang for his fortitude in voicing doubts about the Sanmenxia project decades earlier. But Huang was not the only technical specialist to raise objections about the design of Sanmenxia. Wen Shanzhang, from the Ministry of Water Conservancy and Electric Power, and Ye Yongyi, of the Beijing Water Conservancy Technical Institute, argued for a much lower dam and a smaller reservoir, primarily for flood prevention and secondarily for multiple use. At a June 1957 meeting convened by Zhou Enlai to discuss the Sanmenxia Dam design, technical experts raised these concerns. Shanxi officials stressed the heavy costs of lost farmland and peasant relocation for a province with limited land and financial resources.[127] In all, a dozen or so participants at the conference voiced their concerns over the dam. Reflecting both the intensity of debate and the importance of the project to the Party, Zhou convened three subsequent conferences to settle the dam's design.[128]

The heightened level of debate over the Sanmenxia project by mid-1957 was a reflection of the Hundred Flowers movement, a brief period of relative freedom of expression for technical and aca-

demic professionals in China. Inspired in part by the perceived failure of the High Tide, Mao initiated the movement in late 1956 to
encourage frank assessment by intellectuals of the Party's "work-
style" to help solve "contradictions among the people." On a more
practical basis, the leadership's appeal to "let a hundred flowers
bloom, let a hundred schools of thought contend," was driven by a
desire to convince skeptical intellectuals and technical personnel
that their contributions to building communist modernity were in-
dispensable. By early 1957, moral suasion, coupled with promises of
better work conditions and greater access to Party membership,
eventually persuaded considerable numbers of intellectuals to prof-
fer public critiques of the Party's work-style. The expanded parameters of public discourse persuaded technical specialists like Huang
Wanli and many of his colleagues to more confidently express their
professional opinions on the Sanmenxia project. Implicit in these
opinions was a rejection of Soviet technical advice. After all, it was
Mao who in 1957 encouraged independence from slavish imitation
of the Soviet Union when he stated:

> our scientific workers, having studied Marxist-Leninist theory
> and having learned the Soviet Union's experience in science,
> have made great improvements in both ideology and work . . .
> we unquestionably should persist in this study in the future. It
> should also be pointed out, however, that . . . some comrades
> do not apply their brains to independent thinking when learn-
> ing from the Soviet Union. . . . They dare not say what Soviet
> scientists have not said before, and they do not have the cour-
> age to come forward and approach those things that Soviet sci-
> entists have criticized. When they find out that some issues are
> still controversial and undecided even in the Soviet Union,
> they are at a loss as to what to do. . . . If we only know how
> to imitate, there is no way for us to solve the many scientific
> problems raised in real life which are unique to our great
> motherland.[129]

Technical experts and engineers took advantage of the heretofore-
closed arena of public discourse opened by the Hundred Flowers

movement. The critique offered at the 1957 All-China Water Conservancy Conference and the opinions offered at the June Sanmenxia conference were expressions of a professional identity. The institutionalization of a self-conscious interest group focused on water management began in April 1957, when a number of engineers established the China Water Conservancy Society. The director of the Society was Zhang Hanying, the most senior of a large group of engineers who had been trained during the Nationalist period and who had gained their first professional experiences during this time. Zhang was also an excellent political choice, as he had demonstrated complete fealty to Party policies, despite their twists and turns. Many other leaders of the association shared a similar profile: advanced degrees earned abroad, plus working experience under the Nationalist government. The goals of the Society were "to unify the water conservancy technicians, to develop scholarly research, to exchange scientific and technical experience with various countries in the world, and to study the advanced scientific accomplishments of the Soviet Union, as well as swiftly to catch up with the most advanced scientific and technical standards in the world."[130] The society established branches in a number of major cities. One issue that generated considerable professional comment was on the fundamental orientation of water conservancy, particularly on the North China Plain. Some professionally trained technicians argued for flood control and drainage, while others argued for storage and irrigation. The terms of this debate were implicit in the June 1957 arguments of Huang Wanli and others about the efficacy of a large reservoir on the Yellow River.

The political road negotiated by these technical experts was, of course, full of pitfalls. It is true that their criticism of Soviet advice did not run counter to Mao's perspectives as he reflected on an increasingly acrimonious relationship with the Soviet leaders. However, by the time Zhou Enlai convened the Sanmenxia conference in June 1957, Mao had already concluded that allowing professionals to roam free intellectually was too threatening to Party prerogatives. This, coupled with the growing symbolic power of the Sanmenxia project for the Party, meant the project would go forward, irrespective of opposition by a gang of "rightist" intellectuals or potential problems with the Soviet design.[131]

## Prelude to the Great Leap: The Farmland Irrigation Campaign, 1957–1958

A second Maoist-inspired wave of mass mobilization was launched as Mao became displeased with the pace of industrial and agricultural growth under Soviet-style economic management. This second torrent of mobilization was the clearest attempt yet at establishing a "Chinese path" to communist modernity. This path, with its own technology complex, was foreshadowed by the High Tide of 1955–1956. Decentralization of economic management, valorization of local knowledge, a focus on agricultural growth, and ideological incentives were again promoted. All these factors swung the pendulum back toward what was increasingly becoming a distinctly Maoist road. Although the Party maintained careful control of large projects and sectors deemed strategically critical, projects like Sanmenxia were not immune to the new forces unleashed in late 1957. The more obvious outcome, from the perspective of agriculturalists on the North China Plain, was a surge in mass-organized small-scale water conservancy projects that indelibly altered the regional landscape.

The first step on this new development trajectory was the Anti-Rightist movement in the autumn of 1957. Mao quickly concluded that critique by intellectuals of the Party's work-style, encouraged by the exhortations of the Hundred Flowers movement, had escalated to unacceptable levels. It seemed that some critics were, explicitly or implicitly, questioning the Party's legitimacy to rule. The goal of the Anti-Rightist movement was straightforward—a reassertion of Party control of state and society by resolutely silencing dissent. "Struggle sessions" were organized by Party committees in each work-unit with the express purpose of identifying those who had challenged Party authority in deed, speech, and thought. This process identified over half a million "rightists" accused of anti-Party activities who were "sent down" to the countryside to learn from the peasantry. Under pressure to produce a sufficient quantity of "rightists," authorities construed a host of actions and words as anti-Party. Perhaps easiest to identify as "rightist" attitudes were instances in which criticisms or alternatives were raised to state plans. The engineers and technical specialists who questioned Sanmenxia were obvious candidates for the "rightist" label and for re-education. Among those who questioned Sanmenxia plans in 1956, it is not known who, precisely, was "sent down," but one of the most eminent engineers to be re-educated for challenging the Sanmenxia project was precisely Huang Wanli. In a dubious distinction, Huang was personally "outed" by Mao, who wrote an editorial in *Renmin Ribao* in 1957 castigating Huang for his views.[132] Huang was relieved of his teaching duties at Qinghua, only to be rehabilitated and reinstated after Mao's death in 1976. With intended irony, the central government dispatched Huang to labor at the Miyun Reservoir near Beijing.[133] While the silencing of technical specialists like Huang proceeded in mid-1957, Mao argued, "[To] build socialism, the working class must have its own army of technical cadres and of professors, teachers, scientists, journalists, writers, artists, and Marxist theorists. . . . The revolutionary cause of the working class will not be fully consolidated until this vast new army of working-class intellectuals comes into being."[134]

Mao's resolve to reignite mass mobilization in the countryside was also strengthened by slippage in peasant and cadre commitment to the APCs. During the retreat from High Tide policies (1956–1957), an alarming number of peasants withdrew from the cooperatives. At the same time, Mao also concluded that commitment to socialist agriculture was waning among rural cadres as reports, real or imagined, revealed cadre collusion in concealing harvests to minimize state exaction. Equally important, perhaps more so from the broader perspective of investment capacity, rural agricultural production stagnated in 1957 despite the reintroduction of material incentives after the slowing of the High Tide. Various bottlenecks in the development of heavy industry, engendered by slumping agricultural output, led economic planners to cast about for ways of reigniting economic expansion. Earlier in that same year, discussion emerged on the need to promote the growth of small and medium-size industries that would presumably be less dependent on large capital investment from the state. Indeed, the state found itself with little capacity for increasing investment in any sector in 1957. Capital assistance from the Soviet Union, never large in any case, had dried up, and trade with the Soviet Union had decreased. From the perspective of the more radical elements of the Party leadership, including Mao, the Soviet model (toward which the pendulum had swung once again after the 1955–1956 High Tide) was proving inadequate. Turning around the economic ship required increasing agricultural output as rapidly as possible. Again, the answer to this dilemma was to get more water to the fields, and the way to do this was a massive campaign-style mobilization of rural cadres and peasant labor. Mao proceeded to urge a set of reforms that pushed the developmental pendulum to the left in unprecedented ways.[135]

In September 1957, the CCP Central Committee and the State Council issued guidelines that winter irrigation work should focus on small-scale irrigation projects. The Farmland Irrigation and Fertilizer Movement (Xingxu nongtian shuili he jifei yundong) consciously resurrected the organizational and mobilizing principles of the High Tide of water conservancy by calling for an increase in the nation's irrigated farmland by 100 million *mu*, to match the increase of irrigated farmland in 1955–1956. Because one-third of China's agricultural land remained without irrigation, China had to adopt Mao's injunction to take "irrigation as the lifeblood of agriculture" in order to expand irrigated acreage. More consciously than its predecessor, the Farmland Irrigation Campaign explicitly stressed local management, local financing, and local knowledge. On September 25, a front-page editorial in *Renmin Ribao* titled "Spark an Upsurge in Farmland Irrigation Construction" raised the curtain on the campaign:

A certain technical knowledge is required by small-scale irrigation projects, but as the [technical] demands are not great, financial outlays small, and results achieved rapidly, we can rely on the strength of the masses to execute [small-scale projects]. . . . In terms of fostering technical capacity and disseminating advanced experiences, we must adopt the worksite as our

classroom. . . . In terms of material we must use local resources. . . . At the same time we must unite and promote the
technical knowledge and construction experiences of the
masses. We must sufficiently utilize the strength, material resources, and the wisdom of the masses, and [must] overcome
reliance upon the central government.[136]

Because the social unit responsible for initiating and executing
small-scale projects would be the APC, the editorial stressed the
critical responsibilities of rural cadres to effectively mobilize and
utilize the strength of the peasantry. Past experiences in leading the
masses would assist in these efforts. The editorial continued: "Party
cadres should train themselves to become the working-class water
conservancy expert, the engineer, and technical expert. . . . At the
same time, cadres at each location should pay attention to identifying and mobilizing an experienced local expert to become an
activist for the project."[137]

As laid out in September 1957, the Farmland Irrigation Campaign was the "opening shot" of the Great Leap Forward—a national mobilization effort to dramatically increase China's industrial and agricultural production.[138] On the North China Plain, the
campaign began during the off-season months of October–March
(1957–1958). Beginning in September, when the campaign was pub-
licly announced, a steady stream of injunctions and directives aimed
at the bureaucracy had appeared in the state-run media, building to
a crescendo in December, the height of China's campaign season.
These articles hammered away on three main themes. First, local
Party representatives should understand that small-scale projects
with local investment and experience would be most effective at exploiting water resources. Second, cadres must rectify their thinking
that little investment from the state would generate little achievement. Small-scale projects depended on local resources. And, last,
cadres must recognize the power of the masses to complete small-
scale projects on a massive scale. One editorial began with a simple
rhetorical choice: "[Should we] rely on the state or on the masses?"
The answer was equally straightforward: "If we do not rely on the
power of the masses, then there is no possibility of achieving our
goals of more, faster, better, and cheaper."[139]

牵福渠 (年画)                                                秦文美作

Happiness Canal. (Courtesy of the International Institute of Social History, Amsterdam)

Reports from localities and provinces suggested that a surge of irrigation activity had indeed been sparked by rural cadres. The original target for the 1957–1958 winter season of 43.9 million newly irrigated *mu* set in September was revised to 61.8 million *mu* in October, then to 92 million *mu* in November. By the end of January, Bo Yibo, head of the State Planning Commission and an ardent proponent of the irrigation campaign, reported that over 100 million men and women had brought water to 117 million *mu*. At the beginning of 1957, one in six Chinese were digging irrigation channels and reservoirs.[140]

The Farmland Irrigation Campaign generated two outcomes that served to usher in a broader production movement that would carry the decentralization and voluntarist spirit toward a "great leap." First, the campaign drove the creation of larger and more integrated APCs, reversing the process of organizational atrophy that had struck them during the retreat from the High Tide. Indeed, in Mao's thinking, the next stage of social development of the countryside would be the creation of state farms on the Soviet model that would eventually supersede the APCs.[141] Organizing the massive pool of labor required to complete any single irrigation project of reasonable size transcended the capacity of any single APC. As a consequence, APCs combined their labor and equipment and established a leadership hierarchy that, for all practical purposes, resembled a super-APC. In response to the general parameters laid down by the center, local officials were responsible for organizing and executing irrigation development. Added incentive was provided by the center that encouraged competition among APCs in completing projects. In his detailed analysis of the period, Roderick MacFarquhar persuasively argues, "The seed of the communes sprouted spontaneously during the water conservancy campaign of winter 1957–1958."[142]

The second outcome of the Farmland Irrigation Campaign, the "upsurge in the countryside," was persuasive enough for Mao to conjure mass mobilization as a broader means of social and economic organization to promote rapid growth. In his Preface to the *Sixty Articles on Work Methods*, composed in early 1957, Mao summarized the moment:

> more than 600 million people of our country, led by the Communist Party, clearly see their own future and duties and have exorcized the evil anti-party, anti-people, and anti-socialist wind fanned up by bourgeois rightists. At the same time they have rectified . . . the mistakes and weaknesses rooted in the subjectivism that the party and people have inherited from the old society. The party has become more united, the morale of the people further heightened, and the party-masses relationship greatly improved. We are now witnessing greater activity

and creativity of the popular masses on the production front than we have ever witnessed before. A new high tide of production has risen, and is still rising.[143]

The *Sixty Articles* was a handbook for Party cadres to advance their work in a rural society that had entered a new stage of production. In article 16, Mao for the first time employed the term "great leap forward" in describing this new stage of development. Mao seemed to believe that his vision of a year earlier was coming to fruition, as he stated that it was time "to rally the people . . . in our country to wage a new battle—the battle against nature to develop our economy and culture."[144] The "battle" would focus on extracting the human and material resources necessary for China to leap into communism.

# 5 Creating a Garden on the North China Plain

THE PRESUMED RESULTS of the 1957–1958 Farmland Irrigation Campaign affirmed a set of values and developmental principles championed by the more radical elements of the CCP leadership, while exerting pressure on more moderate central planners to fall in with the new program. Local (versus central) investment, small-scale water storage and irrigation (versus a singular focus on large reservoirs and flood control), locally based experience and knowledge (versus technical expertise), ideological (versus material) incentives were central to the radical technology complex. However, it would be a mistake to conclude that the values and goals associated with the Soviet model disappeared. As we shall see, the continuation of the Sanmenxia project was a testament to the continued patronage of this developmental perspective. But there was a decisive shift in the overall tenor of development that favored a technology complex that may simply be called Maoist. The oscillation between the two complexes reflected deep political divisions among the leadership that had profound consequences for China's people and landscape.

In the first half of 1958, the CCP leadership reflected upon the previous winter's Farmland Irrigation Campaign. Tension continued to exist between radical and moderate camps over what conclusions should be reached from the campaign and how to proceed from them. Officials on the North China Plain called for an extension of mass mobilization as cadres from Henan and Anhui reported on the emergence of large collectives engaged in extensive irrigation projects. The Party leadership took particular note of Chayashan in Henan Province, where a collective grew to include 9,400 households, and of Xushui County in Anhui Province, where a local official recruited 100,000 farmers, dividing them along military lines into battalions and platoons to "attack" one irrigation project after another.[1] At the CCP conference in Nanning (Guangxi Province) in early 1958, Mao embraced the slogan "battle hard for three years to change the face of China," which was first championed by the Anhui CCP Provincial Committee.[2] After a senior Party conclave in Chengdu in March, Mao was persuaded by more radical members of the leadership and by provincial activists that the winter irrigation campaign had proven the transformative enthusiasm and strength of the peasantry. He concluded that "the same techniques should be applied on a wider scale, from the forging of pig iron to the elimination of flies. The water conservancy campaign, in short, became part of the chain of events leading to the Great Leap Forward and the formation of the People's Communes."[3]

In early May 1958, the Second Session of the Eighth Congress of the CCP ratified the "Great Leap Forward" that had "organically" developed beginning in the winter of 1957. In his work report of May 5 to the assembled delegates, Liu Shaoqi, vice-chair of the CCP and Mao's anointed successor, outlined the major currents that had led to the present moment:

> The broad masses of the working people have realized more fully that [their] individual and immediate interests depend on and are bound up with collective and long-term interests; and the happiness of the individual lies in the realization of the lofty socialist ideals of all the people. That is why they [the people] have displayed a heroic communist spirit of self-sacrifice in their work. . . . Comrade Mao Zedong has put forward the slogans "catch up with and outstrip Britain in 15 years," "build socialism by exerting our utmost efforts and pressing ahead consistently to achieve greater, faster, better and more economical results" . . . "and battle hard for three years to bring about a basic change in the features of most areas"—all these calls have quickly gripped the imagination of the huge army of hundreds of millions of working people and have been transformed into an immense material force. There

has emerged in physical labour and other work a high degree of socialist initiative, a surging, militant spirit, a keenness in learning and studying that will not rest short of its aims, [and] a fearless creative spirit. . . . The spring of 1958 witnessed the beginning of a leap forward on every front in our socialist construction. . . . In agriculture the greatest leap took place in the campaign of the co-operative farmers to build irrigation works. From last October to April [of] this year, the irrigated acreage throughout the country increased by 350 million *mou* [*mu*], that is 80 million *mou* more than the total added during the eight years since liberation and 110 million *mou* more than the total acreage brought under irrigation in the years since liberation. . . . In their ceaseless struggle to transform nature, the people are continuously transforming society and themselves. . . . We are now in a great period in the history of our country, the period of development by leaps and bounds. . . . For more than a hundred years our country suffered from the oppression of foreign aggressors which made us backward in many respects. . . . [But] within a very short historical period we shall certainly leave every capitalist country in the world far behind us.[4]

All that remained to communicate the message of the Great Leap was a photo opportunity. On May 25, Mao led senior colleagues to the construction site of the Ming Tombs Reservoir (Shisanling shuiku) outside Beijing to dig dirt. The images introduced, or reintroduced, a Yan'an-style mobilization that blurred the distinction between Party leaders and the masses by emphasizing shared labor and sacrifice. One month later, Zhou Enlai led a phalanx of 500 Central Committee members and government officials on a five-day labor stint at the reservoir, where they lived in a school and labored daily from 3:00 P.M. to 11:00 P.M.[5]

## The Great Leap Forward, 1958–1960

In the second half of 1958, the outlines of the Great Leap Forward became more distinct, and the question of the most appropriate "socialist road" generated two distinct opinions within the leader-

Mao digging at Ming Tombs Reservoir. (© 2006 Sovfoto/Eastfoto. Used under license.)

ship. Although often obscured by the need to present a unified front, the High Tide of 1955 and the subsequent retreat suggested an emerging split between central planners and the proponents of mass mobilization. Mao had carefully negotiated between the two camps, but his tendency was to embrace the enthusiasm and creative spirit of the masses over central planning. With reports of wild success in the water conservancy campaigns of 1957–1958, "the weight of Mao's interests, and of his hopes, had unquestionably shifted toward the masses and the countryside."[6] By all indications, in mid-1958 Mao was "still predisposed to the belief that organizational changes, particularly when combined with resource mobilization campaigns, could provide a dramatic breakthrough to a path of more rapid development."[7]

National industrial strategy continued to emphasize the advance of heavy industry, but during the Great Leap Forward the pace of development was quickened, with a large infusion of capital goods. These capital infusions included imported machinery as well as industrial inputs from rural society (e.g., iron generated by the iconic "backyard furnaces"). But the major organizational novelty of the Great Leap Forward occurred in the countryside, where social reorganization coupled with large-scale exploitation of resources was expected to generate a breakthrough in production. The state would provide a modicum of state funds, but the battle against nature would primarily rely on a tidal release of peasant labor made possible by the organization of the new people's communes. During 1958, not only did Mao's thinking reprise his ambitions of the High Tide, but it also expressed the peasant revolution that preceded 1949. The Great Leap Forward mentality shared an ethos of struggle and sacrifice that "Mao linked explicitly to the past of armed struggle." The Great Leap Forward had an "existential continuity with the spirit of Yan'an."[8]

The organization of the people's communes was a further extension of the amalgamation of APCs that occurred during the Farmland Irrigation Campaign the previous winter. The fundamental supposition was that economies of scale would increase output. Achieving these economies of scale required communal social organization. Again, as with the merging of APCs during the 1957–1958 winter irrigation campaigns, the initial formation of communes was trumpeted as an organic development that arose from the creative spirit of the masses in several locations (notably in Henan and Anhui provinces on the North China Plain). These new phenomena, springing from the productive impulse of the rural sector, were highlighted in the state-run media and served as object lessons to all provincial and rural cadres. Communalization spread through China like wildfire. By the end of 1958, according to state statistics, 99.1 percent of rural households were enrolled in people's communes. Although the varying size of the communes, as well as subsequent local experimentation with form, provided some diversity in the model, all of China's communes had several fundamental features. First, the commune served as the basic administrative unit and provider of social services in the countryside. Household registration, policing, taxation, and health services were all responsibilities of the commune. The commune also provided many of the social welfare functions traditionally managed by the household in China, including child care, elder care, and meal preparation. Second, private property rights were curtailed. Although families continued to retain their individual abodes, formerly private plots, on which households could grow supplement provisions or earn income from private sales at local markets were now held by the commune. Markets were also eliminated. Ownership of household agricultural implements and household property related to personal consumption were largely assumed by the commune. And, finally, income differentials were suppressed as net communal income was distributed, not on the basis of labor inputs, but on a per capita basis that reflected communist ideals.[9]

Although ultimately it led to disaster, the development paradigm of the Great Leap Forward was not irrational. Pooling land would better accommodate new irrigation systems and agricultural mechanization. The assumption of household responsibilities by the commune liberated labor, mainly of women, to contribute to production focused on building irrigation systems and smelting iron. The conversion of surplus labor into capital inputs for agricultural development was one path counseled for developing countries in the postwar period. Indeed, the notion of small-scale, "appropriate technology" would gain a following in the West in response to the sometimes-negative social and environmental impact of large-scale industrial undertakings. Understandably, the Chinese were looking to achieve an economic breakthrough that would be sustainable and appropriate to Chinese conditions. By the end of the first year of the farmland irrigation campaign, the Central Committee reaffirmed the "large scale movement for water conservancy construction" while emphasizing the "three primaries": small-scale projects, storage, and organization and execution by the people's communes.[10]

The massive exploitation of resources during the Great Leap Forward was effected not through industrial technologies but through a purposeful restructuring of rural society that released labor for state-directed ends. Representatives of Party and state power were present at every level of society. This degree of state penetration was unprecedented in Chinese history. The Party had the capacity both to efficiently communicate goals and directives to its constituency and to determine compliance. Yet a pervasive presence of the Party alone is insufficient to explain the effective mobilization of hundreds of millions of peasants to surrender property, to relinquish a strong identity to households, and to trudge to the fields to dig irrigation ditches, ponds, and reservoirs. The mere presence of Party officials and the proximity of the state's military power likely reminded rural inhabitants of the raw coercive power of the state. Rectification campaigns like the Anti-Rightist movement were less than subtle reminders of that power. Still, it remains quite extraordinary that the Party bureaucracy, spread throughout this large society, was able to maintain the threat of coercive power to mobilize rural society as quickly as it did during the Farmland Irrigation Campaign in the winter of 1957–1958 and the subsequent years of the Great Leap Forward. It is hard to overestimate how effective Mao and the Party were in structuring a system of rewards and punishments to coerce compliance down to the lowest levels of the Party bureaucracy. In turn, the rural cadres must be credited with the success of mass mobilization. To be sure, these officials were engaged in forms of mediation between Party dictates and local interests, suggesting a degree of insularity in local society, and the maintenance of traditions from the imperial period. Yet it is impossible not to recognize the transformational social and ecological outcomes generated by this mass mobilization of human labor organized and led by the Party bureaucracy.

## Landscape and Culture in the Great Leap Forward

During the Great Leap the Party-state expended considerable effort in state-run print and other media to create an identity of goals and values between the Party and rural and urban residents as well as within the Party and in state bureaucracies. As with the People's Victory Canal and the 1955 Yellow River Exhibition, the state attempted to carefully manage both traditional and modern symbols and values of landscape. This land had been physically and metaphorically etched by the humiliation of China's modern history, but the landscape simultaneously suggested the regenerative productive potential that must be extracted by human labor. From this point of view, a reengineering of the countryside would impart a new conception of beauty, a beauty that reflected the reemergence of China and the realization of a communist utopia—a region transformed by irrigation channels, reservoirs, ponds, and dams. The intriguing difference between efforts to transform the landscape in the early 1950s and during the Great Leap Forward was the conscious decision in the late 1950s to call upon a technology complex quite different from the Soviet-inspired model. Along with peasant experience and knowledge, decentralization, and local financing, the CCP's mobilization effort was firmly embedded in the history of the Party—a history that included the central act of violent revolution based on mobilization of the countryside.

A flurry of editorials trumpeting the success of the winter's irrigation campaign, and an even greater number announcing the reinstatement of High Tide policies, sanctioned by the Eighth Party Congress (Second Session) in May 1958, unequivocally signaled the victory of radicals in the Party leadership. The beginning of a new irrigation movement was at hand. In "The New Form of the Water Conservancy Movement," published in *Renmin Ribao* in June 1958, Vice-Minister of Water Conservancy and Electric Power Li Baohua offered a detailed exploration of the theoretical and practical bases for the irrigation campaign. Li first offered the previous winter's campaign as a model of emulation for the *shuilihua* (literally, water conservancy-ization) of the country. The overriding theme was that an unprecedented transformation of nature (*ziran mianmao*) could be accomplished only by the mass mobilization of labor. Drawing upon examples, mostly from the North China Plain, Li highlighted localities that engaged in reforestation, land reclamation, and the digging of reservoirs and channels, all of which broke through outmoded practices and represented the leading wave of a great leap in agricultural production. The battle waged by the masses transcended the overly elaborate technical plans of the past by relying on the vibrant experiences of the masses. This kind of mass movement broke through the mysterious perspectives of [modern] technology. "At several locations, the great labor of the masses has revised the laws of hydrology."[11] The "melon and vine" (a system of medium-size reservoirs connected by channels fed by a larger waterway), the grape and vine (small reservoirs connected by channels), networks of channels, and other irrigation breakthroughs had no precedent. As soon as the masses grasped the technique of building dams, they built all types of dams (e.g., earthen, rock fill, brick), they even "dared build dams as high as 50 feet."[12]

Li also drew a distinct parallel with building communism. He wrote, "The movement has smashed the old notion that only those who benefit are responsible. This reflects the people's conscious and heroic socialist spirit. Not only has water conservancy construction transcended the boundaries of household, county, and district, but it is a movement of socialist construction for all people. As the conscious transformation of water conservancy construction, the outcomes of the movement represent a further step in the elevation of communist thinking."[13] These achievements, which were transforming the North China Plain from privation to affluence, could only be attributed to the correct path of the Party, which advanced collectivization, conducted Party rectification, struggled against Rightists, fought waste, and combatted conservatism. "If there was no direct leadership of the Central Committee and Party Committees at all levels, the new pattern of water conservancy construction would not have been possible."[14]

Li then equated the Great Leap Forward with the spirit of Yan'an. In the battle (between the storage/small-scale/mass-based line versus the drainage/large-scale/state line), the confrontation over water management mirrored the debates over development trajectories among the Party's leadership and reflected a blurring of the lines separating Party and peasantry. All were collectively engaged in a battle with nature. The commune-based assaults on the landscape were consonant with the peasant-based localized guerrilla warfare of the Yan'an period. Li expended considerable verbiage in defense of this strategy as he argued that an emphasis on water storage was, in fact, a better way to combat floods (the major concern of those advocating water drainage). An additional attribute of the spirit of Yan'an was experimentalism, an adaptive attitude that shaped the revolution to local conditions. This experimental approach stood in contrast to dogmatism (*jiaotiao zhuyi*), which was formulated and enforced from the center. Li recognized that locally organized small-scale irrigation projects generated apprehension among some technical experts. But he dismissed the objections of the experts as technical dogma contrary to the spirit of local enthusiasm and experience. Li continued his attack on such dogmatic, conservative technicians by decrying their endless compilation of data and their endless discussions. In the realm of irrigation, he said, there was a failure to learn from the experiences of the upsurge in irrigation in 1952 that sprang from the people:

> These experiences have been lost in reams of plans and regulations. Bureaucratization restrained local and mass initiative. The 1952 plan to draw Yellow River water for irrigation in Shandong came to nothing. Those projects that were completed were capital-intensive. But those irrigation projects that drew upon the water supply of the Yellow River last year broke through the myths that restrained the masses [from acting]. If we follow the old precepts, we will never be able to complete large-scale irrigation projects within three years. If we release the bonds placed upon the masses we can complete them in half a year.[15]

Finally, Li issued an injunction to the bureaucracy to be vigilant against the "observing-the-tide-clique" (*guan chao pai*) that waited to see what would happen next and the "oppose-rash-advance-clique" (*fan maojin*) that looked for errors as a pretext for launching a counterassault on the program. In more concrete terms, Li outlined the program's goal: China would have 1.2 billion *mu* of irrigated farmland by 1962—that is, 70 percent of farmland in China.[16]

Chinese media outlets published a stream of formulaic articles that instructed, cajoled, and denounced those opposed to mass mobilization (i.e., planners and technicians). The articles also reaffirmed the heroic spirit of Yan'an. In a lengthy article about Henan

Province, the author of one report rhetorically asked if it was possible to increase irrigated acreage by 20 million *mu*. If the peasants developed an attitude of aggressively attacking *(menggong)* nature, then the entire face of the province could be transformed. In the great upsurge of transforming nature, wave upon wave *(yi lang ganyue yi lang)* of attacks, conducted in the mountains, hills, and plains, would time and again surpass any targets for irrigated land. The report concluded, "The collectivized peasant who threw off the [private] small-holder system [*xiao nong jingji*] will sprint down the path of socialism. Will you [party officials] accompany [the peasants]? Can we dare to proclaim victory? Can we lead the masses to ever greater victories? These are the serious issues that confront the leaders [Party cadres] of the movement. Twenty million *mu* of new irrigated acreage [in Henan Province]? If properly motivated and led, the peasantry can achieve the previously unthinkable 50 million *mu!*"[17]

The article on Henan Province also suggested the existence of a "conversion process," in which the witnessing of such superhuman achievements engendered belief and trust in the Party as it led the masses to communism. The authors stated: "This instance of a 'high tide' of water conservancy is not only an economic movement but also a rich movement of socialism. It [the water conservancy campaign] will cause people to tangibly witness the greatness of the Party." The article recounts the conversion story of Zheng Hai, a "well-to-do middle peasant" from Henan who was seriously afflicted with capitalism: "Members of the collective attempted to rectify his thinking on three occasions, but he remained unconvinced, as he constantly spoke strangely. But the water conservancy movement brought him to socialism. From the complete regulation of his village's water, he saw the brilliant future [*canlan qianjing*] of socialism, [and this vision of the future] transformed him into an activist for water conservancy. Every morning he called for people to start work. Commune members praised him by saying: Zheng Hai used to talk strangely, now he has changed into a 'crowing rooster.'"[18]

There were also well-to-do middle peasants who donated thirty years of savings to support water conservancy construction. The article cited a former middle peasant, who said: "Beat me with a

hundred rods, I still won't leave the commune."[19] Reinforcing the
sense of identity with the Party, simple ditties were propagated by
Party cadres to communicate in direct terms with the peasantry.
For example, an article extolling the movement in Henan included
this song:

> The Communist Party resembles my dad and mum.
> Water conservancy gives boundless blessings.
> Now we eat without relying on Heaven to bestow.
> The peasants act as the old Dragon King.[20]

On the eve of the 1958–1959 Great Leap Forward winter irrigation campaign an editorial in *Renmin Ribao* titled "Set off an Even
Larger High Tide of Farmland Water Conservancy" established
the tone of the movement and the targets for the coming season.
With rural society increasingly regimented into a hierarchy of work
teams, production brigades, and communes, the rhetorical style of
the campaign adopted an appropriately militarized tone. The editorial reviewed the extraordinary accomplishments of the previous
year by attributing success to the 500 million intrepid peasants who
fought bravely under the leadership of the CCP to bring about a
"brilliant military victory" *(huihuang zhan'guo)*. As reported in the
editorial, the numbers were stunning: the previous campaign season (1958–1959) generated an *additional* 480 million *mu* of irrigated
farmland. In one year, more irrigated acreage was added than during the thousands of years before "liberation" (230 million *mu*) and
in the eight years since (290 million *mu*), as impressive as that last
number was historically. More irrigated acreage was added in China
in one year than globally in the previous twenty years (400 million
*mu*, excluding China). "This unprecedented achievement was only
possible by 600 million Chinese laboring under the leadership of
the Chinese Communist Party."[21] This was a "brilliant military victory" not accomplished by heavy industrial earth-moving equipment, but by "a million men with teaspoons."[22]

The directive on water conservancy work for 1959 (issued by the
Central Committee in August 1958) noted that over half of China's
cultivated area was now irrigated, with China's irrigated acreage

comprising one-third of the global total. With the hyperbole common to Great Leap Forward pronouncements, the directive opined that within two years of struggle it was entirely possible that all of China's cultivable area could be irrigated and that all of China's farmland could be "water conservancy-ized."[23] For 1959, the Central Committee established a national goal of irrigating an additional 490 million *mu* (10 million more than the prior year), reclaiming 72.8 million *mu* of waterlogged land, and constructing 96 billion cubic meters of earth and stonework. Because the work for 1959 had already started in August 1958, when the directive was adopted by the Central Committee, there was every expectation that these targets would be surpassed.[24] Of course, the purpose of mass-based irrigation development was to increase agricultural production. Indeed, grain production reached a high of 200 million metric tons in 1958, prompting the comment, "[This] was only the first round on the front-lines of the battle for agricultural production."[25] Grain production, cotton production, and the production of oil-bearing crops would all "smash through production targets" for 1958 based on the continued development of irrigation.[26]

The optimism expressed in the 1959 directive drew a direct line between the mastery of water and the building of communism. It stated: "Our greatest purpose is to gradually advance from socialism to communism. From this perspective, our construction in water conservancy is intended not only to solve the questions of drought and floods but also to progress in controlling climate and remaking nature . . . the only limitation is the ineffective exploitation of water resources."[27] Fulfilling these utopian dreams meant that "every cubic meter of flow, every foot of drop [should] assist socialist construction." The potential limitations of water resources gave rise to a sense that greater interventions were necessary, and indeed possible, to secure this water. Thus, the editorial foresaw the need to impound water on all small rivers as well as to divert southern Chinese water to the north.[28] Indeed, dreams of massive Soviet-style projects were not completely eradicated in the delirium of the Great Leap Forward. An October 1958 article, titled "The Transfer of Southern Water to the North—The Yangtze and Yellow River Join Hands," noted, "[Because] of the expanding development of agriculture and industry along the Yellow River in the past few years, there is a growing sense that the water resources of the river are not adequate."[29]

## Communal Life and the Great Leap Irrigation Campaign

During the winter agricultural off-seasons of 1958–1959 and 1959–1960, it has been estimated that 100 million to 120 million peasants were engaged in farmland water conservancy projects.[30] Labor input for these projects hit a high in 1958–1959, when each member of the rural work force averaged 120 days engaged in digging and hauling. The outcome of communalization was to convert rural seasonal underemployment to seasonal unemployment. Prior to communization, households still attended to sundry household and agricultural chores (e.g., feeding the pigs or gathering manure). Now that these tasks were performed communally, pools of *unemployed* labor were created that could dig irrigation channels and reservoirs. The strategy proved successful, but in the face of the extraordinarily ambitious targets for irrigation development and the demands of the "backyard furnaces," a new phenomenon developed—a labor shortage that eventually affected the fundamental work of agriculture: planting, tending, and harvesting. Thus, the Central Committee's directives for the 1958–1959 conservancy campaign sought to encourage the development of simple tools and other rudimentary labor-saving devices to increase efficiency in the irrigation drive.[31]

During the Great Leap Forward, each household was enrolled in a production team that typically centered on a natural village. Production teams were organized into production brigades within each commune, typically containing 5,000 households, but sometimes as many as 20,000. As noted above, the commune managed production and social services with broad guidance from provincial governments. The regimentation of rural society allowed efficient recruitment for irrigation projects. In some localities, up to 90 percent of able-bodied men and women were engaged in water projects during the 1958–1959 agricultural off-season. Up to one-third of the

workers were women. These numbers reflected the fact that irrigation development was the single largest area of investment for the
communes.[32] Because projects were small in scale and typically local, workers lived at home or boarded with other families. In addition to peasants drafted from the production team, labor units included troops from the People's Liberation Army (PLA). In
agricultural communes near urban areas, episodic labor was supplied by industrial workers, teachers, and others. The need to train
a cadre of "peasant technicians" was theoretically the outcome of
on-the-job training, which was augmented by brief, one-week training courses in surveying, design, and construction. Shandong Province was reported to have trained 1 million peasant experts in
1958.[33] In addition to communal administration and local knowledge, the last of the "three primaries" was small scale. The lack of
investment capital available to local communes required them to
rely on inputs of local material. Thus, "wood, bamboo, tile, and
other material were substituted freely for steel, iron, and cement. . . .
There probably were as many 'formulas' for concrete as there were
construction sites. One such 'formula' called for 70 percent ground
old brick, 25 percent lime, and 5 percent gypsum."[34]

The administration of Great Leap Forward irrigation projects lay
totally with the commune. All projects were also subject to approval
by the commune. Projects at the village level were executed at the
production team level, those that crossed the boundary of two production teams (i.e., two villages) were managed by the brigade, and
so on. This strong sense of self-reliance in local irrigation development was buttressed by three other factors. First, in the ideological
realm, cadres were exhorted to criticize, to work, and to lead, in
order to transcend a conservative peasant culture infused with fa-
talism, familism, and superstition. Cadres were instructed to hammer away at the notion that individual material rewards were secondary to the benefits that would accrue to the collective and to the
nation through hard work. Second, the organization of irrigation
work should be highly militarized. Because brigades were often the
basic administrative unit, "commands" were established on site, and
workers were organized into battalions and companies. Party cadres and members of the communal militia formed "breakthrough"

teams that performed the more dangerous tasks. Such teams were held up for emulation. The presence of the armed communal militias was undoubtedly recognized by the peasantry as a potent expression of state power. The militias were ostensibly organized to fortify military readiness in the aftermath of the Taiwan Strait Crisis in August 1958, when Taiwanese and Chinese military forces clashed. Meanwhile, the threat to the nation and the growing rural militia movement heightened a sense of patriotic fervor. Once again, this fervor served as a mandate for an attack on natural forces in order to strengthen China in a contest with external enemies. We have descriptions of Party cadres fanning this patriotic fervor by leading the peasant workers in song as they marched to irrigation project sites:

> With a hoe in the right hand and a rifle in the left,
> We consider the field the battleground where we use
> Our hoes to attain the 10,000-catty-per-mou [*mu*] target,
> And our rifles in training to guard the fatherland.
> If the American imperialists dare to invade,
> We would definitely annihilate all of them.[35]

The heightened sense of military organization was reflected in the labels for various types of irrigation operations: "general mobilization" was coined for more routine projects; "combat disposition" suggested readiness to engage in large projects; "exterminating combat" connoted overcoming the enemy (i.e., the natural environment); and "surprise attack" meant smaller projects. Military metaphors reinforced the imperatives of coordination, discipline, and, above all, subordination to the Party.[36]

What about the individual peasant? Was there a firm belief in the building of communism that mobilized him or her to dig a quota of 3 m³ of dirt per day? On the one hand, communalization was a frontal attack on the economic and moral cell of Chinese culture—the family. This must have created a pervasive sense of psychological displacement, though there were intense state efforts to introduce the commune as the new "hearth." On the other hand, although perhaps less obvious than in similar transformative efforts in the

Idealized commune. (Courtesy of the International Institute of Social History, Amsterdam)

Soviet Union, there were nevertheless many reminders of the raw power of the state. In addition, there were other powerful moral and economic forces that compelled peasant compliance with state goals. The revolution, the moment that China "stood up," was less than a decade old. And this revolution had been based in the countryside, a revolution for and by the peasantry. For Mao and the Party, land reform was the tangible reward for this broad social base of support. The Party spared little effort in creating and sustaining the myths and symbols of a revolution that welded a partnership between rural society and the Party. The revivification of the Yan'an mythos was a powerful reminder of that historical trajectory. Mao's injunction that "everyone is a soldier" recreated the myth of the Party and the peasant united in resistance to enemies at home and abroad. With a moral world animated by the visions of military victories, and a physical world regimented along military lines, communal life was organized around a set of ritual practices (e.g., the irrigation battalion marching to the irrigation project), and was animated by a creed celebrated in poster, song, and revolutionary discourse that "produced bonds of solidarity without requiring uniformity of belief."[37] Indeed, communalization transformed rural society. But the speed with which it was accomplished not only speaks to the perceived power of the state and the precocious capacity of the Communist Party but also attests to the solidarity of a priestlike Mao with a faithful Party. All these factors reflected the promise of material gains for rural communities on the North China Plain—villages that had now been empowered to transform their historically tenuous ecological setting. And it was on the North China Plain that mass mobilization appeared to be most transformative. Peasant participation in the mass mobilization irrigation drive was paying material benefits as grain production reached a post-1949 high of 200 million metric tons. Unfortunately for long-term rural economic development, these agricultural returns, and whatever level of identity they promoted between Party and rural society, quickly receded. The revolution that had been fought on behalf of the peasantry looked like something else entirely as the Great Leap Forward transitioned from hope to desperation.

## Sanmenxia and the Great Leap Forward

By April 1957, several months before the Anti-Rightist movement set the stage for the Great Leap Forward, construction had begun on Sanmenxia. The project started before the parameters of the dam had been settled. The Party's resolve to build the large-scale capital-intensive project in the political crosswinds of the Great Leap Forward suggests the continued strength of the ideal of "walking on two legs." But it would be incorrect to assume that the mass movement Great Leap Forward did not leave its imprimatur on the Soviet-inspired Sanmenxia project. The Great Leap Forward's strong emphasis on speed and economy affected implementation of the project. At the same time, the towering structure increasingly transcended any contribution to multipurpose Yellow River management. It became a monument to the capacity of the Party to engineer such a massive structure.

The descriptions of the Sanmenxia site and of the ceremonies celebrating the start of construction are snapshots of the normative values and symbols of the Soviet-style technology complex. Several days before the project's opening ceremony, a wonderfully detailed sketch of the worksite and Sanmenxia City was published in *Renmin Ribao*. As construction approached, the area near the train station was filled with every type of equipment. Nearby was a 40,000-square-foot warehouse for machinery and material. Every day the train delivered equipment like the massive 45-ton crane produced by the Dalian Rolling Stock Factory. Similar mechanical "giants" *(juren)* arrived from all over the country. Everywhere was an expanse of machinery where just a year earlier there had been only wilderness. A short distance from the station was the transformation of village into city. Sanmenxia City now had a post office, a bank, and a department store with fabric, medicine, and "cultural products." There was a host of small shops, a public bath, and a clean and hygienic salon "where the girls could get a permanent wave."[38] A club with professional and amateur opera troupes, and a library completed a new cityscape that, compared with the ascetic descriptions of Great Leap Forward irrigation projects of less than a year later, seemed rather bourgeois.

In addition to a celebration of high-modernist industrial machinery and urbanism, other accounts extolled the value of technical knowledge, technical elites, central planning, and Soviet advice. Over 600 old and young technical experts were at the site by April 1957. In addition, "science halls" were established at the construction site to provide technical training for cadres and other personnel.[39] Technical expertise was provided by the 5,000 Soviet technicians involved in the project. An emphasis on blueprints suggested size, structure, planning, and technology. Successful execution required the careful coordination of multiple state agencies to provide equipment, transportation, and workers. An editorial in *Renmin Ribao* titled "Everyone Support Sanmenxia" issued a challenge to ministries and agencies to carefully coordinate contributions to the construction of a project critically important to the nation. Only with this equipment could "the earth be leveled to raise a family" *(ping di qi jia)*—an interesting juxtaposition, intended or otherwise, with the labors of Yü the Great in creating Chinese civilization.[40]

Experienced Party cadres were also needed and were seconded to Sanmenxia from around the country. The lessons from the large reservoir projects on the Huai River system had provided technical precedents and a pool of experienced cadres in large project management. Sanmenxia was a focal point of interest for this experienced group. A project of its size and complexity required all the human and material resources of the nation. The project bound the nation together in an organizational and metaphorical network centered on the lower Yellow River valley. A front-page report on the ceremony marking the start of construction recounted the grandeur and beauty of the physical location, with the three massive rocks midstream, colorful banners, a huge cloth sketch map, a brilliant sun, the surging river, the multihued rocks on the valley walls, and a large banner on the south side of the river, with the couplet: "A cure for the destruction is coming, the Yellow River running clear is at hand." At 12:50 P.M., Liu Zihou, the director of the Sanmenxia Engineering Bureau, gave the order to commence. One after another, the "thunderous explosions on the left bank broke the valley walls into pieces that danced in the air, as dust and smoke swirled in the wind and drifted away."[41]

The blasting of Sanmenxia must have echoed in the halls of the Yellow River Conservancy Commission in Beijing as technical discussions over the dam continued. These debates went on for two full years after the first consultative meeting hosted by Zhou Enlai in June 1957. At that meeting, during the more open atmosphere of the Hundred Flowers movement, a dozen or so professionally trained engineers, as well as representatives from upstream Shaanxi Province, voiced objections to the high-water design of 360 meters initially proposed by the Soviet advisors. What remained contested, even after construction had begun, was the basic function of the dam and reservoir as reflected in the dueling formulations of "store water, settle sediment" (xu shui lan sha) and "store floodwater, discharge sediment" (lan hong pai sha). The first option had been proposed by the Soviet Union and focused on storing water to settle all sediment behind the dam, while utilizing high water levels to facilitate transport, irrigation, and hydroelectric generation. With all sediment settled in the reservoir, water release gates, located relatively high on the dam, would discharge clear water downstream.

Proponents of this plan also argued that there simply was not a better site on the river for a project that would serve as the critical regulating facility for the development of the entire series of dams on the Yellow River (proposed in the Yellow River Plan of 1955). However, opponents of the Soviet design countered with two arguments: first, the major function of the dam should be to regulate downstream flooding, hence, with a larger reservoir, there would be significant seasonal variation in reservoir depth that would necessarily reduce multiple usage (i.e., transport, irrigation, or hydropower). These opponents also argued that the Soviet plan underestimated the rate of sedimentation. The reservoir would simply fill up with sediment too quickly if silt was not flushed downstream. These dual considerations impelled opponents of the original plan to argue for a smaller dam and reservoir, and for lower discharge conduits that would release water plus sediment. The Anti-Rightist movement silenced critics such as Huang Wanli, but technical debate continued within the design teams of the YRCC. Beginning late in 1957, the debates increasingly betrayed provincial interests. Shaanxi officials were most vociferous in their concerns that a large reservoir would back up water in the Wei River (where the Wei merged with the Yellow River, not far from the potential extent of the larger-design Sanmenxia reservoir). Shaanxi officials feared that a backed-up Wei River would inundate the city of Xian.[42]

The project was too important for technical designs to remain incomplete. Zhou Enlai convened three additional conferences to settle the matter. YRCC officials also flew en masse to Moscow for consultations. Blueprints went back and forth between the YRCC and the Lenin Design Institute in Moscow. Finally, in June 1959, the matter was settled: the reservoir depth would be 340 meters until 1967, after which it would increase to 350 meters. In addition, water discharge outlets on the dam would be lowered from 330 meters to 300 meters.

Debate over the design of Sanmenxia was framed by the assumptions and goals of the Great Leap Forward. On the one hand, continued patronage of the project was premised on extending the efficacy of mass-based irrigation projects by better regulating Yellow

Joint Sanmenxia planning meeting in the USSR. From Shuili bu Huanghe shuili weiyuanhui, ed., *Renmin zhili Huanghe liushi nian* (Zhengzhou: Huanghe shuili chubanshe, 2006), 102.

River flows in times of drought and floods. From this perspective, Sanmenxia complemented Great Leap Forward priorities. Although mass mobilization techniques and the Great Leap Forward's celebration of human will over material endowments could energize labor recruitment and discipline, it was simply impossible to build a multipurpose facility as large as Sanmenxia without capital-intensive inputs and a central bureaucratic presence that could effectively procure and manage the human and material resources necessary for such a large project. Thus, in some ways, the Sanmenxia project was protected from the normative standards of the technology complex of the Great Leap Forward by powerful central patronage. In addition, the long-term promise that multipurpose projects held for regional development of industry and agriculture was too alluring to be dismissed. The importance of the project to the Party was suggested by the frequent interventions of Zhou Enlai and by the procession of senior Party leaders to the site.[43] On the other hand, the Great Leap Forward's obsession with water storage may well have contributed to the ultimate decision to retain the thrust of the Soviet plan (which emphasized massive storage of water *and* of silt).

The other Great Leap Forward obsession, speed, also had an impact on the project. Originally scheduled for completion by 1964, the project was largely finished by 1962. The emphasis on speed, combined with economic pressures and a cultural environment that denigrated technical expertise, may have all combined to advance "straightforward," seemingly "commonsensical," and "practical" solutions to complex problems like the behavior of silt. Such Great Leap Forward reductionism likely contributed to the decision to plug the silt discharge tubes on the dam. The decision would have fateful consequences.

The imposition of the Yan'an narrative on the Sanmenxia project during the Great Leap Forward had several results. Great Leap discourse generated a vision of the Sanmenxia work site as the facade of a moral community that sanctioned violence upon the landscape to extract the productive power of water. In 1958, the worksite was said to resemble "a battleground."[44] The objective in this confrontation with nature was "to transform the river that from ancient times has been violent, stubborn and unruly; to tame the river by bending it to the human will."[45] Descriptions of the project during the Great Leap Forward emphasized the capacity of the human will, not the capacity of industrial technologies, to transform the face of Sanmenxia. But the spirit of battle was also said to be embedded in other presumed cultural qualities of the Chinese people such as their industriousness and a capacity to "eat bitterness": "Upon the rocky islands of the gorge are built the mainstays of the massive dam symbolizing the thousands of years of our nation's culture—a foundation powerfully standing amid the unrestrained billows of the river."[46] Vivid pictorial representations of heroism reflected the muscular deeds of model workers, emerging victorious from their struggle against overwhelming odds. For example, armed with pneumatic drill, CCP cadre Wang Jinxiang led his shock brigade of young laborers to the foundation to dig out layers of mud and sediment before conditions changed. At the critical moment, the laborers fought for twenty-four hours without leaving the battlefield of the dam site. "In the truck on the way back to the workers' dormitory, everybody broke out in laughter, [because] except for their two glittering eyes, from head to toe they had become clay figurines [niren]."[47]

Throughout this period, the Sanmenxia project was celebrated in verse and song that invoked revolutionary romanticism. In late 1958 *Renmin Ribao* published several poetic tributes to Sanmenxia, such as:

### WE MUST BATTLE THE YELLOW RIVER
### ABOVE AND BELOW

The boldness of heroic peoples is great,
We must battle the Yellow River above and below.
Blocking the surging water of the Yellow River,
Chopping [off] the scales, claws, and teeth of the wicked
    dragon.[48]

The adoption of the dragon as a symbol of the Yellow River, and of water in a general sense, was consistent with long-held cultural traditions. Transmitted legends of Yü the Great had described the taming of the Yellow River dragon and how this had made the North China Plain safe for Chinese civilization and culture to flourish in relative peace. The appropriation of this myth to serve contemporary goals during the Great Leap Forward, however, added layers of particular violence to this myth. As suggested by the verse above, the construction of the Sanmenxia dam across the Yellow River invited images not only of the taming of the dragon but also of transforming the dragon to an extent that neared complete emasculation. Other rhetorical renderings of Sanmenxia adopted the imagery of a violent, ferocious, stubborn, obstreperous, roaring dragon that had been seized by the throat (the dam across the river) through the efforts of pneumatic drill-wielding shock brigades of workers who subdued the dragon-river and then enlisted the river in service to communist modernity.[49]

The rhetorical attacks on the landscape evoked the restaging of a violent revolution. There was little question of the centrality of the Yellow River valley and the North China Plain to the revolutionary mythos of the Party. And the re-imposition of Yan'an discourse into national discussions during the Great Leap Forward was reflective of the importance of the North China Plain to the Party. From a purely material perspective, one need only look at the Yellow River Plan of 1955 and the farmland irrigation campaigns

of the Great Leap Forward to judge the mythic *and* material connections that this region held for the Party and the state. Indeed,
so important were both dimensions that the landscape had to be
attacked and drastically transformed to create the conditions necessary for the rebirth of Chinese civilization. And this rebirth would
take place on that same stage as the original act of creation (the
myth of Yü the Great) and as the Yan'an revolutionary drama. The
latter image remained powerfully implicit, the former powerfully
explicit.

Beneath a bright autumn moon, the festivities of China's National Day celebration in Tiananmen Square on October 1, 1958,
embodied the spirit of the Great Leap Forward. Over 200,000 participants, men and women, old and young, stood on a grand stage to
celebrate the Great Leap Forward. Volleys of fireworks lit the sky.
Workers from all over Beijing performed their own skits, rhythmic
chants, and "cross-talk" (*duihua;* two-person comic exchanges),
here "versifying about the Great Leap Forward," there moving to
the "red banner dance." From the four sides of the square people
yelled, "We must liberate Taiwan!" and "Socialism is good." Workers from the Beijing Cotton Cloth Factory No. 3 performed a skit
called "Topple the American Wolves." At the center of the square
"people's heroes" were recognized, "among them [were] heroes who
fought the invading American army, [fought the] Japanese devils
[*Riben guizi*], and [fought] the traitorous army of Chiang Kai-shek
[*Jiang zei jun*]."[50] In the small alleys off the main thoroughfares, lan-
terns burned brightly amid unceasing singing. During all the excite-
ment, as described in the next morning's newspapers, the "Sanmenxia
Cantata" was premiered.

The music for "Sanmenxia Cantata" was written by Xi Xianhe,
with lyrics by Guang Weiran. Guang had also written the lyrics to
the "Yellow River Cantata," which in 1958 remained an iconic testament to the Yan'an revolutionary will and to anti-Japanese nationalism. Indeed, a comparison of the two cantatas serves to illustrate
important similarities and subtle distinctions between the Yan'an
and Great Leap Forward periods. The identity of the Yellow River
and North China Plain with the revolution and Party remained the
central thread running through all three periods. The heroic, volun-

tarist, mass-mobilizing essence of the work at Sanmenxia recalled the power of the revolutionary will of the Yan'an period. The idiom—a cantata for a large choir that, in effect, committed and subsumed individual expression to the collective—was never as fresh in the late 1950s as during the Yan'an and Great Leap Forward periods. What was subtly different in the "Sanmenxia Cantata," however, was the singular identification of the Chinese landscape as enemy. The "Yellow River Cantata" celebrated resistance to the Japanese enemy. The landscape was a metaphor for the revolutionary struggle (crossing the turbulent river), and the setting from which Chinese civilization arose. In the "Sanmenxia Cantata," however, this dual character of river and region, both as nurturing parent and as symbol for existential struggle, was subtly reweighted by the purposeful displacement of a composite enemy (the American-Japanese-Chiang Kai-shek threat) with the image of the landscape as foe. With all the mobilizing power of the Party, the training of all guns on the landscape had a remarkably potent transformative effect.[51]

## Great Leap Outcomes: The Mass Irrigation Campaign

The same *Renmin Ribao* article chronicling the joy of the 1958 National Day celebration in Beijing ended by noting: "In the suburban districts, members of people's communes hosted dinner parties where they enjoyed the beautiful full moon and ate mooncakes and fruit supplied free of charge by the commune as they discussed how, beginning tomorrow, they could contribute even more labor to secure an even greater harvest to welcome the tenth National Day [next year]."[52] These dreams proved illusory as grain production declined precipitously after reaching a high of 200 million metric tons in 1958. The grain output for 1959 and 1960 was considerably below the annual average for the years *before* the Great Leap Forward. As production dropped, however, government procurement increased. Thus, the per capita amount of grain retained at the communal level declined to a low of 211 kg in 1960.[53]

Faced with a growing sense of crisis in the countryside, calls from elements of the Party leadership in mid-1959 to relax the Great Leap Forward policies were rejected by Mao. He ridiculed such exhortations as rash, and he cashiered those who suggested retrenchment. Unable to disentangle Great Leap Forward outcomes on the ground from the contentious politics at the top, Mao argued that intensification, not relaxation, of efforts would generate bounteous harvests necessary for China to "leap" into communism. Following the reaffirmation of the Great Leap Forward at the Lushan Conference in June 1959, Liao Luyan, minister of agriculture, articulated the slogan "take grain as the key link." This campaign within a campaign was designed to increase cultivated acreage by 10 percent. As in previous movements to expand production, "grain first" meant bringing more water to irrigate marginal lands. At the same time, the campaign incentivized lake and wetlands reclamation in areas such as Dongting and Poyang Lakes.[54]

The Great Leap Forward sucked the life from rural society. Digging irrigation ditches and reservoirs, round-the-clock tending of backyard furnaces, and farming under the unremitting pressure of faster, cheaper, and better exhausted the physical and human resources of the North China Plain. Grain output declined. At the same time, state procurement of grain increased through a bureaucratic logic that forced compliance from local officials, promoted false reporting to the center, and drove Party leaders to ignore or reject rural distress. Increased state procurement of decreasing grain production resulted in widespread famine. The results of the Great Leap were devastating. Deaths from starvation exceeded 25 million. It seemed that the revolutionary spirit of Yan'an succeeded only in consuming its historical social base of support.

But what of the legacy of the Great Leap Forward on the landscape? Did the Great Leap Forward mass campaigns create an enduring irrigation infrastructure on the North China Plain that generated increased agricultural production *in the long run?* The answer is difficult to clearly discern. There is considerable evidence that the campaign style of the Great Leap Forward, with its overriding emphasis on speed, local knowledge, human labor, and thrift produced irrigation systems that were substandard. Precise judgments are complicated by unreliable data.[55] Perhaps the better question is whether the Great Leap Forward generated significantly increased consumption of water, irrespective of whether these with-

drawals fueled increased production. Here, the answer is an un-
equivocal "yes."

The irrigation campaigns of the late 1950s promoted a developmental framework that contributed to a massive transformation of
China's landscape. The aggressive expansion of irrigation networks,
however inefficient, likely contributed to a *longer-term* capacity to
feed a population that experienced profound growth beginning in
the 1950s. Indeed, some commentators have praised China's ability
to feed itself for most of the post-1949 period—a perspective that,
implicitly or otherwise, respects the social welfare function of state
guarantees of at least minimum sustenance for all Chinese citizens
(again, with the obvious exception of the Great Leap Forward period). At the same time, such guarantees relied on the state's pur-
posefully suppressing demand by rationing food. However, the
meeting of even minimum food needs should not be overlooked.
This accomplishment required intensification of traditional agricultural methods like irrigation, as well as the application of modern agricultural inputs like chemical fertilizers. When one consid-
ers that even this modest agricultural production was accomplished
with a massive drawdown of the limited water resources of the
North China Plain, one can better understand the multiplying effect on water resource demand experienced in China when, for example, the state lifted constraints on demand in the post-Mao era,
when water consumption in the industrial and urban sectors increased
dramatically. In other words, the irrigation campaigns of the Great
Leap Forward may, indeed, have helped ensure China's food self-
sufficiency, but this accomplishment came at a heavy cost to the
water resources of the North China Plain.

Two other analogous Great Leap Forward initiatives may serve as
proxies to explore the outcomes of the mass irrigation campaign.
The Great Leap Forward "campaign to eliminate the four pests"
mobilized communes to attack the enemies of production: flies,
rats, mosquitoes, and sparrows. Banging on anything that could be
banged, bands of commune members harassed birds until they fell
dead from exhaustion. It is difficult to imagine that communal members were not aware of the net benefit of sparrows (which probably
ate more crop-threatening bugs than they did kernels of grain).[56]

The second mass effort during the Great Leap Forward was a soil conservation drive that focused on creating massive swaths of shelterbelts, likely patterned on similar Soviet efforts, to protect farmland from windblown sand from the steppe-desert regions. Reports, however, suggested that only 30 percent of plantings survived.[57] Neglect, plus a significant demand for wood to fuel backyard furnaces, compromised the effort. This pattern of misplaced planning, poor coordination, and unintended consequences was consistent with the outcomes of the mass irrigation campaign.

The Party initially attributed the Great Leap Forward famine to the "three difficult years" (1959–1961), when unusual flood and drought conditions wreaked havoc on agricultural production on the North China Plain. There is evidence to suggest severe localized drought in the Huai River valley of Jiangsu and Anhui. But we also know, and this is widely acknowledged in China, that the famine was generated by a convergence of forces: organizational, political, and environmental. Continued attribution, in whole or in part, of this large-scale famine to too little water or too much water calls into question the efficacy of flood control and irrigation efforts that China initiated after 1949. Here, again, data are unreliable, but they suggest directionality, and when combined with comments from state and Party officials, such data can present a reasonably reliable picture of irrigation on the North China Plain. For example, one is struck by the incongruity of the claims made concerning increased irrigated acreage and reports of declining agricultural production. The state boasted that 4,000 large irrigation systems had been constructed by the end of 1959, and irrigated acreage increased from 16 million to 167 million hectares since 1949.[58] Most of these gains were achieved during the Great Leap Forward. By the end of 1960, irrigated acreage reached 60 percent of all farmland in China (from 30 percent in 1957).[59] On the North China Plain, there was rapid expansion of irrigated land, as water was diverted to the fields from rivers and lakes. Reports stated that irrigated acreage in the Yellow River valley increased from 37,500 hectares in 1957 to 2.1 million hectares at the end of 1959, a fifty-eight-fold increase. At the same time, withdrawals of Yellow River water for irrigation in 1959 increased eighty-three-fold from diversions in 1957 (from 0.16 billion m³ to 13.3 billion m³).[60] However, massive irrigation, along with double-cropping and increased use of organic and inorganic fertilizers, did not increase agricultural production. In fact, production declined. To be sure, declining agricultural output was attributable to a set of material and moral disincentives faced by peasants. At the same time, there is evidence that failure to increase agricultural output was due to the shortcomings of irrigation systems. Furthermore, the mass farmland irrigation campaign may have also threatened heretofore viable farmland, thereby contributing to production declines. In either event, the net result of Great Leap Forward was a massive draw upon surface water resources.

The inefficiency of the new irrigation networks was a consequence of Great Leap Forward priorities. Pressures for haste, a lack of technical expertise, and an overreliance on local planning and management led to irrigation development without increased agricultural output. Promoters of Great Leap Forward policies were not unaware of the potential conflict between mass mobilization and efficiency. Reflecting the arguments between central planners and proponents of mass mobilization, a *Renmin Ribao* editorial framed the issue in this way:

> There are two views on this matter. One is to set up the task of raising construction productivity and the task of raising the standards of construction against each other, holding that "high construction productivity can be achieved only at the expense of the construction standards," and that "one must work slowly in order to do the job properly." This view is not correct. The other view is that high productivity and high standards are two opposing factors in one unity, and that it is entirely possible to bring about both high productivity and high standards of construction. This is a correct view.[61]

Unfortunately, *Renmin Ribao* was wrong. Technical experts were virtually absent in communal projects. Technical expertise, which had been trimmed to follow the winds of the Great Leap Forward by the Anti-Rightist movement, was largely concentrated in regional river commissions like the YRCC. Into the technical void stepped local Party representatives, who were responsible not only for mass mobilization but also for technical direction. They were also under extreme pressure to conclude projects as quickly as possible.[62] Unrealistic mandates from the Party to show results led to inefficient irrigation systems.

The replumbing of the North China Plain created a host of soil problems, including waterlogging and soil alkalization/salinization. The construction of reservoirs and channels obstructed drainage. Pernicious salts and other minerals rose to the surface in saturated soils with devastating effects on soil fertility. By the early 1960s, one estimate calculated that two-thirds of irrigated land in North China was threatened by salinization due to poor drainage. The area of saline farmland in Hebei, Shandong, and Henan rose to an estimated 1.9 million to 3.2 million hectares.[63] Irrigation systems, whose channels were unlined and inadequately packed, lost substantial amounts of water to seepage, further raising water tables. A speaker at the 1960 National People's Congress "complained that 40 to 60 percent of water [withdrawals] was lost through leakage in channels."[64] The People's Victory Canal serves as a concrete example of massive inefficiencies. The system underwent substantial expansion during the Great Leap Forward, but irrigated land was quickly degraded. Saline land increased from 140,000 *mu* in 1957 to 330,000 *mu* by 1962. In that same year, the Henan provincial government closed down most of the irrigation projects that drew water from the Yellow River in order to rehabilitate salinized land.[65]

The mass mobilization generated by the "irrigation panacea" had a paradoxical twist. In spite of the saturation of the North China Plain through irrigation, water shortages were getting worse. Was it possible to sustain large irrigation networks in a region with a limited endowment of water—an endowment that had traditionally precluded irrigation development? During the Great Leap Forward, further pressure on the water supplies of downstream regions of the Yellow River was caused by the development of large irrigation systems in the even drier areas of upstream (Gansu and Ningxia Province and Inner Mongolia Autonomous Region), where it typically "takes four times as much [irrigation] water to grow a field of wheat as it does in the lower [Yellow River] basin."[66] In addition, on the North China Plain, greater numbers of reservoirs and channels significantly increased evaporation.[67] But it was simply the large withdrawals of water for irrigation that pressured supply most egregiously. "So many small ponds had been dug and so many canals and irrigation schemes had been completed that there actually was not enough water available to fill the ponds and irrigate the fields."[68] At its peak in the late 1950s, the People's Victory irrigation system diverted nearly a billion square meters of Yellow River water. Already by 1958, downstream water pressures drove local strategies to secure water. The CCP Central Committee and State Council observed that localities built dams across streams and "bored holes" in embankments to divert water. It was reported that some streams and rivers ceased to flow.[69]

## Great Leap Forward Outcomes: Sanmenxia

By 1960 the technical debates over the height of Sanmenxia had been resolved. In September of that year, 12 diversion outlets were sealed and the reservoir began to impound water. The first Soviet-made turbines arrived in December, shortly before Soviet advisors in China were recalled following the end of the Sino-Soviet partnership. By April 1961 the dam had reached its first-phase design of 353 meters. At the end of 1961, however, a year and a half after the dam began to impound water, before a single kilowatt had been generated, serious problems arose. The speed of sedimentation was well beyond expectations. Silt began to amass upstream as the growing reservoir slowed upstream flow. At the Tongguan hydrologic station, more than 100 kilometers upstream, near the confluence of the Wei and Yellow Rivers, the width of the Yellow River channel had increased 4.4 meters since impounding began in mid-1960. At the same time, the backing up of the Wei River sparked a rise in riparian water tables that promoted widespread salinization. Hydrologists feared that if the reservoir continued to expand, more farmland would be flooded and large numbers of people would have to be relocated. Most importantly, the reduction in the water discharge capacity of the Wei River threatened the city of Xian. The datum that most stunned the Yellow River Commission and Party leaders was that by early 1962 the reservoir had already lost half its capacity of 9.8 billion m³ due to siltation.[70]

The debate over Sanmenxia dragged on for the next two years as the reservoir rapidly silted up. Finally, in 1964, during the retreat from Great Leap Forward policies, Zhou convened a large conference to recommend future directions for Yellow River management. The most pressing item on the agenda was Sanmenxia. Opinions ranged widely. Around this time, Mao "went so far as to suggest that the dam be destroyed by aircraft with bombs if no other option was available in the effort to save Xian."[71] Ultimately, the State Council approved a recommendation made by the conference to revise the operating principle of the dam from "impound water and store silt" (xu shui lan sha) to "retain floodwater and discharge silt" (zhi hong pai sha). This change meant that during the rainy season the discharge gates would be fully open to drain flood crests. Thus downstream areas would again be challenged with heavy water and silt discharge. The dream of having the Yellow River run clear would have to be deferred. Ultimately, a decision was made to reopen four of the twelve water diversion gates, as well as to convert four of the eight penstocks (where turbine fan blades were to have been installed) to discharge outlets and to dig two 900-meter by 11-meter tunnels through the rock cliff on the left side of the dam to serve as additional discharge conduits. The discharge volume of the dam was raised from 3,080 m³/second to 6,000 m³/second to facilitate a greater release of silt. The first overhaul was completed by 1968.[72]

The engineers who drafted the 1955 Yellow River plan had invested Sanmenxia with central importance for the hydraulic stability of the North China Plain. In the original plans, after flow and silt were regulated by the dam, threats to the downstream region would attenuate. Clear water released at Sanmenxia (silt having settled behind the dam) would wash the silt from the downstream bed that had accumulated over decades. The riverbed would cease to rise, and the centuries-old burden of dike building and rebuilding would come to an end. Floods would be averted. With regulated downstream flows, Yellow River water could effectively and reliably be withdrawn for irritation and transport. The onset of the Great Leap Forward did not change the fundamental design and goals of the 1955 plan but added emphasis upon the "three bigs": big diversions, big storage, and big irrigation. Between 1958 and 1960, twenty-two water diversion gates were built into the Yellow River dikes designed to withdraw 3,361 m³/second of water and to irrigate 8.9 million *mu* of land. In the lower valley, Yellow River water was diverted to ten irrigation districts of over 100,000 *mu* each. Thirty-one reservoirs, with a total capacity of 3.3 million m³, were built on the downstream plain, where they withdrew more than 16 million m³ of water from the Yellow River per annum.[73]

The 1964 change to the operating principle of the Sanmenxia Dam, however, generated new problems. Because large amounts of water and silt would again flow downstream, key irrigation projects had to be rethought. According to the 1955 Yellow River plan, downstream flow and silt would be regulated by Sanmenxia to such a degree that it would be possible to erect dams across the lower reaches of the Yellow River to feed three large irrigation projects, two in Henan and one in Shandong. After 1957, the YRCC added three more irrigation storage plans to the original three. Construction was initiated on two of the key projects, at Huayuankou (Henan) and Weishan (Shandong). Both projects were completed in an extraordinarily rapid manner (in the case of Huayuankou, within six months). By December 1962, however, it became increasingly clear that Sanmenxia could no longer function to impound silt. Because of the increasing downstream siltation, the Huayuankou irrigation dam and reservoir were blown up in June 1963. The same fate befell the Weishan project later that year. With the exception of a similar project near Ji'nan (Shandong), where construction was halted, other plans to dam the downstream channel for irrigation never got off the ground because Great Leap Forward policies were abandoned by 1961.[74]

The forces that engendered the Great Leap Forward, some layered atop pre-1958 patterns, some unique to the late 1950s, led to the Sanmenxia debacle. It is difficult to imagine that even at the height of the Great Leap Forward, anyone with a modicum of technical knowledge believed that the river would "run clear," as envisioned in the 1955 plan. But this slogan did reflect the articulation of a goal, however utopian, by political elites that mandated a certain audacity in the formulation of technical plans. A technical plan that nodded to this utopian vision but offered something less stunning or a plan difficult to comprehend in its technical complexity was unlikely to impress Party elites, who felt increasing pressure to modernize rapidly. The Soviet advisors acted as enablers of this technical vision, and the mid-twentieth-century faith in technological systems provided the global context for the movement. However, there was perhaps nothing quite as audacious as the erection of a dam across the Yellow River. After 1955, the degree to which the Party invested the dam with particular meaning became increasingly clear. By then, it was manifest that Sanmenxia had become a powerful symbol of the Party's ability to transform the immutable. At a material level, Sanmenxia would be the fulcrum upon which regulation of the river would depend. It became the key project upon which all the other key projects would depend. However small the odds of success might have been in a different polity and different cultural climate, Great Leap Forward political sensibilities compounded the problem. Any realistic chance for success relied upon dramatically reducing sediment runoff from the loess regions. Planners were well aware of this requirement, but the entire effort, like the Great Leap Forward's drive to increase irrigation and grain production, did not generate the intended results. Fatigued and hungry peasants experienced diminishing returns of mass mobilization and thus saw few direct incentives to work hard. The goals of the 1955 plan for sediment control may have been optimistic, but the targets for soil conservation, including the scope of farmland and the speed of implementation, were even more unrealistic. The silt kept flowing off the hills.

The 1964 conference called by Zhou to review the Sanmenxia project and Yellow River plans marked an important transition from Great Leap Forward policies. Indeed, the retreat from Great Leap Forward policies in general and from mass mobilization for water management in particular was already in effect by 1962. As early as 1958, an announcement that "the collapse of dams of medium-sized and small reservoirs has occurred in quite a few areas" suggested the consequences of Great Leap Forward mass mobilization.[75] Agriculture minister Liao Luyan had already stated in

1960 that "in the ten years since the founding of our country we have already achieved tremendous success in rectifying natural conditions. Nevertheless it is by no means yet practicable to assume complete control over natural condition in the near future."[76] By 1962, calls for "readjustment" (tiaozheng), "consolidation" (gonggu), "replenishment" (chongshi), and the raising of quality standards signaled a retreat from the Great Leap Forward farmland irrigation campaign.[77]

## The 1950s in Historical Context

The drive toward socialist modernity in China after 1949 was a variation of the high-modernist theme that pervaded technological complexes in a wide variety of polities during the second half of the twentieth century. Irrespective of developmental path, either that of central planning or of mass mobilization, the CCP imbibed fundamental assumptions about the primacy of the state to organize and direct knowledge, society, and resources to achieve state-defined goals. As structured by the pervasive assumptions of high modernism, state goals were aimed at maximizing the productive power of water. In this view, water that ran freely to the sea was water wasted. The institutional model for the full exploitation of water was multipurpose development—the construction of regulating structures that could irrigate, generate power, control floods, and aid transportation. All would contribute to the expansion of China's agricultural and industrial production, which would create a modern and affluent economic infrastructure with which to transition to communism.[78]

But there were objective forces that circumscribed the range of choices available to Party leaders to meet these goals. First, the state had limited financial resources with which to pursue capital-intensive projects. The economic landscape of post-1949 China was complicated, to say the least, after decades of war. Second, China was isolated internationally. The best candidate for international partnership was the Soviet Union. However, it quickly became clear that substantial financial assistance would not be forthcoming from Stalin. Soon after 1949, domestic and international realities pointed to the need to "walk on two legs" in order to address ecological distress on the North China Plain. A region that had been the crucible for a successful revolution also held the potential to destabilize revolutionary outcomes. China had labor resources, but it also had Soviet technical advisors who were smitten with central planning, capital intensity, and modern technology. In its first effort to reimpose ecological stability on the North China Plain, the state adopted elements of mass mobilization *and* central planning.

Both development paths employed "brute force technologies."[79] As a combination of industrial technology with state organizational and coercive capacities, "brute force technologies" engaged in massive projects of "geo-engineering" that transformed landscapes into factories of production. These transformations were evidenced by Soviet plans in Central Asia and by mid-twentieth-century projects in the American West. Although China's political and technical elites of the 1930s and 1940s were attracted to "geo-engineering," state capacity to bring these visions to reality coalesced only after the communist victory of 1949. The sorts of muscular technologies intimately associated with brute-force transformations in the Soviet Union and other industrialized countries were reflected in projects such as Sanmenxia, but in the early years of the PRC organizational innovation usually substituted for technological innovation in the mass mobilization irrigation campaigns.

The relationship between the state/Party and rural society after 1949 invites other comparisons to long-running tenets of statecraft. Did the PRC reintroduce state concerns of the imperial period through its programs to sustain ecologically vulnerable regions by recreating the material and moral basis of rural society through redistributive mechanisms?[80] In part, yes. After 1949, land reform and redistributive food policies suggested the broad concern of the Party for its rural social base of support, particularly in regions like the North China Plain, where ecological breakdown generated social and economic conditions that had been conducive to CCP revolutionary mobilization. There was a tension, however, between the Party and its rural base of support, between the Party's identity as a rural organizer of revolution and its urban goal of industrialization. Industrial growth was premised on wringing material resources out of the rural sector, not through capital investment but by intensified labor inputs. During the first decade of communist rule, economic policies were largely premised upon sustaining the rapid growth of urban industrial centers. Although the moral qualities and material contributions of the peasantry continued to be celebrated throughout this period, a decidedly anti-urban/pro-peasant turn climaxed in the late 1950s with the Great Leap Forward. The revival of the Yan'an spirit to the national narrative reinforced the centrality of rural China in general, and the Yellow River valley and North China in particular. Beginning with the Anti-Rightist movement, discredited intellectuals like Huang Wanli were "sent down" to the countryside to learn the revolutionary spirit of the peasantry and to engage in manual labor, often on irrigation projects.[81] The effort to industrialize the countryside with backyard steel–making furnaces, and the relocation of coastal urban industries to the countryside reflected this anti-urban spirit. Thus the massive transformation of the landscape to more fully exploit water resources, particularly as it occurred during the Great Leap Forward, was sanctioned by a moral framework that valorized a revolutionary landscape and a revolutionary rural society.

Central to the leadership's efforts to mobilize support among the Party and peasantry for controlling the waters was the use of metaphors and symbolic representations of landscape and river to show China's historical struggle for national liberation and social revolution. In this sense, the Yellow River served as totem. As the geographical origin of Chinese civilization and as the birthplace of the CCP-led revolution, the river provided a cultural touchstone for national identity. The entire valley was a landscape with which to create epic narratives of a "second creation," namely, the creation of Communist China. The landscape of the second creation was the same landscape where Yü the Great had created the ecological and material conditions for Chinese civilization. Mao was the new Yü. As the old saying went, "When a sage appeared, the river would run clear." As suggested above, "second creation stories" often celebrated rebirth through violence of self against an alien other.[82] The "alien other" in this formulation was a landscape that needed to be rectified to increase production and that simultaneously served as metaphor for internal and external enemies, historical and contemporary.

At the same time, mass mobilization was a way to create an identity of interests between the peasantry and the party. Campaigns mandated shared sacrifice for the benefit of the local community and, by extension, for the nation. Artifacts used in mass mobilization were also imbued with these sorts of symbolic meaning. The wheelbarrow, the pneumatic drill, and earth-tamping tools, for example, were celebrated as technological expressions of local knowledge, egalitarianism, and self-sufficiency. In the hands of peasants imbued with revolutionary ideals, these tools could make "the mountains bow and the rivers yield." At the other end of the scale, large dams such as Sanmenxia were symbols of the power of the Party to organize society and to muster technological know-how to tame nature.

Ultimately, the question is this: How did hydraulic engineering practices affect the ecology of the North China Plain in the post-1949 period? Or, more precisely, what was the legacy of Maoist-period technological choices for the water resources on the North China Plain? Inadequate and unreliable data from this period frustrate any precise evaluation of irrigated acreage, soil conservation, or flood prevention. But it is certain that the use of surface water increased dramatically on the North China Plain after 1949. An abiding faith in engineering solutions, premised either on large-scale industrial technologies or mass-based organization, left a resource debt to post-Mao generations in a region that had struggled with ecological marginality for centuries. Indeed, increased water scarcity on the North China Plain was already apparent to technical and political elites as talk of a large diversion of water from the Yangzte River basin to North China increased by the late 1950s. But there was an additional source of water that remained to be tapped that would ultimately postpone radical engineering solutions. That source was underground.

## Going Underground in the 1960s and 1970s

On July 25, 1966, seventy-three-year-old Chairman Mao went for a swim. Mao appeared, unannounced, on the banks of the Yangtze River near Wuhan to participate in a mass swim held to commemorate a dip that he had taken a decade earlier in the same river. Nine days later, on July 25, a photo of Mao at the river appeared in the state-run media. At the time, Mao was rumored to be in ill health. Indeed, he had maintained a low profile since the debacle of the Great Leap Forward, with which he was identified. The Yangtze River event and its coverage by the Xinhua (New China) News Agency, however, was carefully staged to let people know that Mao was still around, still buoyant as it were, and still capable of arousing mass sentiment. Shortly after the photos of Mao in the Yangtze were published, school children were organized to emulate the Great Helmsman by swimming across the Yangtze. A far more serious audience for the spectacle was the group of senior leaders who had led the retreat from Great Leap Forward policies earlier in the decade. Mao did not like what he was seeing as a result of this retreat. A slide back to bureaucratism, the reintroduction of "pernicious" bourgeois habits and values, and loss of revolutionary fervor suggested to Mao the need to reignite revolutionary mass mobilization. Mao so serenely, seemingly effortlessly, floating upon the river became an iconic image of the tempest of the Cultural Revolution (1966–1976).

The revolutionary fervor of the Cultural Revolution succeeded in eliminating many proponents of centralized Soviet-style economic management. To Mao, these cadres were conservative obstacles to the mobilization of the revolutionary will of the masses. Although the Cultural Revolution can be labeled a political movement, its particular rhetorical style promoted a return to aggressive exploitation of water on the North China Plain through campaign-style initiatives. Already in 1964, Mao had launched a movement called "In agriculture learn from Dazhai" (nongye xue Dazhai), which served as the mantra for rural development throughout the Cultural Revolution and up to the end of the Maoist period. The spirit of Dazhai was "self-reliance"—that is, reliance upon communal resources, human and material, to free productive forces from the constraints of nature.[83] At a practical level, this meant a renewed and sustained thrust to expand irrigation on the North China Plain. But the increasing drawdown of surface water for irrigation, the failure of Sanmenxia to regulate water for multipurpose use, and

Mao swimming in the Yangtze River, 1966. (© 2006 Sovfoto/Eastfoto. Used under license.)

generally dry weather conditions created water stresses that were exemplified by the first desiccation of the Yellow River in the early 1970s. The bottleneck in the water supply turned attention underground. Throughout the 1960s and 1970s, massive exploitation of groundwater granted a critical but ephemeral reprieve to water constraints for agricultural communities on the North China Plain. The post–Great Leap Forward readjustment period did, indeed, scale back the hypercampaign style of mass mobilization efforts, but the focus, throughout the 1960s and 1970s, remained small-scale projects aimed at local needs for irrigation, flood control, and, to a lesser degree, power generation.

The failure of Sanmenxia to function as the lynchpin in the grand dreams of multipurpose development, originally envisioned in the 1955 Yellow River Plan, reinforced the impulse to focus communal efforts on local water management schemes. These local management efforts centered on expanding irrigated agriculture through tapping groundwater sources. The critical new component to the technological matrix that enabled communes to exploit these resources was simple and cheap electrical pumps. The efficiency of these pumps generated such dramatic returns during the 1960s and 1970s that agricultural productivity on the North China Plain became largely dependent upon subsurface water. The relative abundance of groundwater resources during the 1960s and 1970s quieted discussions of water transfers from the Yangtze basin to the North China Plain, but in the longer term, the exploitation of these resources left a critical water debt for agriculture, industry, household, and ecological needs in the post-Mao era (when discussion of large-scale water transfers from South China again arose).

## The Continuing Saga of Sanmenxia and the Yellow River

The reconstruction of Sanmenxia Dam, necessitated by the sedimentation of its reservoir by the early 1960s, took the better part of a decade. Completed in two phases, 1965–1968 and 1970–1973, the dam's function was completely transformed. In the first phase, two silt discharge tunnels were constructed around the dam and eight of the original thirteen penstocks were converted to water/silt discharge tubes. During the second phase, the eight outlets at the base of the dam (that originally facilitated flow-through during construction) were opened to drain water and silt. The remaining four penstocks were lowered, allowing increased discharge that, in turn, shrank the size of the reservoir. One outcome of the redesign was to relieve the dangerously high water levels on the Wei River that threatened Xian. The impact of the new design on the downstream area, the North China Plain, was also significant. Because silt would again flow downstream, it was necessary to organize two additional mass mobilization dike-maintenance campaigns in 1962–1965 and 1978–1985.

Sanmenxia Dam, 1961. (© 2006 Sovfoto/Eastfoto. Used under license.)

The future of downstream Yellow River management in the post–Great Leap Forward period was discussed at the December 1964 meeting called by Zhou Enlai. In part, the meeting provided a forum for renouncing the management style of the Great Leap Forward. In its summation of the genesis of the Sanmenxia quandary, the official report faulted improper planning, insufficient technical data and guidance, and rash advance. The long-time leader of the YRCC Wang Huayun stressed to the assembly that the Yellow River would simply never run clear. The attempt to impound silt behind Sanmenxia was a failure. Wang proposed that the management principle for the Yellow River be to impound silt upstream by soil conservation measures, while sluicing silt down the lower valley. Characterized as an assembly with the "din of a hundred geese" (*bai gu zheng wu*), the conference discussed Wang's proposal as well as others. A central question was what to do with the silt that would continue to flow downstream. Some delegates argued for restoring the original function of Sanmenxia as a massive silt trap, in conjunction with aggressive upstream soil conservation measures. Others argued for a novel conception of utilizing the water and silt by drawing both water and silt from the river to nourish riparian farmland. The debates continued until the developments of the Cultural Revolution preempted further discussion on how to manage silt in the downstream.[84]

Indeed, during much of the Cultural Revolution, there was little centrally managed development of the lower Yellow River valley. The work of the YRCC was denounced for not contributing to production, and its bureaucratic and technical personnel were ridiculed by Cultural Revolutionary activists, usually young Red Guards mobilized by calls from Chairman Mao to "bombard the headquarters" of Party and government bureaucracies staffed by cadres insufficiently imbued with revolutionary values. Hoping to fashion these bureaucracies more responsive to revolutionary goals (as articulated by Mao), activists subjected Party and government leaders at every level to verbal and physical attacks. This "rectification campaign" ultimately removed layers of personnel from virtually every institution in China. By 1970, most cadres of the YRCC had been "sent down" to communes to work in local dike construction or to be re-educated at cadre training schools. Virtually any individual with a modicum of technical training was of particular interest to the Red Guards. For example, Wang Huayun, a loyal follower of the Maoist line through the changing political winds of the 1950s, was labeled "a follower of the capitalist road" *(zou ziben zhuyi daolu de)*, subjected to public denunciations, and forced to sweep roads and scrub toilets.[85] The consequences of the Cultural Revolution for river management on the North China Plain centered on the growing rate of siltation in the Yellow River valley. By and large, this was because land conservation efforts in the midvalley loess highlands ground to a halt. Complicating the state's ability to gather any empirical data on the silt levels in the Yellow River and its tributaries was a complete breakdown of the post-1949 hydrologic monitoring system.

By the early 1970s, after the most intense period of the Cultural Revolution had passed, a number of senior personnel, such as Wang Huayun, were rehabilitated, and the YRCC re-emerged as a functioning government agency. Technical work resumed, and shortly thereafter a renewed trickle of hydrological data suggested the heightened flood threats from the rising riverbed downstream. In late 1975, with strong central government guidance, the downstream provinces organized a third major dike rehabilitation campaign that mobilized millions of commune members to dig and tamp earth along the Yellow River. Beginning in the mid-1970s, renewed soil conservation measures, along with the construction of silt detention basins and development of large-scale dredging capacities, the rate of sedimentation in the downstream bed slowed from the unsustainable levels during the Cultural Revolution.

## Irrigating the Plain

The frenetic mass mobilization efforts of the Great Leap Forward had receded, but large-scale expansion of irrigation continued during the 1960s and 1970s. By and large, this pattern continued to be driven by small-scale projects that were initiated and managed by communal units. Medium-scale projects that extracted water directly from the major rivers like the Yellow River also continued after the Great Leap Forward. But, increasingly, attention turned to rehabilitating surface water irrigation systems and aggressive exploitation of rich groundwater resources.

After the Cultural Revolution, a great deal of effort was expended on repairing irrigation systems on the North China Plain that had been initially developed in haste during the Great Leap Forward. Many of the mass-based irrigation schemes that drew upon Yellow River water, or its tributaries, had to be re-engineered. The core problems were siltation of irrigation systems and inadequate drainage. The proliferation of irrigation channels and reservoirs obliterated existing drainage systems that carried off water during summer storms. At the same time, irrigation gates became clogged with sediment that was originally to have been sequestered upstream by Sanmenxia. The obsession with storing water in countless reservoirs sparked a rise in water tables that, in turn, generated severe salt and alkali problems.[86]

Following the Great Leap Forward and the post–Great Leap Forward retreat, there were two other periods of advance-retreat that initially promoted aggressive mass-based irrigation, only to be followed by consolidation and retrenchment. Before 1949, there were only 12 million *mu* of irrigated farmland in the Yellow River valley, mostly in the midstream region. After 1949, irrigation developed rapidly. Beginning in 1949–1957, an additional 10 million *mu* of irrigated farmland was developed based on rehabilitated pre-1949 irrigation systems and on new systems like the People's Victory Canal. Indeed, in the aftermath of the Sanmenxia debacle, with the collapse of the dream of a silt-free Yellow River, the pace of development quickened in the 1960s for irrigation facilities that drew directly upon surface water. At the same time, the potential benefit of silt to reclaim riparian saline soils was aggressively exploited by the expansion of existing systems like the People's Victory Canal and the construction of new downstream systems.

As a result of accelerated development during the Great Leap Forward, the People's Victory irrigation district grew to over 40 million *mu* (from 5 million *mu* at its inception in 1953). But according to a 1964 report, secondary salinization was so massive that the system was shut down in 1961 to prevent further loss of farmland. Before irrigation began in the early 1950s, the average depth of groundwater in the People's Victory irrigation district was 3–4 meters. By 1961, average groundwater depths rose to just 1.3–1.5 meters below the surface. Salinized land was calculated to be nearly half of all land irrigated in the People's Victory system. Poor harvests reflected increased secondary salinization. In 1961, average annual grain harvest in the region was 1.3 tons per hectare. Harvests increased to 3 tons in 1957, but breakneck extension of the irrigation system during the Great Leap Forward generated salinization and declining agricultural returns. The 1962 harvest totaled only 1.4 tons per hectare. Because of its lack of drainage capacity, the soil was too saturated to produce. Furthermore, the inefficiency of the People's Victory Canal irrigation system was stunning. Only 20–30 percent of water diverted from the river at the headworks reached the fields. Including the water diversion complex at the People's Victory project, there were a total of 22 irrigation diversion gates on the lower course of the Yellow River, with a water withdrawal rate of 3,361 m³/sec. With the data on water usage given by the example of the People's Victory project, one gets a sense of how much water these projects consumed and how meager were the returns.[87]

As illustrated by the People's Victory Canal Project, many of the irrigation projects, either initiated or expanded dramatically during the Great Leap Forward, entered a period of retrenchment after 1961. Large state-managed irrigation districts and smaller communally managed projects were either suspended or re-engineered to improve drainage capacity. Guiding such activities were technical experts who had been rehabilitated and reintegrated into water planning agencies, such as the YRCC. But with the onset of the Cultural Revolution, the technology complex rather quickly swung back to mass mobilization and local knowledge.

Influenced by weather (drought) and politics (the Cultural Revolution), there was a resurgence of surface water irrigation projects beginning in 1965. Large state irrigation districts desperately sought to avoid a repetition of the widespread post–Great Leap Forward secondary salinization problems that had virtually paralyzed agricultural development on the North China Plain. In Henan and Shandong, greater attention was paid to existing and new projects by incorporating drainage and silt retention basins and channels. In 1968, the YRCC announced that irrigation projects that drew upon Yellow River water were again in full swing. Despite the political chaos of the Cultural Revolution, large irrigation districts were developed from the mid- to the late 1960s through the 1970s. By 1972, the total acreage irrigated in the downstream valley reached nearly 12 million *mu*, with annual withdrawal of 8.2 billion m$^3$ of Yellow River water.[88]

During the same period, diversions of Yellow River water also contributed to the mounting demand for water resources in the lower valley. Drought struck the North China Plain in the early 1970s, sparking increased demand for surface water for irrigation and urban water use. The limited capacity of surface water on the North China Plain to continue serving agricultural, industrial, and residential needs had been recognized back in the 1950s, when initial planning for the South-to-North Water Diversion project commenced. But a vivid reminder of these limitations was manifest in the late spring of 1972, when the Yellow River dried up. Downstream agricultural production in Shandong Province was compromised, and several hundred thousand rural residents lacked fresh water. In 1972, 1973, and 1975, the first large-scale water diversions from the Yellow River flowed to Tianjin through the People's Victory Canal. Tianjin, China's third-largest city, is located on the northern edge of the North China Plain. During these drought years, 450 million m$^3$ of water was sent to sustain residential and urban needs. The transfer not only suggested the continued attraction of engineering solutions for hydrological problems but also was a clear suggestion of the tension between urban and rural water use that increased during the post-Mao period as demand from both sectors grew substantially.[89]

The increase in irrigated farmland from the late 1960s to the early 1970s represented a second wave of irrigation development in China. In addition to the rehabilitation of surface irrigation, the critical component of this expansion of irrigation was the introduction and wide dissemination of small-scale pumping technologies that made possible extensive and intensive exploitation of groundwater resources on the North China Plain. By 1963, more than 10,000 pumping stations had already been established in China. Simple pumping technologies complemented the valorization of self-reliance in the late 1960s and early 1970s (i.e., the Dazhai model). These pumps were all manufactured in China and ran on electricity or diesel fuel. They were cheap to make and operate, and their simple mechanisms could be tuned and repaired by cadres from each production team. The pumps replaced the need to convey water, either through human carriage or by conduits, from streams, lakes, and reservoirs to the fields. In addition, water pumped from groundwater sources via tube wells (or pump wells) could more be more precisely timed for application to the fields and required little coordination among collaborating or competing jurisdictions.[90]

Along with the increasing exploitation of domestic oil reserves and the expansion of refining capacity to produce ample supply of diesel, the spread of groundwater pumps was abetted by a robust expansion of small-scale hydroelectric generation facilities. These facilities could be constructed wherever a sufficient fall occurred in natural or built waterways. As such, small hydroelectric generation plants served as adjuncts to irrigation systems, large and small. Virtually all these facilities had a generation capacity of less than 1,000 kilowatts, but many generated much smaller amounts, all the way down to single-digit kilowatts, just enough to run pumps to convey surface or groundwater to the fields. In 1965 China had the capacity to generate approximately 8.5 million horsepower of mechanical and electrical power. This total increased to 50 million horsepower by 1975, two-thirds of which powered pumps for irrigation and drainage.[91] Pumps also were used extensively for surface irrigation projects that heretofore were largely gravity-fed on the lower course of the Yellow River in Henan and Shandong. By the early 1970s, there were more than 80 pumping stations and siphons along the Yellow River.[92]

The application of mechanized pumping began on a modest scale during the Great Leap Forward, but in the period immediately following its end, there was an aggressive expansion of tube-well irrigation on the North China Plain. Along with the development of local energy sources, the remarkable growth of groundwater exploitation had several additional explanations. First, as suggested above, since the 1950s it had been understood that China's surface water resources were limited. Second, the massive problems with salinization that accompanied breakneck development of surface water resources during the 1950s convinced many planners of the benefit of the relative precision and simplicity by which groundwater could be applied to fields. Third, in the aftermath of the Great Leap Forward, there was a renewed promotion of mechanization, technical solutions, and capital investment, coupled with a deemphasis on mass mobilization. And, last, in the security-conscious environment following the Sino-Soviet split of 1960, the development of groundwater resources throughout the country, but particularly in the arid northwest, was necessary to sustain industry and a viable transportation system.[93]

In the early 1960s, a spate of articles in the state-run media, as well as an increasing number of reports from technical conferences, disseminated information on groundwater resources and exploitation. In a long article published in 1961, titled "The Use of Mechanized Wells and Groundwater," the author carefully referred to the pioneering use of groundwater in China centuries earlier in a discussion of the advantages of modern mechanization. Reflecting the political tenor of the period after the Great Leap Forward, this article and others from the early 1960s emphasized the necessity of planning, organization, and coordination within the commune as groundwater extraction increased. Again, seeking some kind of accommodation between the reintroduction of technical planning, centralization, and the continuation of the commune as the foundational social organization in the countryside, the articles explained the utility of local (communal) exploitation of groundwater and told of teams of hydrologists and geologists, who had been dispatched across the North China Plain, to assist communal drilling teams.[94]

During the radical periods of the Cultural Revolution and throughout much of the 1970s, a new set of heroes joined the revolutionary pantheon of the Party. Well-drilling teams, formed at the production team and production brigade level, were the shock troops of the "groundwater campaign" (*dixia shui yundong*) that would transform the whole of the North China Plain into a productive garden. As occurred during the Great Leap Forward water conservancy campaigns, the Cultural Revolution landscape was again rendered a canvass upon which both reactionary and revolutionary values were imprinted. In a fascinating retrospective of the Mancheng County (Hebei) Well-Drilling Team, published in 1972, the political and ideological contortions of the period immediately after the Great Leap Forward through the Cultural Revolution were clearly illustrated. In this telling, the well-drillers of Mancheng County were central characters in a drama that reached its denouement with the successful drilling of thousands of wells. The story began by recalling with the language and image of combat the heroic efforts of the team in 1963, when members struggled with a stuck drill bit. Over the next 103 days, with little sleep, the team battled to extract the bit with burning impatience (*xinji ru fen*), bravery, and the energy of tigers and dragons (*shenglong-huohu*). Subsequently, however, the revisionist, anti-revolutionary spirit of the retrenchment following the end of the Great Leap Forward infected the team's revolutionary identity. Called upon in 1965 to dig tube wells in an area of the county known as East Valley, the team agreed but only on condition of a payment of 8,000 yuan. The team had been "poisoned by the anti-revolutionary revisionist line of Liu Shaoqi" (*Liu Shaoqi fan geming xiuzheng zhuyi luxian de duhai*).[95] Liu Shaoqi was the vice-chairman of the CCP and the chairman of the PRC, but most importantly in the late 1950s he had been designated Mao's successor. Unfortunately for Liu, his tepid support of Great Leap Forward policies and his identification with the retrenchment policies marked him as a member of a group of revisionists that Mao intended to oust from government and Party leadership through the mechanism of the Cultural Revolution. Liu was ousted in 1966 and died in relative isolation in 1969, after mental torment and physical abuse by Red Guards. By April 1969, when the Ninth Party Congress was held, the destruction of the Party and government bureaucracies was deemed complete. In that same year, the Mancheng Well-Drilling

Team had a conversion experience. Heavily criticized for losing their revolutionary and peasant consciousness, as well as for following the "capitalist roader" Liu Shaoqi, the team regained the spirit of class struggle. The team returned to East Valley in the early winter of 1969, facing the bitter cold and wind and fighting bravely around the clock to drill a 120-meter deep well in 40 days. In December 1972, when the story of the Mancheng County Well-Drilling Team was published, the county was suffering from the worst drought in fifty years, for no precipitation had fallen in the preceding seven months. But the revolutionary spirit of the well-drilling team, once lost but rediscovered, allowed the harvest in East Valley to surpass that of the previous year. That year, 50,000 additional *mu* of winter wheat were cultivated over the previous year. Grain "flowed like the Yellow River." All this agricultural productivity was made possible by nearly 2,000 mechanized pumps the team had helped install over the previous ten years (see Table 5.1).[96]

The rapid expansion of irrigated acreage in the North China Plain during the 1960s and 1970s translated into greater food production to sustain a rapidly expanding population. In contrast to the hyperdevelopment of the Great Leap Forward, the expansion of irrigated acreage did not come at the expense of land degradation.

*Table 5.1*  Increase in mechanized irrigation in China

| Province/City | Year | Mechanized pumps |
|---|---|---|
| Hebei | 1965 | Unknown |
| | 1970 | 153,000 |
| | 1976 | 500,000 |
| Henan | 1965 | 20,000 |
| | 1974 | 460,000 |
| Shandong | 1965 | Unknown |
| | 1970 | 69,000 |
| | 1974 | 192,000 |
| Beijing | 1965 | 8,000 |
| | 1976 | 33,000 |

*Data source:* Eduard Vermeer, *Water Conservancy and Irrigation in China: Social, Economic and Agrotechnical Aspects* (Leiden: Leiden University Press, 1977), 193.

Indeed, more carefully planned exploitation of surface and ground-water allowed for the reclamation of saline soils in many parts of the North China Plain. Reclaimed land added to total cultivated acre-age, but more importantly, irrigated lands generated a doubling of agricultural output.

Groundwater exploitation supported the intensification of agri-culture in North China during the remaining years of the Maoist period. A few statistics suffice to give some sense of the impor-tance of groundwater to agricultural production in this period. In 1949, the sum of China's irrigated farmland totaled 240 million *mu* (16 million hectares), or 16 percent of all arable land in the country. Surface water was applied mostly by traditional technol-ogies, including human conveyance, while total installed electric power available for irrigation and drainage was only 97,000 hp (72,330 kw). The second wave of irrigation development, from 1965 to the early 1970s, was mostly focused on the North China Plain, where the sinking of 1.2 million wells by 1973 brought China's total irrigated farmland to 650 million *mu* (43 million hectares), representing nearly half (45 percent) of all farmland in China.[97] Intensification and expansion of irrigation, as well as increased fertilizer use, enabled the North China Plain (and the country as a whole) to realize the mantra of Dazhai self-sufficiency. But the sustainability of such self-sufficiency would be called into question during the rapid modernization drive in the post-Mao era.

## The Red Flag Canal and the Yellow River

As suggested above, the challenge of maintaining and expanding the hydraulic system of the North China Plain was conditioned, in part, by the dynamics of *upstream* development of transregional wa-terways like the Yellow River. The exploitation of these upstream waterways continued apace for the remainder of the Maoist era. Of greatest significance for the long-term waterscape of the North China Plain was the continued re-articulation of Sanmenxia, the construction of upstream multipurpose dams on the Yellow River, and the development of irrigation systems in the midstream of the

Yellow River valley. A closer look at the Red Flag Canal also suggests the continued interplay of the landscape and further definition and redefinition of nation and state.

Construction of the Red Flag Canal began in 1961 and was completed in 1965. The canal diverts water from the Zhang River, a tributary of the Wei River that eventually flows into the Hai River. Situated in a basin at the very edge of the North China Plain, the diversion canal is located at the junction of Shanxi, Henan, and Hebei Provinces. The construction of the Red Flag Canal became a signature event for local self-sufficiency and was nationally celebrated during the Cultural Revolution as a representation of the strength and correctness of Mao Zedong Thought. On April 20, 1966, a grand ceremony, with several hundred thousand participants, was held to commemorate the opening of the canal. On one side of the meeting grounds a red banner proclaimed: "Hold high the great red banner of Mao Zedong Thought"; on the other side was the slogan: "Fight against heaven and earth to re-arrange the rivers and mountains of Lin County."[98] From dawn until noon, speeches from provincial and county representatives, model workers, and a host of Party representatives celebrated the revolutionary spirit of Mao, which was responsible for changing the landscape of this drought-stricken region. Finally, at 12:20 P.M., the diversion gates were opened to fill the main irrigation channels with water. Amid the din of gurgling waters, applause, firecrackers, and gongs and drums, people from villages in Henan and Hebei jumped for joy, shouting "Long live Chairman Mao; long live the Communist Party."[99] Upon seeing the waters of the Zhang River flowing through the canal, an old woman said: "The waters of the Zhang River! Twenty years ago [i.e., 1946], drought forced me to flee. As I passed the Zhang River I drank from it. I would never have believed that today it flows to us."[100]

The Red Flag Canal and irrigation system was celebrated as the epitome of the Dazhai spirit of self-sufficiency. The building of the canal relied mainly on local resources, with 79 percent of costs covered through county, commune, production brigade, and production team funding. All construction material was supplied locally. There was not a single engineer or college student involved in the project, as all technical problems were solved with local experience and knowledge. Farmers became engineers.[101] Numerous images were widely disseminated that showed workers dangling from ropes with pick axes, chipping away the sides of sheer cliffs. These images were also celebrated in films. Among several filmed versions of the construction, the most widely circulated documentary was called "Red Flag Canal" ("Hongqi qu"). The film was distributed nationally and internationally, and chronicled in dramatic fashion the heroic spirit of its builders and the transformative effects of the project on agricultural production.

Operating under the considerable pressures of the "Learn from Dazhai" movement, North China irrigation systems like the Red Flag Canal were developed with remarkable speed. Data that extolled the achievements of the Red Flag Canal system may well have been amplified beyond reality, but there is little reason to doubt that these construction projects were a considerable drain on the water resources of the North China Plain. At the same time, the extraordinary achievement of the Red Flag Canal was suggestive of the type of pressure that local officials faced to rapidly expand irrigated acreage, with little expectation of aid from the central government. In Lin County, the heart of the Red Flag project, a network of secondary irrigation canals, fed by the main trunk of the Red Flag Canal, totaled over 900 km by 1969. This network irrigated 600,000 *mu*, compared to less than 10,000 irrigated *mu* before "Liberation." In 1969, the country reported that its wheat production had tripled from the previous year. By 1973, not only had the Red Flag irrigation network continued to expand, but a Lin County official told a delegation of the Ho Chi Minh Youth Labor League from North Vietnam that yields per *mu* had increased fivefold since 1949 (from 100 *jin* to 500 *jin*).[102] The injunction to be self-reliant—to depend on local resources and knowledge—extended the effort at replumbing the landscape throughout the North China Plain. But as the final stanza of the song "Red Flag Canal" makes clear, beyond the socialist brilliance of the foreground, there lies a developmental approach and technological complex that is thoroughly Maoist in inspiration, all with less than a subtle suggestion of powers approximating those of Yü the Great:

Beyond the brilliance, gazing beyond the highest heavens
The source of the Red Flag Canal is Zhongnanhai [Party
   and government headquarters]
Chairman Mao personally planned, and with his own
   hands opened
Water from Beijing comes
Water from Beijing comes.[103]

In addition to the Red Flag Canal, there were multiple irrigation projects in the upstream sections of waterways that drained through the North China Plain. The largest drawdown of these surface waters occurred in the multiple installations on the midsection of the Yellow River and its tributaries, including the Wei River. Some of these projects were extensions of older irrigation undertakings that had been initiated or refurbished during the Republican era. A few projects were exclusively irrigation works, but some facilities were components of multipurpose installations. The three largest irrigation districts that drew water from the upper and middle Yellow River valley were the Qingtongxia, Hetao, and Weihe Irrigation Districts. All were developed (or were considerably expanded) after 1949; and all these systems reduced downstream flows to the North China Plain.

The Hetao Irrigation District, as one example, is located in the northernmost extreme of the Yellow River bend in Inner Mongolia. Irrigation in this region first began in the Han dynasty with significant expansion in the later imperial period. The environmental conditions of the Great Bend have long been challenging for sedentary agriculture. The region only receives 130–220 mm of precipitation per year, and land-use changes over recent centuries, mainly encompassing the removal of native flora cover, have led to desertification. These ecological conditions are both an explanation for and an outcome of the region's role as a traditional contact zone for the sedentary Han agriculturalists to the south and the non-Han pastoralists to the north. The development and expansion of irrigated agriculture in the imperial period, advanced by a combination of state and non-state forces, was an expression of the greater integration of this region into the Chinese empire.[104] In the late Qing, the irrigated acreage of this area expanded significantly, with the addition of seven new offtake canals on the Yellow River. But the greatest expansion of the system came after 1949. From 1950 to 1960, irrigated farmland in the Inner Mongolian Autonomous Region rose from less than 5 million *mu* to over 20 million *mu*, with over 70 percent of the water sourced from the Yellow River. The sharpest increase in irrigated farmland occurred during the Great Leap, but acute problems with salinity generated a precipitous decline to 10 million *mu* of irrigated land by 1970. Reclamation and better drainage reversed the trend, so that by the end of the Maoist period irrigated agricultural land in Inner Mongolia once again exceeded 15 million *mu*.[105] Similarly, immediately after the Great Leap Forward, some 1.9 million–3.2 million hectares of land was categorized as saline farmland in Hebei, Henan, and Shandong. In all three provinces, construction was halted on many new irrigation systems beginning in 1960. But increased reclamation efforts and a greater number of improved drainage systems meant that by the mid-1970s salinized farmland had decreased to 1.4 million hectares on the North China Plain.[106]

A final example of upstream water resource development that diverted water from the North China Plain was the Qingtongxia Irrigation System in the upper Yellow River valley in Ningxia Province. Initially, the project irrigated over 330,000 hectares of agricultural land. Like the Hetao system, the Qingtongxia system was aggressively expanded from its historical roots in the imperial period. The irrigation system is fed by a multipurpose dam that was completed in 1968. Qingtongxia is one of six dams across the main stem of the Yellow River, including Sanmenxia. These capital-intensive projects were one component of a large system of dams and reservoirs built in the upper and middle valleys of the Yellow, Huai, and Hai Rivers (and their tributaries) between 1949 and 1978. The Qingtongxia irrigation network was representative of the PRC's construction and management of such large systems. It extracted some 4.5 billion m³ of Yellow River water per year, well beyond crop irrigation requirements. Inefficiencies in canal and gate construction required maintaining excessive water levels in the system. A poorly articulated drainage system also meant that a large proportion of excess water in the system was not returned to the Yellow River but instead percolated to the subsurface. The legacy of Maoist-era enthusiasm for irrigation at Qingtongxia, as with virtually all other large and small-scale irrigation projects, was twofold: excessive use of surface-water resources and land degradation through salinization.[107]

## The End of Maoism

Until his final years, Mao continued to indulge his passion for swimming. But his love of floating, now exclusively performed in his carefully regulated pool in the Zhongnanhai compound in Beijing, was no longer employed as a political symbol, except perhaps to fellow Party leaders who were keeping close tabs on his health in anticipation of a leadership scramble upon his death. Mao died in late 1976, and Maoism died shortly afterward. In relatively quick order, Deng Xiaoping emerged as Party leader by 1978 and quickly charted a new developmental path premised on achieving "four modernizations" in agriculture, industry, national defense, and science and technology. First suggested more than a decade earlier by Zhou Enlai in the recovery period after the Great Leap Forward, the "Four Modernizations" represented a signal transition from the technology complex of Maoism. Under Deng's leadership, the pendulum swung dramatically to the promotion of capital-intensive projects, valorization of industrial technology and technical experts, and internationalization.[108] The forces unleashed by the post-Mao reforms dramatically transformed China's economy and society, with profound implications for the nation's water resources.

Surveying the landscape of the North China Plain from the perspective of 1978, one sees a continued commitment to "ordering the waters" on the major waterways of North China: the Yellow, Huai, and Hai Rivers. This commitment to maintain an ecological balance through a combination of tradition and innovation extended the implicit compact between state and society that had been a source of political legitimacy for ruling elites throughout the imperial period. This commitment was reaffirmed by the CCP after 1978. And this will likely hold true for successor Chinese states well into the future. Post-1949 innovations in "ordering the waters" included technology that expressed itself in massive multipurpose dams on the main stems of all the major rivers and their tributaries on the North China Plain. In 1949, China had less than a dozen modern reservoirs. By the end of the Maoist period, the country had close to 2,500 large- and medium-size reservoirs with a total capacity of more than 10 million m$^3$, along with some 80,000 small-scale reservoirs and ponds.[109] The storage of massive amounts of water behind these structures was the key to developing irrigation, hydroelectricity, and improved transport. The herculean human effort to store water was directly related to the second major change on the landscape of the North China Plain that was clearly visible at the end of the Maoist period in 1978. The expansion of irrigation networks between 1949 and 1978, in a region traditionally underserved by irrigation, was indeed a profound transition.

The Maoist experience of water development on the North China Plain defies any simple evaluation. The "walking on two legs" developmental approach seemed to be compatible with a country like China, which had a strong state capacity and an abundance of labor but had limited access to capital or to modern technology for much of the 1949–1978 period.[110] However, the dual developmental approach became compromised, as the "legs" were strongly identified with Soviet-style central planners, on the one hand, and with the advocates of mass-mobilization, on the other. The "logic" of Chinese political struggle resulted in a tendency to swing suddenly and profoundly between developmental strategies, most notably in the swings to the left during the Great Leap Forward and the Cultural Revolution, when the landscape was dramatically transformed through the accelerated development of irrigation networks. From one perspective the hyperdevelopment of the Great Leap irrigation campaigns generated salinity problems that negatively impacted agricultural productivity. From this same critical perspective, one cannot neglect the 30 million to 40 million deaths that occurred during the Great Leap Forward. Crushing pressure to expand farmland irrigation was one component of an invidious set of circumstances leading to quite possibly the greatest famine in human history. The massive transformation of the landscape, either through surges of

"Spring Rains." (Courtesy of the International Institute of Social History, Amsterdam)

mass mobilization, or during more moderate periods, was made possible by a Party that had the capacity to control local society in unprecedented ways. At the same time, the Party expended considerable effort in managing traditional symbols and meanings of water and in reconstructing narratives of China's "century of humiliation" and peasant revolution. The landscape of the North China Plain became the stage on which these narratives were reenacted.

The transformation of the North China landscape represented an opportunity to redress China's humiliation and to extend the revolution. Was anyone listening? Did such attempts to imbue the landscape with historical and revolutionary import mobilize the peasants to engage in replumbing the North China Plain? The question is difficult to answer, but state and Party were engaged in the familiar behavior of political elites in a variety of cultural contexts that have the capacity—the institutional and organization tools at hand as it were—to fashion and refashion narratives that advance the transformation of nature to serve national identity and construction of the state. At the same time, through all the political and social upheaval, through all the deaths, through all the professional and personal lives damaged, if not extinguished during the Anti-Rightist movement, Great Leap Forward, and Cultural Revolution, through all the degradation of large tracts of farmland, through all the efforts to reclaim land damaged by mass-mobilization, China managed to feed itself. China accomplished this by expanding agricultural production in a country that was experiencing a dramatic demographic spike. Indeed, the CCP came to power on the promise of fundamental restructuring of rural society and of the institutions that controlled the production and distribution of food. In order to achieve the dual goal of eliminating famine in the countryside and supporting urban industrial growth, the Party encouraged increased agricultural production by expanding application of water and of restructuring rural society into communal units, while assuming redistributive functions. By 1978, the net gains of agricultural production made possible by irrigation, particularly from groundwater sources, "led to the transfer of foodstuffs from north to south, reversing the south's historical role as the 'breadbasket of China.'"[111] As Lillian Li notes, however, "By controlling the consumption of food at the household and individual level, the PRC has exercised a redistributive role that [the] Qing government never attempted, and it has done so with a thoroughness unprecedented in world history, except perhaps in wartime."[112] Thus, one could persuasively argue that, during the 1950s and 1960s, China's food demand was suppressed to a subsistence level and, more generally, that the manic development experience of 1949–1978 held China back by two decades. One can see in the post-Mao reforms what might have been possible immediately after 1949.

What is perhaps easier to conclude is that the aggressive exploitation of water during the Maoist era further stretched the traditionally scarce waters of the North China Plain. The aggressive development of water resources by a modernizing state is not unique to China's post-1949 history. The adoption of multipurpose river development, for example, was consistent with the paradigm of modernity pursued by polities of all stripes during the twentieth

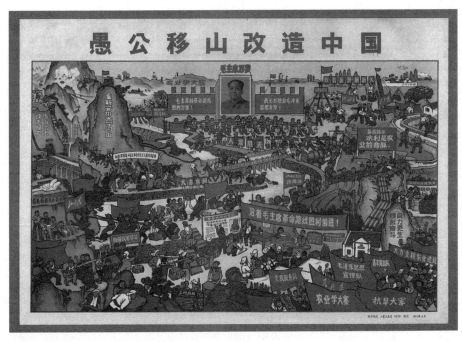

"Transform China in the Spirit of the Foolish Old Man." (Courtesy of the International Institute of Social History, Amsterdam)

century—a management perspective that projected "a powerfully pro-development discourse that saw nature as a resource to be mined or an obstacle to be conquered."[113] In this sense, the damming of rivers and the aggressive irrigation campaigns in China after 1949 was not any unique outcome of "Mao's war on nature"; indeed, it was entirely consistent with a long-standing impulse to "order the waters" and to promote and manage the ecological conditions necessary for the maintenance of Chinese society on the North China Plain. At the same time, however, the Party, particularly in its Maoist phases, had a priority to develop natural resources as quickly as possible and to bring maximum human capital to rapidly rebuild wealth and power in a world that was deemed hostile to the Party and the country. Such thoroughgoing and speedy mobilization of China's water resources necessary to spur agricultural and industrial growth could be achieved only through projects that required minimal capital investment, limited technical knowledge, and that placed little management burden on a massive government bureau-cracy. This Maoist approach had critical consequences for water and land on the North China Plain. In the early 1980s, a researcher at the Chinese Academy of Sciences' Institute of Soil Sciences re-flected on these practices: "In the late 1950s . . . water was diverted blindly from the Huang He [Yellow River] for irrigation in order to solve the drought problem. Drainage systems were not provided. A large number of reservoirs were built in the plain, blocking the natural drainage channels. The canals carried water for long peri-ods, resulting in considerable seepage, which raised the water table. Secondary salinization and swamping of the soil occurred exten-sively. This lesson has driven home to us that the ecosystem of the plain is unstable."[114] Increased consumption was exacerbated by the massive inefficiencies of the water supply system. These inefficien-cies are largely explained by the mass mobilization techniques with which water resources were developed on the North China Plain after 1949. For example, in a study of local irrigation practices in Hebei transmission losses reached as high as 60 percent in the post-Mao period.[115] The following account of the fate of reservoirs on the North China Plain, summarized in the 1990s, serves as a proxy for the quality of construction of water management facilities more generally:

> By 1980, 2,976 dams [in China] had collapsed, including two large-scale dams [the Shimantan and Banqiao dams on the Huai River]. One hundred and seventeen medium-sized, and 2,857 small dams had also collapsed. On average, China wit-nessed 110 collapses per year, with the worst year being 1973, when 554 dams collapsed. The official death toll (not including the Banqiao and Shimantan collapses) resulting from dam fail-ures came to 9,937. Some people say that among the more than 2,000 dam collapses, only 181 involved fatalities but this hardly seems accurate. By 1981, the number of formally recognized dam collapses had risen to 3,200, or roughly 3.7 percent of all dams. According to Ma Shoulong, the chief engineer of the Water Resources Bureau of Henan Province, "The crap from that era [the Great Leap Forward] has not yet been cleaned up."

In 1958, more than 110 dams were built in Henan; by 1966 half of them had collapsed.[116]

In other words, the development of water resources in the Maoist era cast a long shadow over the post-Mao reforms.

# 6 *Managing Legacies, Managing Growth*

THE HOT, DRY AUGUST of 1992 was a particularly unsettling period for the farmers in the North China villages of Baishan and Panyang. On the afternoon of August 5, residents from both villages hurled insults and explosives at one another across the Zhang River, which feeds the Red Flag Canal Irrigation System. That summer, the river trickled more than it flowed. The emotions of the afternoon were fueled by events of the previous night, when seventy Baishan villagers waded into the river to erect a barrage to divert water illicitly to their fields in Hebei. Upon hearing of the treachery, Panyang villagers assembled to drive the dam builders away. With combatants arrayed on each bank of the river, the exchange commenced. Two days later, villagers from Baishan crossed the river to the Henan side and dynamited an irrigation canal that watered Panyang. The rest of the month proceeded in similar fashion, with periodic outbreaks of hostilities. Then, in the predawn hours of August 22, explosions along the Red Flag Canal sent a cascade of water roaring down a 90-meter slope, straight to the fields and houses of Panyang Village.[1]

According to a correspondent from *Nongmin Ribao* (Farmers Daily), who wrote a front-page story on the conflict a couple of weeks later, Party and government leaders were stunned by the acute disturbance. Yet some locals informed the reporter that conflict had been going on for twenty years; others said thirty. The farmers also acknowledged that the rancor had escalated during the past several years. Indeed, before 1988, when tensions were perhaps more latent, villagers seemed to get along reasonably well. There were dozens of intermarriages between the villages. The previous Panyang CCP branch secretary was the uncle of the current Baishan CCP branch secretary, and the director of the Baishan Rural Collective was the nephew of the Panyang Village CCP branch Party secretary. But water ran thicker than blood. Violence between the two villages continued throughout the decade with considerable loss of life and property.[2]

The exchange of mortar and locally fashioned bombs across the Zhang River in 1992 was both destination and point of departure—that is, the conflict over water resources along the Zhang River and the Red Flag Canal lay along a historical continuum that connected past and present and that suggested the parameters of a future trajectory. More concretely, escalating contestation over water resources on the North China Plain during the 1990s was a consequence of a historical arc that combined climate, geography, and social forces. Looking back over the events of 1992, a 2004 article in a regional water management journal provided critical context with the observation that violence over Zhang River waters had intensified by the early 1990s. According to the article, in the early 1960s, when the Red Flag Canal was being constructed, siphoning water from the Zhang River seemed reasonable, as water was plentiful. At that time the annual flow of the Zhang River was 5 billion m³. Although the completed Red Flag Canal diverted 70 percent of the river's flow, there was still plenty of water in the Zhang to service downstream needs. But after the 1980s the annual flow of the Zhang River decreased to 1 billion m³. Upstream withdrawals for irrigation and local industry intensified downstream competition for water, like that between Baishan and Panyang Villages. Compounding the struggle over a dwindling supply of river water was drought. During the 1990s, the region suffered the worst drought in forty years. In the downstream areas, the Zhang River ran dry. The upper reaches of the Red Flag Canal in Shanxi Province also dried up on eight different occasions. Ecological stress generated economic and social instability. A local official from Baishan explained the desperation felt by farmers as they attempted to build the dam to divert water to their fields. The construction was done during the critical days of the growing cycle. Because of drought, hundreds of *mu* astride the river had no access to water. All the peasants had a stake in the availability of water. The local official commented, "Who wouldn't be anxious?" Discussions with Pan-yang villagers expressed similar economic as well as social dependence upon water. They said that the water from the Red Flag Canal was their "lifeblood," and that they had "sacrificed over 100 lives" to build the canal.[3]

## Challenges in the Reform Period

The story of this intervillage dispute over water resources was a relatively early indicator of water stress on the North China Plain that grew progressively acute under the extraordinary demand generated by the economic growth of the post-Mao reform era (1978–). The story of water on the North China Plain for much of the twentieth century, particularly after 1949, centers on technological and organizational innovations that generated greater hydrologic stability for the major river systems (through flood control), and that exploited the productive powers of surface and subsurface water resources. Beginning in 1978, these "fixes" to the traditional challenges of floods and famine also provided the resource base to support the massive post-Mao restructuring of agriculture, industry, and urban life. The pace and degree of these transformations, however, suggested that the technological and sociological solutions to the long-term water challenges of the North China Plain were temporary. A sense of "ecological crisis," so manifest during the late imperial and Republican periods, reappeared during the post-Mao period as consumption and degradation of water resources complicated allocation decisions and posed existential issues for the Party—a party that was, and remains, largely dependent on its capacity to engineer economic opportunities for its constituencies.[4]

The economic reforms of the post-Mao era had a profound impact on the waterscape of a region that had been re-engineered during the Maoist period. Rural reforms erased communal obligations by granting farmers long-term leases on land and greater discretion over what to cultivate and where to market. Shortly thereafter, the restructuring of the state-owned industrial sector and the introduction of market forces generated a manufacturing boom and a large-scale migration to urban centers. China's gross domestic product (GDP) grew by over 10 percent, year after year, and sobering images of fouled water and sobering scientific data showed that ecological limits had been reached. Engineering solutions, such as large-scale interbasin water transfers, the likes of which had not been seen in China since the peak of Soviet influence, were now necessary to sustain some semblance of ecological balance on the North China Plain. A survey of data sketches a compelling image of water constraints. Total available water resources dropped from 41.9 billion m³/year during 1956–1979 to 32.5 billion m³/year during 1980–2009. On a per capita basis, the decline of water resources was more stunning. As a consequence of demographic growth on the North China Plain (whose population increased from 57 million in 1980 to 133 million in 2009), per capita water availability plunged from 735 m³/year in 1952 to 302 m³/year in 2009. In comparison, total national water resources per capita were 2,079 m³/year in 2009 (the global per capita 2009 average was over 6,225 m³/year). As per capita supply declined, overall consumption of water skyrocketed. Household demand in 2009 increased by 144 percent over 1980. This will continue as total water consumption is expected to increase from 41 billion m³/year in 2009 to 46.2 billion m³/year in 2020.[5]

Compounding the tenuous supply–demand balance of water was the emergence of two additional problems: pollution and global warming. In 2009, 25 percent of Yellow River system water did not meet the Grade V standard (i.e., it was unfit for human consumption or agricultural use). Similar conditions pertained to the two other major river systems on the North China Plain: the Huai River (17 percent) and the Hai River (42 percent). At the same time, a 2010 publication by several government and research organizations in China forecasted a 27 percent decline of glacial volume in the Qinghai-Tibet Plateau and the Himalayas, the source of many of China's rivers, as well as transboundary arteries such as the Salween, Ganges, Mekong, Ayeryawady, Brahmaputra, and Indus Rivers. Since the 1970s, increasing pressures on surface water led to the full tilt pumping of groundwater on the North China Plain.[6]

By 1997, an accumulation of evidence suggested critical problems in water allocation. Unsettling reports of localized violence like that in the Red Flag Canal Irrigation District, embarrassing national and international reporting on the putrefaction of the Huai River, stunning images of the dried-up Yellow River, and the doubts of influential international observers about China's capacity to feed itself all combined in the late 1990s to focus attention on the North China Plain. This "perfect storm" sparked the state to allocate substantial funds to research organizations like the Chinese Academy of Sciences to better understand the forces contributing to water scarcity and water degradation.[7] On the demand side, population growth was a foundational driver. From 2000 to 2005, in just five years, the key age bracket between the ages of fifteen and sixty-four increased from 856 million to 922 million, "a staggering increase" of 66 million in the prime of their years of consumption.[8] But demographic expansion was only one part of the story.

Beginning in 2000, much of the government-funded research studies focused on the Yellow River as proxy for the region's water challenges. The fundamental question addressed by all these studies was whether the decreased flow of water and the downstream dry-up of the Yellow River were caused by changes in climate, human activities, or a combination of both. Although there were observable changes in temperature and precipitation over the previous fifty years in China, virtually every study concluded that aggressive water resource development since 1949 was "the most important factor in causing the frequent zero-discharge episodes on the main stem [of the Yellow River] in the lower basin."[9] At the same time, the potential impact of climate forces on the water resources of North China suggested an even more uncertain future. Thus, from the perspective of the new century, water stress on the North China Plain was exacerbated by transformations in China's economy and society that pushed per capita consumption ever higher during the post-Mao period.[10]

## Competing Demands

Remarkable economic growth in China over the past several decades has caused an increase in water consumption across all sectors. Given the scarce endowment of water on the North China

Plain, burgeoning urban and industrial demand has complicated the calculus of allocation. On the one hand, China's modernization drive has depended on continued industrial and commercial growth and associated processes of urban expansion. On the other hand, the imperative to expand agricultural production to meet China's food demand continues to motivate state guarantees of sufficient water supply to agriculture. These multiple objectives have produced an allocation system that is suspended somewhere between the state and the market. Demand for water as input to productive processes is largely driven by market demand for the goods and services generated in all three sectors. Supply, however, is largely regulated by pricing mechanisms established by the state. The fundamental rationale for such a system has critical economic and political dimensions as the state seeks to negotiate a delicate supply profile that satisfies important social constituencies. On the North China Plain, a critically important sector to sustain is agriculture—an enduring state commitment reaching well back into the imperial period. And yet, state guarantees of water supply in China, again particularly to rural constituencies, have promoted inefficiencies that compound the struggle to effectively allocate scarce water resources across all sectors of the economy.

As in most regions around the globe, in China agriculture is the largest consumer of water. Although agricultural water use on the North China Plain declined from a high of 84 percent during 1980–1999 to 75 percent in 2007, absolute consumption has continued to increase.[11] Robust demand is a function of increasing pressure on the North China Plain to feed the country. The legacy of China's "century of humiliation" at the hands of the international system has continued to inform a national priority to maintain grain self-sufficiency. During the most intense mass mobilization movements of the Maoist period, the slogan of local self-sufficiency permeated the drives to increase agricultural production through expansion of irrigation and fertilizer use. The "Grain-first" and "Learn from Dazhai" movements during the Great Leap Forward of the late 1950s and the 1960s were perhaps the most explicit rendering of such desires for self-reliance in an international environment deemed threatening to China. This broad sentiment has persisted during the post-Mao era, perpetuating the need to maintain and expand water supplies to meet the country's food needs. Thus, although the percentage share of water consumption by the agricultural sector has indeed gradually declined in the post-Mao era, the intensification of agriculture has grown as a consequence of reforms in the agricultural sector that began in 1980.[12]

The centerpiece of these reforms was a package of initiatives collectively termed the "household responsibility system." The fundamental thrust of the program, which was championed by Deng Xiaoping, was a reworking of the immediate post–Great Leap Forward retrenchment policies designed to restore agricultural production. The household responsibility system returned greater autonomy to the household. Individual households were required to fulfill production contracts made with the state (at state-set procurement prices), but farmers were free to sell any production beyond that quota at local markets, which appeared virtually overnight after 1980. The introduction of market demand established variable prices for agricultural output that encouraged farmers to cultivate higher value-added products, such as fruits and vegetables. The outcomes of rural reforms included a quick dissolution of communes in the countryside. Rural households were granted long-term leases on land and given increased autonomy over production and marketing decisions. Producing for the market and not the state, farmers responded by intensifying agricultural production by greater application of water and fertilizer.[13]

The changing structure of agriculture in China, and the broader effects of post-1978 economic reform, gave farmers powerful incentives to increase agricultural output. From a national perspective, the fact that the North China Plain was already China's largest expanse of agricultural land suggested that, if Party leaders continued to insist on food self-sufficiency, the region would inevitably play a major role in the broader intensification of agriculture. During 1995–2005, the area sown to crops in the Yellow River basin grew by 5.8 percent. At the same time, in the seven southern coastal provinces of China cultivated acreage declined by 4.6 percent.[14] In other words, agricultural production has been growing in precisely the region that is among the driest in the country. As a consequence, greater exploitation of water resources was necessary to sustain this expansion. In a reversal of food production patterns that pertained for most of the late imperial period, the reform period has witnessed the emergence of North China as the granary for the south. Data from the reform period (1995–2005) clearly reflects the intensification of agriculture on the North China Plain. By 2000, the North China Plain produced 60 percent of China's wheat and 40 percent of its corn on 22 percent of the country's land resources and with 4 percent of its water resources.[15] This type of agriculture depended on irrigation. Thus, during the reform period, irrigation in North China increased by nearly 40 percent. At the same time, water withdrawals for irrigation exceeded rainfall by 200–300 percent, meaning that growth in irrigated agriculture was sustained only by further extraction of groundwater resources at rates that have widely been recognized as unsustainable.[16] To cite one set of data, in the western portions of the North China Plain, shallow groundwater tables declined from depths of about 1 meter in the 1950s to more than 20 meters in the 1980s and to 30 meters in the 1990s.[17] The threats to China's capacity to remain self-sufficient in food revived the impulse to address the water imbalance between North and South China by massive water diversions from the Yangtze River to the North China Plain. But as the geographer Michael Webber points out, "The paradoxical implication of this shift in agricultural production into North China is that nearly 10% of the water used in agriculture in the North is employed in producing food that is exported to South China . . . the net effect is the annual export of virtual water embodied in food from the water-scarce North to the water-abundant South, in volumes that exceed those proposed for the South-North [water] transfer."[18]

A critical issue relevant to water, and to the capacity of China's agriculture to feed its population, is the amount of land under cultivation. Over the past several decades, considerable attention has been given to the amount of agricultural land lost to urban and industrial expansion. There is little doubt that farmland conversion has happened in the rapidly urbanizing and industrializing areas of East and Southeast China and in the Yangtze River valley. But the range of assembled data on China's total cultivated acreage reflects some ambiguity.[19] For example, one study highlights the stunning amount of agricultural land that has been submerged by reservoirs alone since 1949: "Between 1957 and 1990, China lost an area of arable land equivalent to the total cropland of France, Germany, Denmark, and the Netherlands."[20] But net loss (or gain) to agricultural output from the construction of reservoirs is unclear. Farmland was submerged, but gains from irrigation made possible from these reservoirs may have more than compensated for the withdrawal of agricultural land.

There is likely a suite of forces generated by economic reform in China that has indeed eroded the country's total cultivated land, including urban and industrial expansion, deforestation, erosion, and land pollution. That significant tracts of land have been withdrawn from agricultural production was clearly suggested by the resolve of state and Party leadership to stabilize cultivated acreage. At the Seventeenth Party Congress, held in 2007, Premier Wen Jiabao articulated a goal of maintaining 120 million hectares of arable land. This "red line" was calculated to ensure food self-sufficiency.[21] The salient point is that China must squeeze more food out of a static (or declining) amount of agricultural land if it is to meet the demands of an expanding population with increasing per capita consumption. Coaxing more agricultural output from the North China Plain requires more water to the fields.

The changing structure of urban food demand driven by rising per capita income has also had a potent effect on agricultural water use. A rapid increase in the consumption of animal protein has resulted in a decline in direct consumption of cereals and grains. Between 1952 and 1992, annual per capita consumption of animal protein increased threefold, from 11 kg to 38 kg. Most of that increase came after the onset of reform. By 1999 annual per capita consumption of animal protein had increased to 65.3 kg.[22] In 1952, only 3.1 percent of energy needs came from animal protein; in 1992, this figure was 19 percent.[23] Raising animal-based foods is not water efficient: "In China, it takes 2,400–12,600 litres of water to produce a kilogram of meat, whereas a kilogram of cereal needs only 800–1,300 litres. The recent rise in meat consumption has pushed China's annual per capita water requirement for food production up by a factor of 3.4 from 255 metres in 1961 to 860 cubic metres in 2003. Compared with China's population growth by a factor of 1.9 over the same period, this suggests that dietary change is making a high demand on water resources."[24] In addition to growing large quantities of feed grain to sustain meat production, farmers on the North China Plain also began switching to more water-intensive crops like fruits, vegetables, nuts, and dairy products to satisfy changing urban dietary preferences.

Because of the expanding market governing food production and consumption decisions, the North China Plain became increasingly critical to China's food security. The millions of production decisions that collectively generated an increasing draw on scarce resources were not made by administrative fiat. Premised on cost–benefit calculations, farmers were making rational decisions based on comparative advantage. And these decisions were calculated on a supply of water that was exempt from the increasing marketization of the Chinese economy. The price of water was set by administrative fiat— and water was cheap. As long as inexpensive supplies were available, there was little incentive to use water efficiently. The massively inefficient systems constructed during the Maoist period continued to be inefficient in the post-Mao period. Irrigation flooding techniques (as opposed to more efficient drip irrigation methods), leaky irrigation channels (more than half of water transmission is lost to percolation), and evaporation from the large and small reservoirs that dotted the landscape all perpetuated increased water losses as water utilization rates increased. While agricultural efficiency increased on North China Plain from 1.52 tons of grain per acre in 1953 to 4.16 in 1984, the efficiency of water use decreased (i.e., proportional losses of water increased as consumption grew).[25] Water use in China necessary to generate 10,000 yuan worth of production was 329 m$^3$ in 2006—six times greater than in the United States.[26]

The urban base of the North China Plain on the eve of the post-Mao reforms was small compared to that in other regions in China. However, during the reform period, urban centers like Beijing and Tianjin, as well as medium-size cities like Shijiazhuang (Hebei) and Zhengzhou (Henan), have experienced tremendous growth. As suggested above, the water resource base of the North China Plain has been affected not only by the forces of reform within the region but also by increased demand in the country as a whole. Thus, rather than focusing solely on northern China, it is perhaps more instructive to examine Beijing, Tianjin, and other cities on the North China Plain in the context of urban expansion in China generally. At the beginning of the reform period, China's urban population was officially 80 million people, but by 2002 it had increased sixfold to 490 million, far outstripping the national population growth rate.[27] Cities in the North China Plain experienced similar growth rates, as urban industrial and commercial expansion drew workers from conditions of underemployment in rural areas. Per capita GDP in Shanghai increased by a factor of sixteen from 1978 to 2002.[28] The consumption dynamics of urban populations created greater pressures on urban water allocations. Indeed, urban water use has expanded more rapidly in the reform era than any other sector, increasing by more than 10 percent per year during the 1990s.[29] The potential for Chinese household water consumption to increase further, in both urban and rural areas, is suggested by an average per capita consumption in China well below that in the United States and Japan. Figures for 2009 show that water consumption in urban China was 212 liters per person per day. If demand trajectories approximate patterns established in Japan and the United States, where per capita consumption is 350 liters per day for both countries, there is plenty more room for growth in household water consumption on China's horizon.[30]

In addition to foundational changes in rural and urban economies, Deng's Four Modernizations also transformed China's industrial infrastructure.[31] The boom in industrial production in China, which followed a transition from an emphasis on heavy industry to light industry, generated a substantial increase in water usage. Nationally, industrial water use expanded from 10 percent of all water consumption in 1980 to 23 percent to 2006.[32] However, the level of industrial growth on the North China Plain was slower than the national average. By the 1990s, industrial use of water comprised less than 10 percent of all demand for water on the North China Plain. Still, any growth posed challenges to allocating China's limited resources. One particular threat to the surface waters of the North China Plain, however, was the development of Township and Village Enterprises (TVEs) in rural areas. It was largely TVEs that created a new stress on water resources on the North China Plain, namely, pollution.

## Pollution of Surface Waters

In the early stages of post-Mao reform, the formation of rural industrial enterprises was encouraged in order to engage surplus labor in rural China. TVEs sprang up throughout the North China Plain and served to slow migration to urban areas. Immediately after 1980, most TVEs were owned collectively by residents at the township or village level. By the mid-1990s, many TVEs were controlled by local governments or private interests. The growth of these small-scale industrial enterprises was stunning. By the early 1990s, there were 1.3 million TVEs in China, which accounted for 30 percent of all industrial production. By 2000 this figure had jumped to 47 percent and accounted for 48 percent of China's exports.[33] At this same time, TVEs absorbed 25 percent of the rural labor force. By 2003, over 120 million peasants had left agricultural work for employment in over 22 million TVEs in the country.[34] In short, TVEs were critical to the development of China's economy during the reform period. Calculated by export value in 1995, the top five product sectors of TVEs were apparel, textiles, leather products, knitted products, and toys.[35] TVEs used considerable amounts of water, but their most critical impact on water resources was pollution. Largely unregulated, and engaged in highly polluting industries like leather tanning, TVEs rendered significant areas of surface water and shallow groundwater sources too polluted for human and agricultural use.

During the reform period, degradation of water resources by rural industrial development was exacerbated by the increasing use of inorganic agricultural fertilizers. Indeed, along with further exploitation of water resources, the other driver of agricultural growth after 1949, particularly during the reform period, was the intensified application of inorganic fertilizers. Consumption of nitrogen fertilizer increased worldwide from 2.865 trillion grams (Tg) in 1998 to 24.8 Tg in 1998, accounting for 30 percent of global usage.[36] China was a major contributor to this global trend. Between 1980 and 2002, inorganic fertilizer use on the North China Plain increased by 260 percent.[37] A local study of Ningjin County in Shandong Province concluded that without heavy dependence on fertilizer, even with a suite of other "green revolution" inputs (including expanded irrigation and high-yielding seeds), the county simply could not have met its goal of increasing food production to match demographic expansion. On an average annualized basis, Ningjin County increased per capita grain production by 14 kg a year from 1949 to 1999, more than meeting the needs of a rapidly expanding population.[38] However, the county paid a longer-term price for the yields afforded by aggressive use of chemical fertilizers: contamination of surface and groundwater. Local variation was high, but overall nitrogen runoff to rivers, lakes, and ponds, ranged from 2 to 5 percent of all nitrogen applied. Loss from leaching (with most chemicals eventually going to groundwater aquifers) ranged from 4 to 19 percent of applied nitrogen in the Beijing region, according to a 1998 study.[39] The authors of the local study of irrigation and fertilizer inputs in Ningjin County concluded that at current rates of water utilization (mostly from groundwater) and of nitrogen fertilizer seepage, the sorts of harvests the county experienced in the past two decades would "unlikely be sustainable for a long period."[40]

Perhaps the most egregious and visible outcome of the cocktail of industrial pollutants and agricultural fertilizers on the North China Plain was reflected in the Huai River valley. Despite the enormous post-1949 investment of human and material capital for flood control and irrigation networks, the region remained one of the poorest in China. By the onset of the reform era there were over 5,000 reservoirs and 4,000 sluicing facilities. These facilities were designed for multipurpose use, but they mostly hydrated the dry agricultural soil of the valley.[41] Because of increased water demand spurred by economic reforms, water utilization rates on the main stem of the Huai River exceeded 70 percent, leading to desiccation of the river channel by the early 1990s.[42] At the same time, massive consumption of water contributed to industrial growth. During 1984–2003 "the total industrial output of the region [the Huai River valley] grew by a factor of nearly 30, while per capita GDP grew by a factor of 6.8."[43] Most of this industrial growth was generated by TVEs. Papermaking, brewing, tanning, and tobacco and food-processing facilities generated goods that found receptive domestic and international markets. Unfortunately, these industries used water inefficiently and discharged large quantities of pollutants. For example, locally owned paper mills used 360 percent more water than state-owned mills.[44] Throughout the reform period, the vast majority of TVEs discharged untreated wastewater directly into lakes, rivers, and streams. According to the State Environmental Protection Agency, in 2002 only 24 percent of China's 62 billion tons of wastewater was treated. TVEs alone discharged over 10 billion tons of wastewater.[45]

The cumulative effects of industrial and agricultural runoff became apparent to downstream denizens of the Huai River valley in northern Anhui and Jiangsu Provinces. In February 1989 the most significant incident to date occurred when seriously polluted water behind the Bengbu (Anhui) water gate was released. This water formed a 60-km-long toxic plume that flowed downstream, threatening the water supply of several million people. All withdrawals from the river were stopped.[46] The fouling of the Huai River became the focal point for increasing local and central government concerns over the longer-term impact of environmental degradation. Lacking effective administrative leverage to enforce environmental regulations at the local level, the Party sanctioned the dissemination of images and stories of ecological deterioration by the state-run media as a means of pressuring local officials to enforce environmental standards upon TVEs. But this effort largely flew in the face of the Four Modernizations, itself a mass movement of sorts, which was encouraged by moral exhortations from the central government and sustained by the promise of material rewards to individuals and communities from economic modernization. From 1980 to 1995, the Huai River became the most polluted waterway in China. According to China's system of surface water-quality classification, the percentage of Huai flow classified as Class IV or worse (no direct contact with humans) increased from 45 percent in 1980 to more than 88 percent in 1995.[47]

During the late 1990s, state-run media also began to highlight health concerns from polluted waterways on the North China plain. For example, an investigative report conducted and aired by China's largest television network, China Central Television (CCTV), focused on Huangmengying, a village located on the largest tributary of the Huai River. In its report, CCTV highlighted the fact that of 204 people who died in the village from 1990 to 2004, 105 died of cancer, a mortality rate 60 percent higher than before 1990, when pollution in the tributary was still relatively low. In the aftermath of the investigative report, local and provincial public health officials confirmed that such high rates of cancer were entirely consistent with levels of pollution in village drinking water. Huangmengying and other small villages in the Huai River basin quickly became known as "cancer villages." Another report highlighted the mortality rates in Yangzhuangxiang, a small village that was located on a tributary of the Huai River in Anhui Province, where mortality rates reached 13 per 1,000, with 80 percent of all deaths attributed to liver cancer. Along the same river, one village reported 40 cases of esophageal cancer among 280 households.[48] Although the environmental condition of surface waters in North China was particularly egregious, a large body of data on the deleterious health effects of degraded water emerged for the country as a whole. An explicit example was the finding that waterborne illnesses were the leading cause of death among children under age five.[49]

## The Yellow River Gone Dry

The jolting images and accounts of Huai River pollution, coupled with the questions about China's food and national security, were punctuated by the unparalleled 1997 desiccation of the Yellow River. As the dry-up reached 700 km inland, downstream communities were deprived of their principal source of water. Contributing additional meaning to the dry-up was that it occurred in a waterway that to many continued to represent "the mother of Chinese civilization." Although the Yellow River had experienced frequent and varying degrees of desiccation since the 1970s, the particularly dry years of the mid-1990s coincided with the unprecedented exploitation of upstream, midstream, and downstream water resources (see Table 6.1). A *Renmin Ribao* article in May 1997 provided a brief historical overview and analysis of the desiccation of the Yellow River. The article noted that the Yellow River experienced regular annual dry-ups in 1970s. In 1972 the Lijin hydrological station in Shandong Province, some 140 km inland, recorded its first dry-up, which lasted for fifteen days. Since 1990, dry-ups had occurred every year and had shared several distinct characteristics. First, the duration of each annual dry-up increased each year, reaching 236 days in 1997. Second, the dry-ups extended increasingly far upstream. In 1997, the desiccated riverbed had reached over 700 km inland from the coast, approaching the large industrial city of Zhengzhou in Henan Province. And finally, annual desiccation occurred earlier each year. In the past, dry-ups typically happened in May or June, but during the late 1990s they commenced in February or March.[50]

A spate of articles in *Renmin Ribao* and other media outlets began to frame water constraints in the broader context of sustainable development. Because desiccation occurred in the downstream provinces of Shandong and Henan, there was considerable lamenting of the lack of basin-wide coordination of storage and extraction of Yellow River resources. Virtually every account focused on the primacy of integrated river management that could better coordinate allocation, pricing, and water conservation practices. In a series of articles in 1997, Jiang Chunyun, a member of the CCP Political Bureau (Politburo) and vice-director of the State Council, called for centralized control over the water resources of the Yellow River valley. His comments were directed at provincial governments that largely made allocation decisions independent of downstream concerns.[51] The articles quoting Jiang and other officials that appeared in the official press following the 1997 dry-up introduced into broader public discourse concepts of ecological balance, sustainable development, and comprehensive basin-wide management of water resources. Addressing these problems would imply greater central bureaucratic presence, as well as more consolidated bureaucratic power, for water allocation decisions. The Yellow River dry-ups of the 1990s were the opening acts of a struggle between the center and the provinces to manage the water resources of the North China Plain.[52]

The progressively serious Yellow River dry-ups added further incentive for state-mobilized research efforts to study the causes and consequences of river desiccation in the broader context of the water resource challenges of the North China Plain. Scholarly and policy discussions largely centered on the anthropogenic and natural climatic forces responsible for North China's water crisis. Although there were differing opinions on the relative primacy of each factor, the balance of this research concluded that exploitation of water resources since 1949 critically conditioned current realities. Scholars such as Xu Jiongxin collected and collated hydrological data that showed the North China Plain had become a "high-frequency zone" of river desiccation. For example, during the 1980s, four of the Hai River's main tributaries dried up each year for an average of 200 days. Northern tributaries of the Huai River experienced an average of 100–200 desiccation days per year. Even midstream tributaries of the Yellow River, including the Wei and Zhang Rivers, had dried up 100–300 days every year during the 1980s. From a historical perspective, the river dry-ups in the 1980s and 1990s on the North China Plain were an anomaly. "From 1809 B.C. to 1949 A.D., no-flow states on the Yellow River were documented for [only] ten years in the Yellow River Stem."[53] Desiccation occurred on average only once every 376 years.

Xu and a host of other scholars concluded that the fundamental reason for these events was the large-scale construction of reservoirs on the mid- and upstream reaches of these waterways.[54] Such analyses identified the fragile ecological foundation upon which large-scale Maoist exploitation of water resources on the North China Plain had been built and also recognized the pressures upon upstream surface waters that fed the North China Plain. Much of this research also pointed to the relatively dry nature of the post-Mao period, particularly the 1990s. Yet, even in the face of low precipitation, "There is no doubt that under natural conditions, even in the high-frequency zone [for dry-ups] . . . river desiccation would rarely occur."[55] Xu concluded:

Only after many reservoirs were constructed in the upper drainage basins did river desiccation become frequent. . . . Since the late 1950s, many reservoirs have been built in the drainage basins of the Yellow and Haihe rivers, and huge quantities of water have been diverted from the river for irrigation every year, so that the runoff entering the lower reaches of the river has declined sharply. The total storage capacity of reservoirs built in the Haihe, Yellow, Huaihe, Liaohe rivers accommodates 84, 73, 72, and 82% of the total annual runoff, respectively, which are the highest of the seven major river systems in China. . . . Therefore it can be thought that water diversion is the major factor for runoff reduction and for river desiccation, and the effect of meteorological factors is of secondary importance.[56]

By the time of the Yellow River dry-up in 1997 large and small retention reservoirs fed irrigation systems that watered some 7.5 million acres in the Yellow River valley (up from 800,000 acres in 1949). There were 3,147 dams (storing 57.4 billion $m^3$ of water), 4,500 diversion works (containing 29.5 billion $m^3$ of water), 29,200 river pumps (capable of extracting 4.8 billion $m^3$ of water), and 875,000 groundwater wells in the Yellow River valley.[57] By 2000, twelve large dams had been built (or were being constructed) on the main course of the Yellow River. These dams had an effective storage capacity of 35.56 billion $m^3$. In the context of an already precarious ecological setting, all this diversion and pumping activity sucked the waterways of the North China Plain dry. By 1993, annual diversions from the Yellow River were 27.6 billion $m^3$, four times that of 1949. Between 80 and 90 percent of the flow of Hai River tributaries was drawn off by the early 1990s. Similar numbers pertained to Huai River tributaries.[58]

The consequences of river desiccation included deterioration of the aquatic environment in the estuary region; aggravation of silt deposition; threat to downstream supply for industrial, residential, and agricultural sectors; and intrusion of seawater in the estuary region.[59] Already by the 1970s and 1980s, some 70,000 square hectares (hm²) of agricultural land had no access to Yellow

*Table 6.1*   Yellow River dry-ups

| Year | Occurrences | Duration (total) | Length (km) |
|------|-------------|------------------|-------------|
| 1972 | 3  | 19  | 310 |
| 1973 | 0  | 0   | 0   |
| 1974 | 2  | 20  | 316 |
| 1975 | 2  | 13  | 278 |
| 1976 | 1  | 8   | 166 |
| 1977 | 0  | 0   | 0   |
| 1978 | 4  | 5   | 104 |
| 1979 | 2  | 21  | 278 |
| 1980 | 3  | 8   | 104 |
| 1981 | 5  | 36  | 662 |
| 1982 | 1  | 10  | 278 |
| 1983 | 1  | 5   | 104 |
| 1984 | 0  | 0   | 0   |
| 1985 | 0  | 0   | 0   |
| 1986 | 0  | 0   | 0   |
| 1987 | 2  | 17  | 216 |
| 1988 | 2  | 5   | 150 |
| 1989 | 3  | 24  | 277 |
| 1990 | 0  | 0   | 0   |
| 1991 | 2  | 16  | 131 |
| 1992 | 5  | 83  | 303 |
| 1993 | 5  | 60  | 278 |
| 1994 | 4  | 74  | 308 |
| 1995 | 3  | 122 | 683 |
| 1996 | 7  | 136 | 579 |
| 1997 | 13 | 226 | 704 |
| 1998 | 16 | 142 | 142 |
| 1999 | 6  | 42  | 42  |

*Data source:* Liu Changming and Jun Xia, "Water Problems and Hydrological Research on the Yellow River and the Hai and Huai River Basins in China," *Hydrological Processes* 18:12 (2004): 2199–2200.

River irrigation water in Henan and Shandong Provinces. In the 1990s, this area increased to half a million hm², generating agricultural losses of 4 billion yuan (roughly US$30 billion) per year. In the Shengli oilfields of Shandong, Yellow River water was unavailable for water injection that sustained production in this aging but critical source of domestic production. Desiccation threatened the nutritional complex of the waters in the estuary, mouth, and littoral areas of the Yellow River that sustained the spawning grounds for wild fish and a large aquaculture industry. And finally, urban water supplies to cities such as Dezhou and Dongyin were halved during the dry-ups of the late 1990s.[60]

## The Growth of an Environmental Consciousness

By the late 1990s, communities on the North China Plain were engaged with the challenge of increasingly limited and polluted water supplies. These realities generated protest. The growth of an environmental consciousness represented, at least initially, a coalition of communities that faced dire ecological conditions, as well as other rural and urban constituencies that were increasingly receptive to notions of sustainability. Ultimately, this coalition came to challenge the "clean up later" orientation as evidence mounted that suggested the negative impacts of water scarcity and pollution on economic vitality and human health.

In the absence of bureaucratic capacity to police polluters, the state quietly encouraged the dissemination of alternative levers. As suggested earlier, during the late 1990s, a form of ecojournalism was tacitly encouraged to shine an incriminating light on local polluters. In addition, call-in radio programs encouraged listeners to report on pollution problems in their neighborhoods. National media outlets like CCTV reported on pollution and other resource problems with relative freedom. A notable example of such investigative journalism was the report on the "cancer villages" in the Huai River valley. But the state was potentially playing a dangerous game. With the political disintegration of the Soviet Union in 1991 providing an object lesson, state and Party leaders were well aware of the role that environmental protest had in burgeoning political movements in the Soviet Union and Eastern Europe. Therefore, China's central government set limits to what it deemed acceptable protest. Still, there were spaces in Chinese society where protest occurred with increasing frequency. One of these spaces was in rural China. With quite remarkable transparency, the state reported that the number of water "disputes" rose in China from 16,747 cases in 1986 to 94,405 in 2004.[61] The scale of these disputes ranged in size and severity, from the interprovincial pitched battles between communities across the Zhang River, discussed earlier, to smaller protests over polluted wells.

In this atmosphere of limited state capacity, there also emerged individuals and institutions that garnered national and international reputations as advocates for environmental health in China. A rather disparate range of actors, including Liang Congjie, Ma Jun, and Dai Qing, negotiated the tenuous line separating sanctioned environmental activism from unsanctioned criticism of the state. Liang Congjie (1932–2010) was a historian and teacher who founded Friends of Nature in 1994, based on his admiration of Greenpeace. Friends of Nature was certainly less confrontational than Greenpeace but occasionally pushed the envelope, as when it produced stunning video evidence, subsequently aired by CCTV, of local officials illegally engaged in cutting virgin forest stands. In the 1990s, Ma Jun was an investigative journalist for the *South China Morning Post* who became increasingly interested in environmental corruption in China. Ma wrote the landmark book *China's Water Crisis* published in China in 2004 and translated into several languages. Ma also developed a website called the China Water Pollution Map (*Zhongguo shui wuran ditu*) that gathered national data on water pollution. Ma was included in *Time* magazine's list of the 100 most influential people for 2006.[62] Last, Dai Qing was a reporter for *Guangming Ribao* (Guangming Daily), who in the late 1980s came to question the wisdom of the proposed Three Gorges Dam (Sanxia) project on the Yangtze River. Dai compiled two volumes of essays, *Yangtze! Yangtze!* and *The River Dragon Has Come*, both of which highlighted the environmental and human costs of the Three Gorges Dam. The original Chinese edition of *Yangtze! Yangtze!* was banned in China immediately after the 1989 Tiananmen student protests, which Dai supported. And *The River Dragon Has Come* met resistance from the censors and was published outside China only several years later. Dai's support of the 1989 student movement landed her in jail for ten months. Dai continues to live in China and to speak out on environmental issues.[63]

Environmental activism and activists in China occupied an ambiguous space between the state and civil society. Liang, Ma, and Dai pioneered a form of environmental activism in China by challenging heretofore unchallengeable state and party dictates. At the same time, these activists could be successful only by understanding the limits of their activism. In addition, it may not be a coincidence that Liang and Dai had significant sources of political capital. Liang was the grandson of Liang Qichao, a noted intellectual of the early twentieth century. Dai was an orphaned daughter of a revolutionary hero, later adopted by one of the "ten marshals" of the People's Liberation Army (Ye Jianying). But as Dai learned in 1989, the space for environmental activism was circumscribed. When her activism extended to student demands in 1989, she landed in jail. This same type of ambiguous space was replicated in the wave of environmental nongovernmental organizations (NGOs) that sprang up in China beginning in the 1990s, many of them modeled on Liang's Friends of Nature. All NGOs must be registered with the state and receive formal state sanction to function. Thus, in effect, they are GONGOs (government nongovernment organizations). International nongovernment organizations (INGOs) also began to operate in China during the 1990s. The legal basis for INGO operation in China also remained tenuous, and many INGOs opted to register as "foreign foundations." Yet the number of INGOs had increased by 2000, as many were able to make accommodations with local governments that were vested with responsibility for enforcing environmental regulations.[64] Despite the ambiguous footprint of environmental activism in China, it is indisputable that a growing environmental consciousness has been sparked by the heightened resource challenges of the reform period. From the tens of thousands of local disputes to the emergence of national environmental advocates who challenged state plans and bureaucratic practices to the proliferation of environmental NGOs in China, the state and Party became well aware of the potential social costs of resource depletion and degradation in China.

## The Dilemma for State and Party

In one of the final scenes of the film "Yellow Earth" ("Huang tudi," 1984), Cuiqiao, a North China peasant seeking to escape the oppression of rural society, drowns as she attempts to cross the Yellow River on a journey to the CCP base camp in Yan'an. The scene unfolds as Cuiqiao is bidding an emotional farewell to her younger brother with an injunction to care for their widowed father. As she begins to pole her way across the river, Cuiqiao sings a traditional folk song revised by Communist activists to celebrate the Party and peasantry. As she disappears into the foggy riverscape, Cuiqiao sings:

> The hammer, the sickle and the scythe,
> For workers and peasants shall build a new life.
> The piebald cock flies over the wall,
> The Communist . . .

As her boat disappears into the fog, Cuiqiao's voice suddenly ceases, with the last verse incomplete. The camera lingers on the roiling waters of the Yellow River then slowly pans to the moon. Cuiqiao has drowned—interred by and in the Yellow River. The complete final verse, sung in its entirety earlier in the film, went "The Communist Party shall save us all."[65]

Directed by Chen Kaige and photographed by Zhang Yimou, both of whom became celebrated filmmakers, the film was the first Chinese production to receive critical international acclaim during the post-Mao period. The film suggested the need to reappraise the Maoist era. The final scene of Cuiqiao's ill-fated attempt to cross the Yellow River stands in stark contrast to the heroic crossing of the Yellow River by the boatmen in the "Yellow River Cantata." The sophisticated treatment of the dilemmas faced by Cuiqiao elicited a plethora of responses in China, ranging from tacit agreement with the indictment of Maoist China to harsh condemnation of the filmmakers' critique of the CCP.[66]

"Yellow Earth" was part of the reappraisal of the Maoist era that accompanied China's market reforms and "opening" to the outside world after Deng Xiaoping assumed power in 1978. The arrest of the Gang of Four and subsequent pronouncements on the mistakes of the Maoist era (e.g., the Great Leap Forward and the Cultural Revolution) were grafted upon an official line that affirmed the fundamental correctness of the revolution led by Mao. This clumsy attempt to legitimize continued Party control reflected serious policy debates among the leadership. Throughout this period, debates centered on the extent and speed of the reforms and were particularly sharp in the mid- to late 1980s as the leadership felt its way along the path of reform. Partisans of more rapid liberalization coalesced around the leaders Zhao Ziyang and Hu Yaobang and included intellectuals such as Fang Lizhi and the university students who expressed their feelings in demonstrations in 1985 and 1986, and most visibly in 1989. Partisans of slower-paced reform who gathered around Li Peng, recognized the potential threat to social order that wide-scale reforms might engender. Indeed, the CCP regarded with alarm the extent to which *perestroika* was unleashing forces beyond the control of Soviet leaders.

These debates implicated the landscape. In June 1988, CCTV aired a six-part documentary called "River Elegy" ("Heshang") that advanced a critique of Chinese society. Adapted from a text of the same name, the documentary employed liberal use of Yellow River images to symbolize a feudal, authoritarian, self-centered, and scientifically bankrupt society that had fallen irretrievably behind Western civilizations. Contrasting with the muddy, yellow waters were images of the blue ocean that represented the cosmopolitan values and perspectives of a seafaring Western world. The authors implicitly charged that rule by the CCP and, by extension, the conservative opponents of deep reform in the post-Mao period, were complicit in maintaining traditional patterns of a peasant, landlocked, bureaucratic society that, despite its aspirations, failed to produce "wealth and power" for China.

Part 5 of the documentary, titled "Sorrow," focused on the Yellow River as a metaphor for China's "ultrastable" pattern of cultural and political reproduction. Dynasties fall and dikes rupture, but each is repaired on the model of its predecessor. "Isn't the thousand-*li* dike that squeezes the raging Yellow River a wonderful symbol of our monolithic social structure? ... After the breach of the Yellow River dike, it was repaired again and the people waited for the next breach. Why have we always been trapped in this ever-repeating cycle of destiny?"[67] The scriptwriters' reading of the Yellow River and the oft-flooded North China Plain as a landscape representative of feudal society—that is, as "China's sorrow"—was entirely consistent with the manipulation of this narrative by Mao and Party leaders after 1949. At the same time, however, the filmmakers were explicit in their judgment of the Mao era:

> Since Liberation [1949], we have heightened and widened the great dike three times, which has guaranteed nearly forty years of calm water—almost a miracle in the modern history of the Yellow River. . . . For centuries the long-cherished wish of the Chinese people has been to clear up the Yellow River—it is like a dream that never disappears. New China once placed all its hopes on the Sanmenxia Dam. In 1955 vice premier Deng Zihui announced . . . to all the representatives of the People's Congress that, after the completion of Sanmenxia Dam, they and the entire nation would be able to see in the lower reaches of the Yellow River what they had been dreaming about for ages: the Yellow River will be clear! At the solemn moment thirty-two years ago, the Chinese seemed still to believe in the old saying, "When a sage appears, the Yellow River will be clear." . . . But the Yellow River did not run clear. The raging mud silted up Sanmenxia Dam, and the backwater spillover into the Jing and Wei rivers, flooding the rich Guanzhong Plain. Thus the age-old national dream vanished once again like a bubble in the swirling yellow water.[68]

The last imagery evokes Cuiqiao's tragic quest for "liberation" in Yellow River.

The use of the Yellow River valley as metaphor of the past was a motif of Chinese political culture during the Mao period. But the "land of sorrow" referenced in "River Elegy" was a landscape firmly identified with Maoist China. The imagery of the ocean suggested that China needed to be incorporated into the "blue civilizations" of the modern world order. At the same time, the manipulation of the landscape served a narrative directed not at China's peasantry but at China's modern intellectuals. In contrast to the Maoist rhetoric focused on the peasantry, Part 6 of the documentary ("Blueness") celebrated the transformative potential of China's intellectuals:

> Chinese history did not create a bourgeoisie for the Chinese people which [sic] could hasten the victory of science and democracy; Chinese culture did not nurture a sense of citizenship. On the contrary, it taught a subject mentality. A subject mentality can only produce obedient people who meekly submit to oppression on the one hand and madmen who act recklessly on the other. But History did give the Chinese people an entirely unique group: its intellectuals. . . . Their talents can be manipulated by others, their wills can be twisted, their souls emasculated, their backbones bent, and their flesh destroyed. And yet, they hold in their hands the weapon to destroy ignorance and superstition. It is they who can conduct a direct dialogue with the "sea-faring" civilization; it is they who can channel the "blue" sweetwater [sic] spring of science and democracy to our yellow earth![69]

The rhetorical weight of "River Elegy" was considerable. Broadcast to an estimated 600 million people in China with access to television sets, the film's castigation of China's traditional culture, indictment of Maoism, and assertive embrace of scientific and democratic ideals aired during the increasingly heated national debate about the nature and speed of reforms during the late 1980s. In the Party leadership, Wang Zhen, a revolutionary commander and one of the "Eight Elders of the Chinese Communist Party," labeled "River Elegy" "cultural nihilism," criticized the film's apparent embrace of wholesale Westernization, and denounced the documentary's critique of socialism and of the CCP. On the other side of the debate, Zhao Ziyang, then general secretary of the CCP, was reported to have promoted the script among aggressive reformers like himself.[70] There is little doubt that "River Elegy" contributed to sensibilities leading to the Tiananmen demonstrations of 1989 that gave voice to frustrations over government corruption, the pace of reform, and the attenuated role of China's urban intellectuals in that reform. The subsequent government crackdown of the 1989 demonstrations closed the curtain on "River Elegy." The documentary and publication of the script were banned in China, and Su Xiaokang was implicated as a supporter of the student movement. Su escaped China in the wave of arrests following the crackdown and ultimately settled in Princeton, New Jersey.

The irony of Su's call for the negation of the past to serve cultural enlightenment is that, on many levels, his aspirations for China were deeply embedded in tradition. The documentary's fundamental purpose (reform) was infused with the traditional role of educated elites (literati); with a utilitarian obsession with history as servant to the present (even as a negation of the present); with the unity and hegemony of imperial China; with a resort to "cultural fetishism" (Yellow River, Great Wall); and with the dichotomy facing all of China's modern reformers—the superiority of the past (the glory of China's civilization) versus the inferiority of the present (China's modern "humiliation"). In his assessment of Su and his coterie of cultural critics, Wang Jing wrote, "[This] modern elite are all dreaming the same dream as their forebears of the dynastic past— wealth, power, and hegemony. . . . What *He shang* [Heshang] forcefully calls for—the agenda of enlightenment (*qimeng*) by means of cultural introspection is time and again eclipsed by a stronger, perhaps unconscious drive for what it rejects—the revival of the glorious past in the future. . . . History is both the dream and the nightmare from which neither the Chinese people nor the intellectuals themselves have awakened."[71]

The crackdown on the Tiananmen demonstrators and the subsequent attacks on cultural critics such as Su seemed to settle intra-Party debates on the future of reform. More aggressive modernizers like Zhao were cashiered, and public debate was stifled. The pace of reform slowed, a consequence of policy decisions and of a reduction in foreign investment. Three years later, however, the reform program was kickstarted by a visit to South China by Deng Xiaoping. In a series of talks in Shenzhen and Zhuhai (two Special Economic Zones [SEZs], where restrictions on foreign investments were intentionally relaxed on an experimental basis), Deng repudiated conservative policies of the post-Tiananmen period. The state-run media gave full coverage to the trip. At a talk to Shenzhen officials, Deng instructed, "You should be bolder in carrying out reform and opening up, dare to make experiments and [you] should not act as women with bound feet. An important experience of Shenzhen is the courage to make breakthroughs."[72]

Deng's trip was a defining moment for two reasons. First, it signaled to Party leaders that the official line had returned to supporting more aggressive economic reform that centered on the restructuring of state-owned industries. Second, Deng's destination was significant. As laboratories for a variety of novel investment policies, Shenzhen and the SEZs were at the leading edge of reform. During the 1980s, Shenzhen was transformed from a rather insignificant appendage of Guangzhou into a boomtown. Its liberal investment regime and its proximity to Hong Kong quickly transformed Shenzhen into a hotbed of capitalism. Shenzhen held out the promise of riches for hundreds of thousands of young Chinese who migrated to the city. Shenzhen boomed, as did China's GDP growth rates after policy adjustments in the aftermath of Deng's visit. Fueled by foreign direct investment and export manufacturing, China's coastal cities like Shenzhen, Guangzhou, Xiamen, and Shanghai became shining examples of "China modern."

Coastal and southern China became the idealized landscape of the new China. At the same time, the sundering of the trinity of land, peasant, and Party centered on North China gathered momentum during the 1990s. History was re-examined. During the Maoist period, "The People's Republic of China . . . presented itself as the heir of a Han people who had come together millennia earlier in the north China plain of the Yellow River valley, built a great civilization, fought to preserve it, and expanded over the centuries by civilizing barbarian invaders."[73] Archaeological evidence, largely suppressed in deference to Maoist discourse of the revolutionary north, was widely disseminated in the reform era. These newer findings suggested multiple geographical nodes, including South and Southwest China, for the genesis of several Chinese "civilizations." The negation of Mao's revolutionary north was accompanied by the rise of a national identity that celebrated the modern development of China. The re-creation of China's millennia of florescence was now posited in economic reform and took place in

China's geographical points of exchange with the "blue" oceans (i.e., the coastal regions of East-Central and South China). The explicit rejection of an exclusive Maoist identity de-centered the Yellow River and the North China Plain from the national narrative. This was increasingly apparent by the reassessment of China's modern history, its relationship with imperialist powers, and the celebration of Ming dynasty voyages of discovery.[74]

At one level, the post-Mao period echoed the withdrawal of the state from water management on the North China Plain in the aftermath of the Opium Wars in the mid-nineteenth century. Just a few decades later, the breakdown of the hydraulic system of the North China Plain, in the late nineteenth century, was a painful reminder of the state commitment made centuries earlier to maintain an ecological equilibrium on the North China Plain. A century later, in the late twentieth century, the state was once again reminded of its historical commitments to North China by the Yellow River dry-ups, mounting evidence of unsustainable water use, and increasing discord over allocation. Although the revolutionary narrative centered on the Yellow River and the North China Plain had been challenged, and the means and ends to "wealth and power" were now located in the modern coastal setting, the traditional state commitment to "ordering the waters" of the North China Plain could not be neglected.

One could argue that from the vantage point of the late Maoist period the ecological constraints of the late imperial period had been pierced. The multipurpose use of water developed by the Maoist technology complex generated electricity to industrialize the countryside, watered fields to grow more grain, and sustained a surge in population growth. The water resource balance left by the Maoist period was, however, more akin to a second mortgage against ecological realities that could not be fundamentally transformed. Although the developmental paradigm changed dramatically with the onset of the reform era, as reflected in the celebration of the modern urban landscape, this alternative path to "wealth and power" still depended upon water. The state could not abandon a traditional commitment, sustained during the Maoist period, to maintain social and economic stability in a region with exceptionally high population densities and a key food-producing region. There has been little evidence during the reform period that China has weakened its resolve to limit its exposure to international grain markets. At the same time, the logic of economic growth inevitably increased water demand by industrial and urban constituencies. Together, these commitments to maintain economic growth have been directly tied to politically legitimacy. The life of the Party (in both its literal and figurative senses) was subject to the same pressures experienced by imperial states to manage the thin ecological margin on the North China Plain. By 1997, with the Yellow River dry-up, it had become increasingly clear that the state and Party could escape neither ecological history nor the state's historical commitment to the North China Plain. The second mortgage on the water resources of the region was coming due. The state was forced to refocus attention on the water resources of the North China Plain.[75]

## Debating China's Water Problem

By the end of the 1990s, mounting evidence of a water dilemma sparked considerable debate within and beyond China concerning the country's capacity to sustain its modernization drive.[76] A particularly provocative assessment of this future was penned by Lester Brown of the Worldwatch Institute. Published in 1995, *Who Will Feed China? Wake-Up Call for a Small Planet* argued that China's development trajectory could not be sustained. Brown argued that in order to feed a growing population in light of limited endowments of water and land, China would be forced to tap international grain markets. The inflationary effect sparked by Chinese grain purchases would price poorer regions of the world out of these markets, thereby aggravating global poverty. Brown envisioned a scenario in which China's grain import needs would range from 207 million tons to 369 million tons by 2030. Yet increasing land withdrawn from production and stagnating crop yields would generate a 20 percent decline in agricultural output by 2030. Cultivated land would decline by an average of 1 percent per year, which would result in 50 percent decline in farmland by 2040. At the same time, there was little prospect for increasing grain yields as per hectare output approximated yields in Japan.[77] The immediate effect of his thesis was to place China's resource challenges at the very forefront of domestic and international debate about China's reemergence as a global power.[78]

A number of global and domestic observers challenged the assumptions of "China's food problem." Soon after the publication of *Who Will Feed China?*, scholars criticized the official data upon which Brown based his arguments. Vaclav Smil, Richard Edmonds, and Robert Ash noted that data from the State Statistical Bureau (now called the National Bureau of Statistics) and other government agencies (which showed national cultivated acreage falling below 100 million hectares) were incorrect. These scholars pointed out that government land data were conditioned by bureaucratic practices in which local officials underreported agricultural lands and newly reclaimed lands in order to mitigate local tax burdens. Although recognizing that industrial development and urban development encroached upon agricultural lands during the 1980s and 1990s, Edmonds and Ash noted that China still had 130 million to 140 million hectares of agricultural land.[79] Other analysts, such as Robert Paarlberg of the Harvard Center for International Affairs, argued that not only were Brown's assumptions incorrect but that China had very little to fear from entering international grain markets. Paarlberg and others who placed great faith in markets argued that farmers who had switched to higher value crops were simply responding to market signals that would make best use of China's scarce land and water resources. He also noted the potential to increase yields from China's agriculture as suggested by historical trajectories in the United States, Australia, South Korea, and Japan. Furthermore, even if China entered global markets, the United States, Australia, and Argentina still had plenty of surplus capacity to meet increased demand from China. Paarlberg concluded that instead of a 20 percent decline in Chinese grain production by 2030, there would be a significant increase in wheat, maize, and rice production. Other institutions, such as the International Food Policy Research Institution, concurred.[80]

In China, the official line on the country's presumed food crisis was that "China can be basically self-sufficient in grain while *suitably* increasing imports."[81] There were, however, inescapable concerns. Particularly worrisome were data showing that grain production was not keeping pace with population growth. In the three decades after 1949, grain output increased by 2.89 percent per annum, outpacing annual population growth of 1.91 percent. After 1984, however, grain output increased only an average 1.21 percent per year, below annual population growth of 1.36 percent.[82] At the same time, the state projected a per capita decline in farmland from 0.08 hectare in 1994 to 0.06 hectare by 2030 as China's population reached 1.56 billion people.[83] Other data emerged in the first decade or so of the post-Mao era that added concerns over China's ability to feed itself. Decreasing per capita water supply, sectoral competition for water, and the inefficiency of agricultural water use added to a growing sense of insecurity over China's agricultural prospects. In the context of these broader developments, a nationwide grain shock occurred in 1993 and 1994, as reports of rice shortages in the south, some accurate, some inflated by rampant speculation, began to spread to markets in the north, precipitating dramatic increases in the price of food throughout China.[84]

Other responses to Brown's article in China reflected fears of American "containment" of China. Adopting such a perspective, a number of Chinese articles reacted strongly to the suggestion that China's rise would upset a global order dominated by American interests. At the same time, the very title of Brown's work, *Who Will Feed China?*, suggested a variation on nineteenth- and early twentieth-century imperialist paternalism. An article published in 1997 in the *Journal of Chinese Geography* reflected these sentiments:

> Mr. Brown gives the impression that trade in foodstuffs is not a component of international trade, but is charity. . . . The economic development of a country with food import needs will spark increased demand that will benefit international trade. Such a simple fact is ignored by Brown. If there are those that think this is a distortion of Mr. Brown's arguments, then a simple analysis of the title of Mr. Brown's book, *Who Will Feed*

*China?*, will make this clear. The word "feed" implies giving food to [as a charitable act]. This is not a question of who will "feed somebody" or who will "be fed by somebody." In our contemporary era there is no country that would simply feed another country or a country that would simply be fed by another. China and the United States are no exceptions. In the first year after the revolution of 1949, China's grain production was only 1,035 kilograms per hectare. Per capita production was only 210 kilograms. In that very difficult year did any foreign scholar or politician ask *Who Will Feed China?* [Now] With the capacity of the Chinese people to provide clothing to warm, and food to eat one's fill, the living standard of the people has steadily increased. Why is the problem of *Who Will Feed China?* raised in such times? Perhaps some people prefer that the Chinese people starve than have a developed economy. To make an analogy, given the fact that the United States imports 25–30 percent of its energy needs, perhaps we might ask, *Who Will Warm the United States?* We believe this is a question that would make Mr. Brown uncomfortable. Indeed, there is a rather large circle of us who feel uncomfortable with Mr. Brown's question.[85]

Notwithstanding such reactions, there was widespread agreement among Chinese researchers and commentators on prescriptive measures to address China's "food problem." Such measures included raising productivity in marginal areas, mitigating the limiting factors on agricultural production (e.g., floods, salinization, water, and soil erosion), breeding and disseminating new strains to increase yields, improving the utilization of fertilizers, implementing best practices in agriculture, and using water resources more efficiently.[86]

## Addressing North China's Water Constraints

During the past twenty years, the state has formulated an increasingly diverse array of policy options designed to allocate China's resources in a manner that accommodates continued economic expansion. To be sure, a strong impulse toward continued faith in "engineering solutions" remained powerful among elements in the central leadership and in China's water management establishment. At the same time, however, a reaction set in against the sort of gigantism that was characteristic of mid-twentieth-century development patterns in North America, Europe, and the Soviet Union (i.e., maximizing supply). Chinese willingness to question this "gigantism" development pattern suggested alternative policy options centering on governance and demand-side management. Examined in their entirety, however, policy responses to North China's water challenge remain a mélange of different priorities. They have not been intentionally ambiguous nor have they been the response of an ill-informed set of actors; rather, the responses have been the net result of disparate challenges from different constituencies and of a fragmented bureaucracy whose diverse components serve a range of constituencies.[87] Generally speaking, there have been three major thrusts during the reform period to address North China's water dilemma: administrative centralization of allocation and pollution control, continued fealty to engineering solutions (to increase supply), and structural adjustments to water infrastructure and markets (to manage demand). Perhaps we might think of this hybrid approach as a variant of "walking on three legs" or "water management with Chinese characteristics." In either case, whether more systematically envisioned or the result of "muddling along," China has managed to maintain agricultural productivity while increasing industrial and urban expansion. But the costs of these projects, the prospects for the health of local communities, and the sustainability of current practices remain in question.[88]

During the 1990s, commentators in China focused on the lack of unified administration of water resources along the country's major waterways and on the North China Plain. At issue was how to coordinate storage and release of reservoir waters to ensure seasonal demand for irrigation, hydroelectric generation, and flood control. As described by the eminent Chinese water expert Liu Changming in 2001, "There is a lack of unified control regarding the operation of large reservoirs and engineering works that collect water. Hydraulic engineering projects constructed along the Yellow River play a key role in the socioeconomic development of northwest and north China. However, different government agencies control their engineering projects and facilities, and the management of water resources is not well coordinated, rendering it impossible to operate the river basin as a single unit."[89] An early attempt at resolving increasing allocation conflicts among riparian regions occurred in the mid-1980s, when the State Council (after five years of contentious negotiations with provincial governments) announced a program to allocate the average flow of 58 billion m³ among the eleven riparian provinces. A total of 37 billion m³ was allocated among the provinces, with the remaining 20 billion m³ intended for ecological flows to ensure sediment flush and breakdown of pollutants in the lower reaches. Because of further reduction in the total flow of the river, as well as egregious violations of the allocation program (most notably by the Inner Mongolia Autonomous Region and Shandong Province), the agreement required immediate revision.[90] The resolve of the central government was bolstered by a 1998 report published by the Geology Bureau of the Chinese Academy of Sciences that concluded that low downstream discharge was leading to accelerated channel accretion. Thus, the chances of a Yellow River course change were the highest since 1949.[91]

The explanation for the threat, however, suggested a case of unintended consequences and underlined the delicate and precarious ecological pendulum of the North China Plain. On the one hand, water impoundment projects in the upstream and midstream since 1949 did, indeed, generate a significant decline in absolute quantities of silt flowing to the downstream. The average annual sediment discharge from 1990 to 1999 was $389.9 \times 10^6$ tons/year, less than one-third the silt discharge of the 1950s.[92] The problem, however, was that a corresponding reduction of flow to the downstream meant that the entire silt load, still high despite real reductions, settled in the downstream bed. In 1998, the State Council revised the original allocation program by imposing reductions of provincial quotas by a fixed percentage based on the reduction of annual river flow. On the other hand, the government mandated that the YRCC prevent future dry-ups by maintaining a minimum runoff of 30–50 m³/second at the river's estuary in Shandong Province.[93]

In addition to reform of China's water governance, a number of structural recommendations were also forwarded to Chinese policy makers, including measures centered on environmental protection, soil and water conservation, pricing mechanisms, and infrastructure.[94] But implementing these measures was another matter entirely. A closer examination of three potential reforms that were intensively discussed around 2000—namely, pollution control, pricing reform, and interbasin transfers—reveals the complexities of implementing policy prescriptions to address water scarcity on the North China Plain.

The challenge that the central government faced in instituting a centrally mandated pollution control regime was presaged by earlier difficulties in convincing Yellow River riparian provinces to abide by allocation agreements in 1985 and 1998. Although the provinces eventually signed on, compliance remained partial. The fundamental source of this challenge to centralized administrative authority, a factor that would likewise mitigate attempts to implement national pollution controls, was a compromise that the Party had made with provincial and local governments early in the reform period (i.e., soon after 1978). The devolution of economic management from the center to the localities was premised on increased allocative efficiency as local and provincial officials more effectively mobilized local resources and engaged directly in domestic and international trade networks. The Party instituted a system of fiscal decentralization that allowed managers of local branches of formerly state-owned conglomerates to retain a portion of their profits. At the same time, local governments could exact taxes from these enterprises.[95] Although the contribution of decentralization to China's modern economic growth is debatable, one indisputable outcome was the increased autonomy of local officials. The strengthening of local autonomy circumscribed the capacity of the state to impose centrally mandated measures such as pollution control. The case of the Huai River cleanup effort is perhaps the most illustrative example of central ambitions that went unfulfilled.[96]

In response to reports of catastrophic pollution in the Huai River during the late 1980s and early 1990s, the State Council adopted the Huai River Pollution Control Plan in 1994. The plan was the first basin-wide pollution abatement effort in China designed to show the commitment and capacity of the state to guarantee ecological health in the face of rapid economic development. The plan outlined a set of goals to be achieved in a two-stage process over ten years. Central to the plan were two broad-based goals: by 1997 all industrial effluents would meet state wastewater standards, with chemical oxygen demand (COD) reduced from 1.5 million tons (in 1993) to 0.89 million tons, and by 2000 total COD was to be further reduced to 0.368 million tons, allowing the main stem and all major tributaries that supply urban consumers to reach Grade III quality (i.e., suitable for human consumption). The state placed the Huai River Pollution Control Plan at the top of its environmental agenda, and it quickly became a symbol, both domestically and internationally, of the efficacy of the state in protecting its citizens. In order to meet the pollution abatement goals, the state embarked on a three-pronged approach: a massive campaign, to be completed by June 1996, to shut down all small-scale polluting industries in the Huai River valley (typically operated as TVEs), a campaign to install industrial wastewater treatment apparatus in factories and workshops by the end of 1997, and construction of 52 municipal wastewater treatments in cities along the Huai and its main tributaries.[97]

By 2001 the chief administrator for the State Environmental Protection Administration proclaimed that the quality of Huai River water had reached Grade III standards and that 70 percent of water in tributaries of the Huai had reached similar quality. But within a few years, there was increasing evidence that the victory was short-lived. In 2003, the Huai River Conservancy Commission reported that in early summer over 80 percent of monitoring stations reported water quality of Grade V or worse. Then in July 2004, flooding in the upstream sections of the river again generated a 150-kilometer band of toxic water that flowed downstream with devastating effect on river life. Shortly thereafter, Xinhua News Agency reported that the "ten-year effort in the Huai basin ended in vain; water quality [had] returned to its starting point."[98]

The failure of the Huai River Pollution Control Plan was the most visible example of a fundamental dynamic of China's broad struggle to effectively regulate environmental conditions during the post-Mao era. China has developed a well-articulated set of environmental guidelines, but lacks the administrative authority to coerce local governments to comply with these mandates. Indeed, the state appeared to be committed to managing the environmental consequences of robust economic growth relatively early in the post-Mao period. The initial signal to such a commitment was the enactment of China's first Environmental Protection Law in 1979. In 1982, Article 26 of the new Constitution of the PRC noted that the state would "protect and improve the environment in which people live and [protect and improve] the ecological environment."[99] Shortly thereafter, in 1983, the then–vice-premier Li Peng announced that environmental protection, along with family planning, was a "fundamental state policy."[100] In 1987 the National Environmental Protection Agency was created under the State Council and was given cabinet-level status. By 1990 a comprehensive Environmental Protection Law was promulgated based on the three principles of "putting protection first and combining prevention with control," "making the instigator of pollution responsible for treating it," and "intensifying environmental management."[101] The new law also set up a system of pollution discharge permits, to be administered by local environmental protection bureaus that were invested with authority to collect fees and impose fines, as well as the ultimate power to close down offending industries. Thus, "by 1996 China had set up a far-reaching network of 8,400 environmental agencies, bureaux and offices at a variety of administrative levels, from province through city, county and township, employing a staff of 96,000 people, and had drafted an array of (on paper) tough regulations concerning environmental protection."[102]

With the power they had accrued through administrative and fiscal decentralization at the beginning of the reform period, local governments were largely able to fend off the intrusion of the center. Local government officials, who were vested with the authority to manage local environmental protection bureaus, were typically more interested, from both a financial and a political standpoint, in economic performance. The structure of rewards and punishments for local Party officials was, first and foremost, designed to valorize economic growth. Investments in environmental protection, to the degree that they had a negative impact on profits, were largely a secondary concern. Thus, even where state pressure led to the installation of pollution control equipment by local companies, subsequent investment in maintenance was often eschewed.[103]

In addition to what, on paper, was a reasonably articulated environmental regulatory program, the country worked to develop a comprehensive water law that defined water rights and created an institutional structure over the water sector. The 1988 Water Law attempted to consolidate all agencies involved in water administration into a single entity: the Ministry of Water Resources (MWR). By statute, the MWR makes water policy and formulates strategic plans for flood control and pollution control. In addition, the ministry is responsible for resolution of water use conflicts. The MWR also governs China's major river basin organizations (e.g., the YRCC). In 2002, the government also divided the country into some 13,000 water classification zones that reported on the environmental conditions of some 5,737 rivers and 980 lakes and reservoirs. The State Environmental Protection Administration (successor to the National Environmental Protection Agency; in 2008 upgraded to the Ministry of Environmental Protection) has put out regular bulletins reporting the conditions in these zones.[104] The major impetus for (and the greatest challenge to) China's new water laws was the need to better define water rights and to empower the central government to allocate water along the length of major river valleys like the Yellow River. Indeed, the continuing power of the provinces to frustrate the central state was foreshadowed by the challenges that the YRCC had experienced in the 1990s in attempting to implement water allocation programs. Realizing the strategic role of water in continued economic growth, the state amended the Water Law in 2002. The amended law made a concerted attempt to incorporate concepts of integrated water resource management like water rights, administration, allocation, pricing, and supply and demand. China's amended Water Law remained ambiguous on a number of fronts that, combined with the erosion of central power, resulted in frustration in the implementation of the Huai River plan and the allocation program for the Yellow River.[105]

However, the bottleneck in China's water administration was not only the outcome of a legal framework and an institutional structure that promoted a significant degree of ambiguity. The campaign-like drive for reform after 1980 promoted a "clean up later" mentality often premised on the industrializing experiences of the West in the nineteenth and twentieth centuries. But this developmental outlook was not shared by all, as many communities continued to be plagued by water supply problems.

The fundamental question that faced the state and Party throughout the reform period was how to balance water as an economic good and as a social good. The fundamental shift of developmental model from Maoism to the market reforms of Deng Xiaoping was accompanied by different assumptions about resource management. The logic of economic growth that increasingly relied upon the market valuation of goods, services, and resources implied that water allocation be determined by demand. Ultimately, the structure of water demand would shift to higher value-added production for consumers who could afford to pay the full cost of (and perhaps a premium cost for) water. These consumers would include industrial users and urban residents with increasing disposable income. Thus, water subject to market pricing would be used efficiently in support of industrial development and urban expansion. Few states around the globe, however, treat water exclusively as a commodity. The price of water rarely recovers supply costs. States subsidize water use because access to water is viewed as a critical social need. China has been no different from most polities around the world that subscribe to the notion that inadequate access to clean water holds the potential for social and political instability.

During the reform period, the state and Party have been forced to negotiate a balance between long-running commitments to rural society and the desire for economic growth. The weight of commitments to the countryside, sensitivity to threats to Party monopoly of political authority, and continued concerns about food security all combine to persuade the state to supply water in sufficient volumes to avoid social discontent. The same social and political considerations lead the state to provide adequate resources for urban residents. Rapid urbanization during the reform period also brought to light unresolved allocation issues that existed between urban and rural consumption. The focus of the Party, which has been concerned about threats to its legitimacy, particularly since the 1989 demonstrations, has been to engineer sufficient GDP growth to enable the national economy to absorb the social dislocation of economic restructuring. In addition to energy supplies, a major input for this economic growth has been water. Water allocation to the modern economic sector, including pricing mechanisms to promote higher value-added use, has been a critical component of China's water dilemma. Furthermore, compounding the issue of physical supply, and implicating virtually every element of the reform effort, has been pollution.

China's fealty to large-scale engineering solutions to address water supply constraints is a direct legacy of large Soviet-style interventions during the Maoist period. This preference was consistent with twentieth-century modernizing visions around the globe. At the same time, regulation of the environment, particularly on the North China Plain, followed a long tradition of Chinese statecraft and was a central idea of the founding myth of Yü the Great. In style and substance, the reform period was heir to Maoist perspectives and practices. China's most internationally recognized engineering project of the reform period, the Three Gorges Dam on the Yangtze River, suggests a continuity that binds the modernist visions of Sun Yat-sen, through the Maoist period when engineering plans were conceived, to the reform period, when the state invested massive financial and moral assets in the project. In many cultural contexts, from the twentieth century to the present—and perhaps even more so in China with its rich history of large public works intimately tied to political legitimacy—the organizational abilities of the state and its capacity to marshal the latest science and technology in service to the nation have created a human-built environment with questionable long-term consequences. Although dam-building and other large replumbing projects reached their zenith in industrial countries during the 1960s and 1970s, the strength of China's traditions in managing water, coupled with the fact that by the onset of the reform period the country still had massive water resources yet to be developed, particularly in the Tibetan-Himalayan plateau region, led China's state leaders to aggressively dam the rivers. One must also consider the fact that China's leadership remained heavily influenced by the technical dreams of the

Maoist period. For example, Li Peng (who served as premier from 1987 to 1998) trained at the Moscow Hydropower Department of the Moscow Electrodynamics Academy from 1948 to 1954. During the reform period, Li was typically identified as an ardent supporter of the Three Gorges project. Eighteen of twenty-four Politburo members of the Fifteenth Central Committee (1997–2002) were engineers, and a majority of the 344 members of the Central Committee were trained as engineers. At the very top of the government and party hierarchies were four professionally trained engineers: Jiang Zemin (electrical engineering), Zhu Rongji (electrical engineering), Hu Jintao (hydraulic engineering) and Wen Jiabao (geomechanics).[106]

The iconic engineering response to water scarcity on the North China Plain during the reform period has been the South-to-North Water Diversion project. First conceived in the early 1950s, the project was resuscitated in the 1990s as it became increasingly clear that water resources on the North China Plain could not sustain agricultural, industrial, and urban needs. Large-scale intraregional water transfers were implemented in the early 1970s when shortages in Tianjin were addressed by emergency diversions from the Yellow River via the People's Victory Canal, the Wei River, and the Grand Canal. In the early 1980s, two additional transfers of Yellow River water to Tianjin were completed. Shortly thereafter, a permanent diversion channel from the Panjiakou Reservoir on the Luan River to Tianjin was constructed. Water was also diverted to Tangshan, Qinhuangdao, and Qingdao.[107]

The feasibility of long-distance water transfers was confirmed by these experiences in the 1970s and 1980s. The original discussion during the 1950s of large-scale diversions from the Yangtze River water to the North China Plain was embedded in the immediate experiences of the Soviet Union that, as early as the 1930s, was sketching out massive diversion plans to redirect "surplus waters."[108] At the same time, the notion of interbasin waterways called to mind the engineering feat of the Grand Canal. The South-to-North project was consistent with the imperial system's abiding interest in maintaining social stability on the ecologically marginal North China Plain through state redistribution of southern resources.

Both projects were expressions of maintaining imperial and national unity by transcending the South–North hydrologic divide.[109]

In terms of the amount of earth moved, the South-to-North Water Diversion project will likely be the single largest engineering undertaking since the construction of the Grand Canal. As formally adopted by the State Council in late 2002, the project will transfer water to Northwest China and the North China Plain via three routes, will cost $64 billion (twice the cost of the Three Gorges Dam), and is projected to take fifty years to complete. When all three routes are completed, 44.8 billion m³ of water (11.8 trillion gallons) will be delivered from the Yangtze River basin to North China. The formal approval of the project in 2002 inspired rhetorical flourishes that evoked the enthusiasm for large-scale engineering projects of the Maoist period. *China's Water Conservancy Report* recalled Mao's observation as he toured the Yellow River with Wang Huayun in 1952: "In the early stages of New China, Chairman Mao Zedong raised the magnificent [*hongwei*] notion that 'Since the water of the south are plentiful, and the waters in the north scarce, if it is possible it's okay to borrow a little.' . . . The South-to-North Water Diversion project is the realization of the optimal development of a twenty-first-century water resource mega-engineering project for our nation."[110]

The *China Economic Report* noted that "approval of the South-to-North Water Diversion project is the correct crystallization of the difficult work of three generations of Party leaders and technical personnel." The article quotes a water management official: "Connecting the four rivers of the Yangtze, Huai, Yellow, and Hai, the three diversion canals will form a 'four horizontals and three verticals' pattern [*si heng, san zong*]." The project would rectify the country's water imbalance. It is hard to imagine a more eloquent articulation of the twentieth-century, high-modernist dreams to efficiently deliver resources to the turbines and fields of the nation.[111]

The clear predilection of Party leadership for engineering solutions to the water challenges on the North China Plain during the reform period was both an implicit rejection of *and* an affirmation of the Maoist development plan. Mass mobilization and purely ideological inducements were out. Although the reform drive reflected certain qualities of a mass movement, and the state and Party appealed to nationalist sentiment, the other leg of the Maoist paradigm, the high-modernist application of a technological complex that included state power and the application of industrial technologies came to dominate the reform period. In the end, the technocrats, who had been alternatively championed and castigated during the Maoist period, had seemingly won.

But such a narrative is too simplistic. Despite its hegemony, the engineering approach was increasingly challenged during the reform period, not only by activists like Dai Qing and Liang Congjie but also by engineers themselves, for whom disagreements (some emanating from the 1950s) had yet to be settled. Now that he had greater public space to engage in debate, Huang Wanli re-emerged from professional purgatory not only to defend his original critique of the Sanmenxia Dam at the beginning of the Great Leap Forward but also to question current projects, such as the Three Gorges Dam on the Yangtze River. Huang was one of the last "rightists" to be rehabilitated, and he was reinstated to the faculty of Qinghua University in Beijing in 1980 at the age of sixty-nine. In the last decades of his life, Huang also spoke out against the South-to-North Water Diversion project. In general, Huang's opposition was framed by what he noted were unintended environmental consequences of large projects. His activism was a double-edged sword for the state in the reform period. On the one hand, his opposition to the Sanmenxia project served to discredit Maoist recklessness. On the other hand, continued Party patronage of engineering projects made leaders wary of Huang's criticism of large-scale interventions.[112]

Huang became a celebrated figure to many in China's expanding environmental movement during the reform period. Although the state attempted to carefully regulate expressions and actions of this movement in the 1990s, an open debate emerged on the future of Sanmenxia that in many ways served as a proxy for the more contentious Three Gorges Project. Sanmenxia Dam became a testament to failed dreams. It produced some electricity but was useless in regulating seasonal flows and controlling silt. A seminal event in the widening discourse on environmental management was a television interview with Zhang Guangdou, one of China's leading hydraulic engineers, who had a Ph.D. from the University of California at Berkeley (1936). Zhang had been involved in the design of major dams for decades, including Sanmenxia and the Three Gorges. On Beijing TV in late January 2004, Zhang admitted that Sanmenxia had been a mistake. He insisted that the Soviet design was to blame. At the same time, he argued that large dams were still critical to managing China's water resources.

> *Zeng Tao [journalist]:* Prof. Zhang, do you think any of the dams you have been associated with have encountered problems or, frankly, been failures?
>
> *Zhang Guangdou:* Sanmenxia was a mistake, and I am not the only one to say that . . . but we could do nothing about it because we had to heed the Soviet experts . . .
>
> *Zeng:* We've heard a lot from the press about the pros and cons of Sanmenxia. Wen Shanzhang has been mentioned as opposing Sanmenxia, and also Huang Wanli, who was treated badly because of his strong and persistent opposition to the dam.
>
> *Zhang:* It was Huang Wanli who opposed Sanmenxia. . . . It was obviously a mistake to label Huang a "rightist."
>
> *Zeng:* Prof. Zhang, you became involved with the Three Gorges project more than half a century ago. Is the Three Gorges Dam the project you love the most and have worked on the most?
>
> *Zhang:* Very much so.[113]

A year earlier, in 2003, Zhang Guangdou even suggested that Sanmenxia Dam be abandoned. Zhang was quoted in a *Renmin Ribao* retrospective on Mao-era dams that signaled intensifying debate over the efficacy of large-scale engineering projects. In part, however, the debate over Sanmenxia was possible only after the completion in 2000 of a second dam across the downstream Yellow River at Xiaolangdi. The new dam assumed functions for the downstream channel, namely, silt retention and flushing, hydroelectric generation, and provision of irrigation water. The continued attraction to engineering solutions was reinforced at the Twentieth Congress of the International Commission on Large Dams, held in Beijing in September 2000. Vice Premier Wen Jiabao was quoted as stating, "The Chinese Government will continue to give priority to the development of water recourses in the course of national economic development . . . to constantly raise the level of development and dam construction."[114]

The Chinese debate over engineering solutions was joined by international networks focused on sustainability and with a domestic activist community that counted among its ranks urban educated elites. From this informal coalition, which included bureaucratic elements, particularly the State Environmental Protection Administration, a water management agenda emerged that argued against solutions aimed solely at increasing water supply.[115] The South-to-North Water Diversion project became a target of considerable attention from this informal coalition. The project received the full backing of the state and celebratory rhetoric in the state-run media, but concerns were raised by technical specialists and by environmental activists about the long-term environmental consequences of river diversion. Indeed, by 2010, as construction was progressing on the eastern and central routes, project implementation slowed because of increasing concerns about the cost, relocation of communities, and environmental consequences. Unofficial estimates suggested that the project's cost would far exceed the official cost of $64 billion. The scale of population resettlement on the central route, which extended 800 miles from the Danjiangkou reservoir on the Han River (a Yangtze tributary) to Beijing, alone was estimated at 350,000 persons. The environmental concerns raised since approval of the project in 2002 centered on two key issues: the impact on the Yangtze River estuary region from a reduction in discharge; and pollution in the Yangtze River from rural and urban wastewater sources, particularly along the eastern route, which would require a large investment to mitigate before potable water could reach consumers on the North China Plain. The accumulated concerns persuaded the government to slow down project investment—a significant disappointment because it was hoped that water diversions could commence in time for the 2008 Beijing Olympics. A serious drought on the North China Plain in 2010–2011, however, caused the government to speed construction. By mid-2011, the project

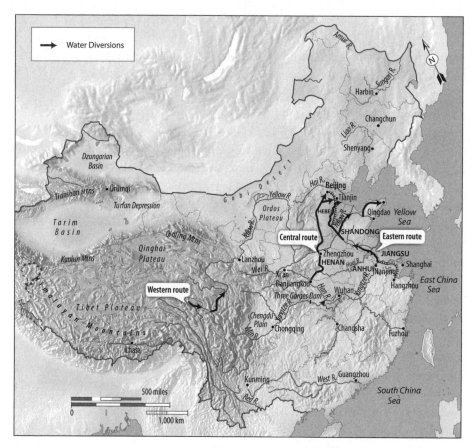

South-to-North Water Diversion project

began to supply water to midpoints along the eastern and central routes. The eastern route, which utilizes significant sections of the old Grand Canal, began delivering water to Jiangsu, Ahnui, and Shandong provinces by the end of 2013, with plans to begin delivering water to Tianjin by 2014. The western route, which traverses far more complex geography, officially remained under further study.[116]

Interviewed in mid-2011 for a *New York Times* article on the South-to-North project, the environmental activist Dai Qing was quoted as saying, "When the water comes to Beijing, there is the danger of the water not being safe to drink. . . . I think this project is a product of the totalitarian regime in Beijing as it seeks to take the resources away from others."[117] Dai went on to suggest the need for alternative management strategies, such as water conservation.

Although more carefully couched in rhetoric less offensive to the CCP, ideas that shifted the focus from ecological interventions to adjustments in how water was consumed were increasingly discussed in scientific, activist, and government circles by the mid-1990s. One broad impulse was the introduction of market-based solutions to environmental management, to replace Maoist "command and control" policies. A progressive taxation system designed to force polluters to internalize pollution costs (by exacting fees for emissions), a market for tradable pollution permits, private resource ownership, and a system of punitive fines were advocated by domestic and international policy actors. Indeed, some of these initiatives were promoted by the central government, but the relative ineffectiveness of the measures—pollution fines perhaps the most obvious example—led some to argue that the liberalization of resource markets had simply not gone far enough. The potential market reform that created the greatest attention was pricing. The state invested considerable effort in studying the ramifications of adjusting the structure of water pricing to promote efficient use and allocation of water to value-added economic activities (including urban residential use).[118]

The greatest focus of restructuring water demand has been the agricultural sector. Farming consumes 75 percent of all water on the North China Plain. By 2000, the extensive farmland irrigation efforts begun in the 1950s brought half of all cropland in the region under irrigation, with some three-quarters of total output coming from irrigated land. But much of this irrigation infrastructure was inefficient. Large losses of water occurred through evaporation and seepage in unlined channels. Central authorities have urged provincial and local officials to incentivize efforts to introduce drip irrigation and to line irrigation networks to prevent seepage. However, reforming the price structure of surface and subsurface irrigation water has been the object of most debate. China's price structure was in many ways similar to that in other developed and developing countries that delivered water at prices well under cost. Some analysts, however, argued that China had the capacity to increase cost to a level sufficient to motivate water conservation. In Henan, for example, the average cost of water supply in 2000 was 0.08 yuan/m$^3$

(slightly more than US$0.01), but water authorities recouped only half that cost from irrigators. Water itself, either surface or sub-surface, was free. In an influential study published by Chinese and Australian scholars, however, farmer responses to price increases in North China showed little gain in water efficiency. Their find-ings suggested that response to price signals would be enhanced if sweeping institutional changes were made in the water sector, in-cluding devolution of authority from irrigation districts to water-user associations and giving greater water rights and autonomy to individual farming enterprises.[119]

China's leaders are caught between the complex forces of long-term and short-term ecological realities, on the one hand, and long-term and short-term state commitments, on the other. Political leaders continue to be confronted with the unfavorable biophysical setting of the North China Plain: low rainfall punctuated by intense summer storms, sediment-laden rivers, and poor soils. In order to maintain its own power and the unity of the nation, the imperial state embarked on public works projects to perpetuate the ecologi-cal conditions necessary to support the reproduction of rural soci-ety. These projects included large river defense works and the construction of the Grand Canal. These state commitments made possible considerable agricultural production and fostered high population density. The perpetual challenge for the state in the imperial period was to make the investments necessary to maintain this human-built environment. At times, the state failed, most nota-bly from the late-nineteenth to the mid-twentieth century, when the North China Plain became "the land of famine." Twentieth-century efforts to replace the imperial state included the historical mandate to restore ecological stability to the North China Plain. Armed with new industrial technologies, political ideologies, and organizational tools borrowed from the West, Japan, and the Soviet Union, the traditional commitment to "manage the waters" was deepened to incorporate the modernist vision of multiple use for water resources, in order to drive agricultural and industrial expan-sion. As heir to these traditional commitments and as an enthusias-tic proponent of the modernist vision, the CCP of the Maoist era left a legacy to reformist leaders that included a vastly altered water ecology. The reformist government, too, was to follow the long tradition of ecological intervention and to understand the politically legitimizing power of ordering the waters. The post-Mao state was receptive to technological complexes that included, among other legacies, a commitment to large-scale engineering projects. Maoist China also bequeathed institutional arrangements and practices that have shaped management of resources in the contemporary era. A fractured bureaucratic structure and the power of local Party cadres, for example, complicate water allocation decisions and pollution control efforts. An additional component of the challenge facing the management of China's water sector is the traditional commitment made to rural society by the imperial state, renewed in radical terms by the CCP in 1949 and perpetuated, in effect, by China's enduring commitment to food security. All these forces—some traditional, some variations on traditional themes, and some new—condition the leadership's response to what is widely considered a worsening water environment on the North China Plain. To take perhaps the two most radical policy options available to address China's water conundrum, one end of this spectrum would be to argue, from a comparative advantage perspective, that China should abandon its ecological management of the North China Plain and should allow the rivers to meander at will and that China should allow the market to set the terms of economic activity, while buying its grain on the international market from low-cost producers like the United States and Australia. China cannot do this—or at least its perception of national self-interest, as informed by historical commitments, will not allow this option. The other end of the spectrum is for the state to reassume total control of allocation. Such a response would irretrievably compromise the legitimacy of the Party and the government. It is somewhere in the middle of the spectrum that Chinese leaders must negotiate a set of policies that effectively allocates resources to a variety of constituencies.

## National and Global Implications of China's Water Crisis

As twenty-first-century China has grappled with policy responses to the dilemma of the North China Plain, there is every indication to suggest that water degradation and allocation issues have only intensified. Writing about a visit to the People's Victory Canal irrigation project in 2006, a journalist observed:

> I went to the People's Victory Irrigation Project. It is just over the river from the Huayuankou dike, where the river flows out onto the North China Plain. The project is a fifty-year-old Communist totem, one of Chairman Mao's first public works [constructed] to feed his people. Its canals irrigate an area about the size of Greater London, where a million people live and work. But things are in a sad state [on the North China Plain]. The main distribution canal is badly polluted with foam from a local paper factory. Much of the equipment has not been replaced since the fifties. The original iron sluice gate that the "great helmsman" Mao opened more than half a century ago is still in place. And, worse, the water is giving out. In the old days, the project was literally awash in water. It sucked almost 800,000 acre-feet from the river each year. . . . No such luck today. "We are always short of water now," director Wang Lizheng said with a sigh, in between nervous puffs on his People's Victory cigarette. "We are only allowed to take half the water from the river as we used to."[120]

The fate of the People's Victory Canal project is a microcosm of the North China Plain, as the state struggles to allocate scarce water resources to help sustain economic growth. And China itself is a microcosm of global water problems. On a per capita basis, there is sufficient water around the globe to meet human demand. The critical problem, of course, is distribution. The north–south hydrologic divide in China mirrors global inequities in water availability. Pollution also hits those regions harder that have limited water resources. At the same time, global demand is dominated by agricultural irrigation, while urban and industrial uses increasingly compete with the rural sector for water. What are the potential internal and external consequences of China's particular water dilemmas? How might resource constraints affect China's economic performance? These are fundamental questions, and the political and social stability of China rests on how they are answered. Given the

increasing ties that bind global economic health with China's economic health, water constraints in China have the potential to shape
global economic exchange. Considering that China's population
numbers some 1.4 billion, what happens politically in China is a
global issue. That these questions of national and global stability rest
on a historically marginal ecological foundation is, of course, rather
remarkable. But what demands our further attention is the compounding effect of global climate change. Climate change is, indeed,
the wildcard that looms large in Chinese policy and scientific circles,
and is one factor that has driven the continued fixation on supply
as the state aggressively pursues massive engineering projects on the
Himalayan and Tibetan plateaus. What, then, might be the national
and global consequences of China's water dilemma over the next
couple of decades?

Water resource constraints will be a key component of China's
internal political discourse. Water scarcity and its impact on economic opportunity could develop into an important fault-line for
the nation's political stability. Over the past two decades, environmental transformations have introduced new values and actors into
China's water sector, which has been monopolized by the state since
1949. The human, material, and cultural costs of all water projects—
from local, small-scale water development to massive projects such
as the Three Gorges Dam and the South-to-North Water Diversion project—have galvanized the Chinese and have brought about
a growing sensitivity to the environmental consequences of breakneck economic development. From activists like Dai Qing and Ma
Jun and organizations like Friends of Nature, to local groups react-
ing to problems in their backyards, the state and the Party are faced
with interest groups that can articulate opposition to national policies and to local conditions (rural and urban) that resonate across
China. Although the Party has occasionally tacitly encouraged
moderate environmental activism and reporting to help the central
government enforce environmental mandates on recalcitrant local
governments and enterprises, it is keenly aware of the role played by
environmental protesters as the leading edge of a broader opposi-
tional agenda to political authority in the Soviet Union and Eastern
bloc states.

As discussed earlier, data released by the state in 2006 and 2007 on the number of public protests and disturbances stunned many observers, in China and abroad. A large number of these protests were generated by grievances over water supply. Water disputes increased from 16,747 in 1986 to 94,405 in 2004.[121] From the protests in 2000, when thousands of angry Shandong farmers responded to an inadequate water supply by diverting reservoir water that had been allocated to Beijing, to the violent clashes between several thousands of villagers, relocated for the South-to-North Water Diversion project, and local officials over inadequate conditions and compensation, many of these protests began in rural areas. Access to sufficient quantities of clean water is one component of China's growing urban–rural divide. Furthermore, the water available to rural users is often polluted from agricultural runoff or from insufficiently regulated rural enterprises. Certainly, urban constituencies are not immune to water problems, as evinced by the headline-grabbing spill of 100 tons of benzene into the Songhua River in 2006. Rural Chinese, however, lack institutional structures to cope with the economic and health consequences of water degradation. Public health systems in rural China are hard-pressed to manage waterborne diseases such as diarrhea, which is the leading cause of death among rural children under five years of age. Frustration over water quantity and quality is one component of a "perfect storm" that, combined with other factors (e.g., environmental issues, corruption, income disparities, and a fraying social safety net), could "present a unifying focal point for dissent that crosses geographic, cultural, socio-economic and political lines."[122]

Perhaps the single greatest concern related to water in China is the continued capacity of China to feed itself. With just 7 percent of the world's arable land, China attempts to feed roughly one-quarter of the world's population. Maintaining grain self-sufficiency requires maintaining enough land with access to clean surface and subsurface water supplies. Because available grain production for human consumption is stressed by land and water constraints, and production and consumption patterns influenced by the post-reform domestic market, the challenge for China to meet its oft-stated goal of food self-sufficiency is to maximize grain production by stabilizing agricultural acreage and by maintaining access to clean irrigation water. As late as 2008, Chinese farmers produced 95 percent of the country's staple agricultural products, but the pressures of demographic growth as well as industrial and urban expansion continue apace.

The historical experience of water allocation in developing regions suggests a clear pattern in which agricultural interests usually lose out to urban and industrial constituencies. The political imperative to placate a potentially more politically activist urban population and the desire to allocate material (and human) resources toward greater value-added economic outputs are the primary determinants of this allocation equation. Competing demands in the industrial and urban sector have placed further stress on Chinese agriculture to increase food supply. The critical issue is how to use water better in the agricultural sector. China's irrigation infrastructure has significant capacity to increase water efficiency, but ambiguous property rights and a fractured administrative structure are significant obstacles to the implementation of demand management policies. At the same time, pollution limits access to clean surface and groundwater sources. The pressures on China's agricultural economy have generated a pernicious cycle that contributes to degraded water resources. Intensification of fertilizer use, a critical input to increasing agricultural yields, has at the same time generated pollution of farmland irrigation sources and waterways. As reported by Renmin University in Beijing, China produced roughly 24 percent of world grain output, but "its use of fertilizer accounted for more than 35 percent of total global consumption," suggesting the significant intensification of agricultural cultivation. The report goes on to note that China's grain production had increased more than eightfold from the 1960s, while use of nitrogen fertilizers had surged by about 55 times.[123]

An additional stressor to China's water balance arises from the energy–water nexus. China has aggressively invested in alternative and renewable sources of energy. These measures have been a response to a variety of projections that China is expected to consume 14 million barrels per day (bpd) of petroleum by 2030.[124] In addition to concerns about oil import dependence, mounting environmental concerns are also driving Chinese investments in clean energy. The International Energy Agency forecasts that China will emit nearly 30 percent of the world's energy-related carbon dioxide emissions by 2030, and "use of vehicles is projected to expand from roughly 24 million vehicles now [2006] to 90–140 million by 2020 . . . increasing the demand for transportation energy from 33% of total Chinese petroleum use to about 57%."[125] Investments in ethanol production to mitigate pollution and relieve supply bottlenecks have been one response to the expansion of China's transportation sector. Although a variety of vegetable feedstocks have been discussed as fuel substitutes, corn has emerged as "a more attractive feedstock for making ethanol in the medium and long term due to . . . [its] extensive availability in North China."[126] In late 2007, the state announced plans to meet roughly 10 percent of its national gasoline demand with biofuels by 2020. This would represent a 400 percent increase in biofuel production. Meeting this target would require a 26 percent increase in corn production. Such an increase would involve massive additional inputs of water. To produce "one liter of maize-based ethanol in China requires 6 times more irrigation water than in the United States, and over 25 times more than in Brazil."[127] At the same time, the aggressive development of shale-gas reserves in the United States has energized hopes for similar development in China.[128] China holds the world's largest reserves of shale gas. Similar with the story of biofuels development, however, producing natural gas from shale requires enormous amounts of water. The competing water demands implicit in the goals of food and energy independence, as well as cleaner air, will force China's leadership to make a set of choices with wide-ranging implications for China's domestic and foreign policy.

China's impact on global energy markets has been widely discussed, but there has been comparatively little discussion of how China's concerns over water and food security might shape interstate relations, particularly in Southeast Asia and sub-Saharan Africa. What are the potential implications of China's water and food security concerns for the global community? One option available to China is to secure access to global resources through direct ownership or long-term leases of foreign land. During the past decade, China has engaged in such investment, though this sort of "resource diplomacy" has aroused mixed sentiment. Although global food demand is projected to increase 50 percent by 2030, many parts of the world have the capacity to increase production to meet this growth. However, agricultural subsidies in many countries will continue to suppress market signals. These subsidies, a regular feature in both developed and developing regions, are a response to sensitivities toward the concerns of politically volatile urban populations and the desire to redistribute wealth from agriculture to expanding industrial sectors. At the same time, in its analysis of how the world "will have to juggle competing and conflicting energy security and food security concerns," the National Intelligence Council argues that in grain surplus areas, such as the United States, Canada, Argentina, and Australia, "demand for biofuels— enhanced by government subsidies—will claim larger areas of cropland and greater volumes of irrigation water. . . . This 'fuel farming' tradeoff, coupled with periodic export controls among Asian producers and rising demand for protein among growing middle classes worldwide, will force grain prices in the global market to fluctuate at levels above today's highs."[129] Lester Brown's warnings about China's impact on global food markets two decades ago echo in familiar tones.

China reflects these global dynamics. Similar to its response in the realm of oil, China has sought to ameliorate domestic agricultural production bottlenecks, deflect reliance on international markets, and minimize future price volatility by investing in agricultural land and production abroad. These efforts have mixed consequences for global interests. Chinese purchases on international corn and soybean markets will benefit grain-producing regions such as in the American Midwest and West. At the same time, China's investments in agricultural regions of Africa and Southeast Asia, as part of its "going abroad" strategy, will likely pose challenges to the self-interests of a variety of polities.

In 2009, Niu Dun, deputy minister of agriculture, stated that China preferred to be self-sufficient and would not aggressively pursue land deals. By that time, China had leased over 1.5 million hectares of farmland abroad, second only to South Korea.[130] China invested in farmland in Kazakhstan, Laos, Australia, the Philippines, Mexico, and the African countries of Mozambique, Zimbabwe, and Tanzania. In addition, there are an estimated twenty-three Chinese-owned farms in Zambia.[131] The motivation for such investments has been to secure farmland for production of biofuel crops like sugar, cassava, and sorghum, as well as for the direct exportation of food to the Chinese market. China was no outlier in these kinds of investment. There was an investment flurry in the first decade of the twenty-first century from a variety of sources (like government and private investment funds) from a host of countries. The International Food Policy Research Institute (IFPRI) estimated in 2009 that "15 to 20 million hectares of farmland have been subject to negotiations or transactions over the last few years."[132] One analyst "estimated that by the end of 2008, China, South Korea, the United Arab Emirates, Japan, and Saudi Arabia controlled over 7.6 million cultivable hectares overseas, more than five times the usable agricultural surface of Belgium."[133]

The fact that an estimated 15 million to 20 million hectares of global farmland have been subject to negotiations or transactions in the past several years has provoked charges of a "new colonialism." Much of this investment is sponsored by states that seek to bypass world markets, in order to secure grain for consumption and biofuel feedstocks. Critics argue that foreign land acquisitions create conditions for continued economic immiseration in South and Southeast Asia and in African countries. At the same time, many developing countries see foreign investment in land as a method for gaining needed technology, knowledge of advanced practices, and employment. As many have argued, global movement of agricultural commodities is really trade in water. Thus, it is no coincidence that many of the most aggressive government-sponsored investments in agricultural farmland emanate from regions that are water poor. And China's "resource diplomacy" is generally conducted solely on economic terms, little encumbered with issues such as human rights or terrorism. This has generated much discussion in U.S. policy circles that China should become a "responsible stakeholder" in global affairs. Significant current and potential farmland investment by China occurs in areas in Africa and Southeast Asia that are important to U.S. security and economic interests.

The investment needed to alleviate pollution and to increase water supply for agricultural, urban, and industrial users will have a negative impact on growth rates in China. Growth rates of the sort that China has experienced over the past two decades are likely sufficient to absorb these types of state expenditures, but state outlays for engineering projects such as the Three Gorges Dam and the South-to-North Water Diversion project are unusually large. Tax receipts to state coffers may also be negatively affected as water scarcity, in turn, lowers economic productivity. Along with pollution abatement and massive project costs, the additional economic and financial burdens of large health-care costs and lost workdays because of waterborne diseases add up to a collective bill for China's economic health that is not insignificant. If, indeed, China's growth rates decline due to various internal and external factors, the postponement of these investments may perpetuate or aggravate adverse ecological conditions that will continue to provide a flashpoint for social and political grievances.

China's water resource challenges are increasingly linked to food quality and health issues in and outside China. China exports significant quantities of food products. Many of the concerns over China's food exports have centered on fish and fish products. The importance of farm-raised fish has become increasingly critical to global food production. Stocks of consumable ocean fish have declined substantially in the face of the ravenous global demand for seafood. China has responded to this demand, producing 117 billion pounds of seafood in 2006, some 70 percent of all farm-raised seafood in the world. Some of this is consumed domestically, but, at the same time, China has become the number one exporter of fish in the world. Agricultural runoff, municipal waste, and industrial effluents have all been sources of water contamination of China's freshwater fisheries. In mid-2010 *Renmin Ribao* reported a particularly serious incident in Fujian Province, where toxic waste from a local copper mine resulted in the death of 1,890 tons of fish. In many instances, waste from fish-producing ponds is recycled into local water systems. Further compounding the problem is the use of antibiotics in fish feed to maintain the health of fish in contaminated waters. These drugs concentrate in the muscle tissue of fish, leaving potentially carcinogenic residue.

A second global health problem related to China's water scarcity is the development and spread of zoonotic diseases. Outbreaks of severe acute respiratory syndrome (SARS) and avian flu in the past decade have shined a light on the transmission of viruses from animals to humans. The emergence of pandemic diseases is likely to occur in regions with a high population density and "close association between human and animals."[134] Water shortages are among the outcomes of environmental change that may be critical to the development of pandemic diseases. Water shortages have intensified the human–animal interface, providing the environmental conditions for the spread of zoonotic diseases.[135] China fits this profile of ecological stress, high population densities, and close human–animal interaction that could foster the emergence of pandemic diseases that ignore national borders.

Water stresses may also continue to generate tensions in the bureaucratic fabric of China. The allocation of scarce water resources falls within the administrative bailiwick of a dizzying number of government ministries and agencies. Urbanization and industrial development have led to increased pressures on China's delicate water balance. Decision making processes within this fractured bureaucracy are notoriously characterized by competing interests and mistrust. After decisions are made, implementation is again shaped by central bureaucratic constituencies and the interests of provincial and local governments, which have proved remarkably immune to central government mandates. To cite one fundamental problem of allocation between rural and urban constituents: The Ministry of Water Resources reports that 60 percent of China's cities face water shortages, while Beijing (on the North China Plain) has access to one-third of the world average per capita supply. In rural sectors, it is estimated that 500 million residents are exposed to contaminated drinking water.[136] There are myriad central, provincial, and municipal agencies involved in resolving these water allocation and water degradation issues. The Ministry of Water Resources, the Ministry of Agriculture, the Ministry of Industry, and the Ministry of

Environmental Protection are just a few of the government agencies, and at only the central level, that have critical interests in water administration. Bureaucratic constituencies fracture and coalesce around water policy. In the same spirit, provincial-level compacts regulating water withdrawals from interprovincial waterways are notoriously contentious and lack effective oversight mechanisms. The status of a 1999 water accord among riparian Yellow River provinces was already considered precarious shortly after the ink dried.[137] Thus water constraints hold the potential to create disequilibria within a state administrative structure already stressed by bureaucratic fault lines.

Climate change is the potential game changer in the millennia-long struggle to manage the scarce resources of the North China Plain. Recession of Himalayan glaciers and of the snowpack on the Tibetan/Qinghai Plateau will have serious consequences for China's rivers that feed and water the high population density in eastern China but that also sustain the population and economies of South and Southeast Asia. The arterial connection between the high plateau and the North China Plain is the Yellow River. State leaders are well aware of China's vulnerabilities to the effects of climate change. At the same time, China has been aggressively developing the water resources of the Himalayan/Tibetan Plateau.

During the past decade, the state channeled substantial money to research institutions like the Chinese Academy of Sciences to forecast the potential consequences of climate change. Of particular concern in this research agenda is the fate of precipitation, glaciers, and snowpack on the Tibet-Qinghai Plateau. The melt from glaciers and annual snowfall from the region feed rivers that serve 47 percent of the world's people.[138] There is little agreement on the precise outcomes of climate change, but a growing body of Chinese and international research suggests that the Himalayan region will be substantially affected by rising temperatures. Greater runoff will initially generate increased flows, which will augment water supplies but, at the same time, will potentially increase flood risk. Over the long term, however, runoff will decrease and other potential consequences of climate change, such as reduced precipitation in the Yellow River valley and North China Plain, will intensify water scarcity. According to a 2007 Chinese study, at currently observed rates of melt, Himalayan glaciers could decline by one-third by 2050 and half by 2090. The anticipated loss of water resources would have a negative impact on China's food production. According to Tang Huajun, the deputy dean of the Chinese Academy of Agricultural Sciences, "A 5 to 10 percent crop loss is foreseeable by 2030 if climate change continues."[139] *Renmin Ribao*, which quoted Tang, continues, "The impact of climate change, coupled with arable land loss and water shortages, will cause a bigger grain production fluctuation and pose a threat to reaching output targets. . . . China, which recorded a grain output of 530.8 million tons in 2009, plans to increase output to 550 million tons by 2020 to ensure grain security for the world's most populous country. China is likely to face an inadequate food supply by 2030 and its overall food production could fall by 23 percent by 2050."[140] In addition to responses such as accelerating use of genetically modified drought-resistant grains, China will more aggressively increase reservoir capacity on transnational waterways in Southwest China.

An important reason for China to exploit Himalayan water resources is the need to address regional economic imbalances by developing the economy of western China. State leaders see hydroelectric generation as a source of cheap energy to develop Tibet and the southwestern provinces, as well as a way to send electricity to industrial centers in Guangdong Province. Both private capital markets and power generation corporations with significant state ties have capitalized on improved transportation infrastructure and technical capacities to target Himalayan rivers for the development of hydroelectric facilities.[141] Aside from their relative distance from population and industrial centers, one could not find better prospects for power generation than these rivers. Rising high out of the Tibetan Plateau, the steep gradient of these rivers provides hydroelectric potential unmatched in China. As of 2013, over 200 dams were under construction or in planning stages in Southwest China.[142]

What will the consequences be for regional economic and political stability as China aggressively develops transboundary water resources? Asia's nine largest rivers originate on the Tibetan Plateau.

Rivers from this region sustain the lives of 1.3 billion people in South and Southeast Asia.[143] For example, over 50 percent of the Brahmaputra River flows through China, but the vast majority of use occurs downstream in India, Nepal, and Bangladesh. The U.S. National Intelligence Council concludes, "With water becoming more scarce in several regions, cooperation over changing water resources is likely to be increasingly difficult within and between states, straining regional relations."[144] Transboundary rivers such as the Mekong (Lancang) and the Salween (Nujiang) are the lifelines of Southeast Asia. For example, the Mekong runs through China, Burma (Myanmar), Laos, Thailand, Cambodia, and Vietnam. In the lower basin, 60 million people rely on aquatic food sources for 80 percent of their protein needs. The International Rivers Network reports that, in addition to aggressive plans for dams by Laos, Vietnam, and other downstream countries, China is building a cascade of eight dams in the upper reaches of the Mekong River, four of them completed, including the massive Xiaowan Dam.[145]

China controls the "water tower of Asia." With such an awesome hold over the resource lifeline of the region, the country is faced with critical decisions about how to wield that power. On the one hand, China's sensitivities to resource dependency leads it to be "one of only three UN member-countries to reject the notion that states have the right not to be adversely affected by activities of upstream countries. Beijing asserts complete sovereignty over resources within its boundaries."[146] However, unilateral development of transboundary waterways will come at the high cost of alienating China's neighbors, one of them, India, in possession of an advanced military capability. Managing its transboundary rivers affords China an opportunity to engage in regional development forums. One regional association with which China has had an ambiguous relationship is the Mekong River Commission (MRC), created in 1957 by riparian countries to consult on issues of common interest in river development. China is not a formal member of the MRC but has made halting steps to share upstream flow and rainfall data with its members. However, it has largely been resistant to consultations involving upstream reservoir management or development plans.

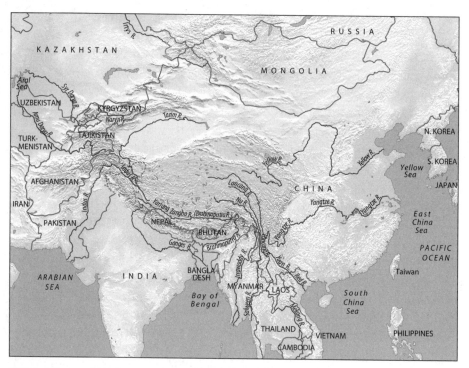

"Water Tower of Asia"

In short, China's water resource challenges, most acutely re-flected on the North China Plain, implicate national, regional, and global security dimensions. Internally, water constraints will continue to affect economic growth in China. The health of the global economy depends on a vibrant Chinese economy. The potential consequences of water scarcity, combined with health concerns engendered by polluted water, can have an impact on political stability. Environmental frustration can be one component of a suite of grievances that can collectively contribute to political instability. A politically unstable China unquestionably translates into an entire host of concerns over global security. At the regional and international levels, global climate change will likely condition China's relations with South and Southeast Asian countries, as the challenges of managing these relationships are accentuated by the diminishing water resources of the Tibetan–Himalayan region. Incorporating China into governance organizations such as the MRC can help stabilize this region. Globally, continued economic expansion, coupled with demographic growth and climate change, will intensify water resource constraints and may cause China to take a more aggressive posture in international grain markets and investment in agricultural resources like land. Finally, China's water quality and quantity issues are directly connected with global food safety and health issues. These dynamics suggest that China and the international community have mutual interests in managing China's water challenges.

# Notes

## Introduction

1. Zhang Guanghui et al., "Huabei pingyuan shui ziyuan jique qinshi yu yinyuan," *Dili kexue yu huanjing bao* 33:2 (June 2011): 172–176; Wang Shiqin et al., "Huabei pingyuan qianceng dixiashui shuiwei dongtai bianhua," *Dili xubao* 63:5 (May 2008): 462–472; Ministry of Water Resources of the People's Republic of China, "China's Glaciers in Danger of Drastic Shrinkage: Report," October 9, 2010, available at www.mwr.gov.cn/english/news/201010/t20101009 _238173.html; Ministry of Water Resources of the People's Republic of China, "Ground Water Tapping Out in North China," April 29, 2010, available at www.mwr.gov.cn/english/news/201004/t20100429_201821.html; Ministry of Water Resources of the People's Republic of China, "Report on the State of the Environment, 2011," July 15, 2013, available at http://english.mep.gov.cn /standards_reports/soe/soe2011/; for water quality classifications, see Ministry of Environmental Protection, "Dibiao shui huanjing zhiliao biaozhun," 1999, available at http://english.mep.gov.cn/standards_reports/standards/water_en vironment/quality_standard/200710/W020061027509896672057.pdf.

2. Notable examples include Judith Shapiro, *Mao's War against Nature: Politics and the Environment in Revolutionary China* (New York: Cambridge University Press, 2001); and Elizabeth Economy, *The River Runs Black: The Environmental Challenge to China's Future* (Ithaca: Cornell University Press, 2004).

3. See Lester Brown, *Who Will Feed China?* (New York: W. W. Norton, 1995). It should be noted that data employed by Brown were called into question; see Jerry McBeath and Jenifer Huang McBeath, "Environmental Stressors and Food Security in China," *Journal of Chinese Political Sciences* 14:49 (2009): 49–80.

4. For additional case studies, see Robert Hathaway and Michael Wills, eds., *Managing New Security Challenges in Asia* (Washington, DC: Johns Hopkins University Press and Woodrow Wilson Center Press, 2014).

5. Arnold Pacey, *Technology in World Civilization: A Thousand-Year History* (Cambridge: MIT Press, 1990).

6. Mark Elvin, "The Environmental Impact of Economic Development," in Brantly Womach, ed., *China's Rise in Historical Perspective* (Lanham, MD: Rowman and Littlefield, 2010), 151–170.

CHAPTER 1  *On the Ecological Margins*

1. For these and other data, see Frank Leeming, *The Changing Geography of China* (Oxford: Blackwell, 1993), 11; and data compiled by the Food and Agricultural Organization of the United Nations, available at www.fao.org/nr /water/aquastat/countries_regions/china/index.stm.

2. Liu Changming and Jun Xia, "Water Problems and Hydrological Research on the Yellow River and the Hai and Huai River Basins in China," *Hydrological Processes* 18:12 (2004): 2197–2210.

3. Ma Jing, Arjen Hoekstra et al., "Virtual versus Real Water Transfers in China," *Philosophical Transactions of the Royal Society B* 361 (2006): 838; also see Liu Junguo and Huub Savenije, "Food Consumption Patterns and Their Effect on Water Requirement in China," *Hydrology and Earth System Sciences* 2:3 (2008): 887–898.

4. Clifton W. Pannell and Laurence J. C. Ma, *China: The Geography of Development and Modernization* (London: Edward Arnold, 1983), 18–19; David N. Keightley, "The Environment of Ancient China," in Michael Loewe and Edward L. Shaughnessy, eds., *The Cambridge History of Ancient China* (Cambridge: Cambridge University Press, 1999), 30–31. For a comprehensive set of historical maps, see Tan Qixiang, ed., *Zhongguo lishi ditu ji*, vols. 1–9, 3rd ed. (Beijing: Zhongguo ditu chubanshe, 1996).

5. J. S. Lee, *The Geology of China* (London: Thomas Murby, 1939).

6. See Gou Xiaohua et al., "Streamflow Variations of the Yellow River over the Past 593 Years in Western China Reconstructed from Tree Rings," *Water Resources Research* 343 (2007): 4179–4186.

7. George Babcock Cressey, *China's Geographic Foundations: A Survey of Its Land and Its People* (New York: McGraw Hill, 1934), 184.

8. Joseph Needham, *Science and Civilization in China*, vol. 1, secs. 1–7 (Cambridge: Cambridge University Press, 1954), 68–69.

9. Ibid., 58.

10. Yu Liansheng, "The Huanghe (Yellow) River: A Review of Its Development, Characteristics, and Future Management Issues," *Continental Shelf Research* 22:3 (February 2002): 391.

11. Keightley, "The Environment of Ancient China," 32–33.

12. Pannel and Ma, *China: The Geography*, 33.

13. Fred Pearce, *When the Rivers Run Dry* (Boston: Beacon Press, 2006), 116.

14. Xu Jiongxin, "A Study of Long Term Environmental Effects of River Regulation on the Yellow River of China in Historical Perspective," *Geografiska Annaler*, Series A, Physical Geography 12 (1968): 61–62.

15. Zhang Jiachang and Lin Zhiguang, *Zhongguo qihou* (Shanghai: Shanghai kexue jishu chubanshe, 1985).

16. For these data and more see Liu and Xia, "Water Problems and Hydrological Research," 2201.

17. Leeming, *The Changing Geography*, 16–17.

18. Keightley, "The Environment of Ancient China," 33.

19. Chang Kwang-Chih, "China on the Eve of the Historical Period," in Loewe and Shaughnessy, *The Cambridge History of Ancient China*, 43.

20. Keightley, "The Environment of Ancient China," 34–35.

21. Cressey, *China's Geographic Foundations*, 162.

22. Ibid., 90.

23. See Tan Minghong, Li Xiubin et al., "Urban Land Expansion and Arable Land Loss in China—A Case Study of Beijing–Tianjin–Hebei Region," *Land Use Policy* 22:3 (July 2005): 187–196.

24. Ibid., 169.

25. Alistair Borthwick, "Is the Yellow River Sustainable?," available at www.soue.org.uk.

26. Shen Caiming et al., "Characteristics of Anomalous Precipitation Events over Eastern China during the Past Five Centuries," *Climate Dynamics* 31 (2008): 471–472.

27. Shen Caiming et al., "Exceptional Drought Events over Eastern China during the Last Five Centuries," *Climatic Change* 85:3–4 (December 2007): 453–471.

28. Shen et al., "Characteristics of Anomalous Precipitation," 456–466; Gou Xiaohua et al., "Streamflow Variations of the Yellow River over the Past 593 Years in Western China Reconstructed from Tree Rings," *Water Resources Research* 343 (2007): 4179.

29. Shen et al., "Exceptional Drought Events," 464–465.

30. Fei Xiaotong, *Small Towns in China: Functions, Problems and Prospects* (Beijing: New World Press, 1986), 155.

31. Mark Elvin, *The Retreat of the Elephant: An Environmental History of China* (New Haven: Yale University Press, 2006), 469.

32. Shen et al., "Exceptional Drought Events," 460.

CHAPTER 2 *Management and Mismanagement in the Imperial Period*

1. Vaclav Smil, "Controlling the River," *Geographical Review* 45:1 (Winter 2001): 253.

2. Chang Kwang-Chih, "China on the Eve of the Historical Period," in Michael Loewe and Edward L. Shaughnessy, eds., *The Cambridge History of Ancient China* (Cambridge: Cambridge University Press, 1999), 55.

3. Ibid., 72.

4. Derk Bodde, "Myths of Ancient China," in Samuel Noah Kramer, ed., *Mythologies of the Ancient World* (Garden City, NY: Doubleday, 1961), 367–408.

5. Mark Edward Lewis, *The Flood Myths of Early China* (Albany: State University of New York Press, 2006), 5.

6. Ibid., 16.

7. Ibid., 19.

8. Ibid., 51.

9. Ibid., 15.

10. Karl Wittfogel, *Oriental Despotism: A Comparative Study of Total Power* (New Haven: Yale University Press, 1957); for Wittfogel retrospective, see Maurice Meisner, "The Despotism of Concepts: Wittfogel and Marx on China," *China Quarterly,* 16 (November–December 1963): 99–112; Frederick Mote, "The Growth of Chinese Despotism: A Critique of Wittfogel's Oriental Despotism as Applied to China," *Oriens Extremis* 8:1 (1961): 1–41.

11. J. R. McNeill, "China's Environment in World Perspective," in Mark Elvin and Liu Ts'ui-jung, eds., *Sediments of Time* (Cambridge: Cambridge University Press, 1998), 36.

12. For a succinct rendering of the Zhengguo Canal, see Joseph Needham, *Science and Civilization in China,* vol. 4, pt. 3, *Physics and Physical Technology: Civil Engineering and Nautics* (Cambridge: Cambridge University Press, 1971), 285. See also Yao Hanyuan, *Huanghe shuili shi yanjiu* (Zhengzhou: Huanghe shuili chubanshe, 2003), 590–598. For contemporary interpretive issues, see Huang Zhangling, "Zhengguo qu yuguan mianji de xingcheng ji qi daxiao wenti," in Zhongguo shuili xuehui shuili shi yanjiuhui, ed., *Zhongguo jindai shuili shi lunwenji* (Nanjing: Hehai daxue chubanshe, 1992), 122–124.

13. Dai Zizhuang, *Zhongguo shuili shuping* (Taibei: Mingwen shuju, 1990), 61.

14. Needham, *Science and Civilization in China,* vol. 4, pt, 3, 296.

15. Lynn White, Jr., "The Historical Roots of Our Ecologic Crisis," *Science* 155 (1967): 1203–1207; for a scholarly take on China, see Rhoads Murphy, "Man and Nature in China," *Modern Asian Studies* 4:1 (1967): 313–333.

16. Murphy, "Man and Nature," 315.

17. McNeil, "China's Environment in World Perspective," 37.

18. Peter Perdue, "A Chinese View of Technology and Nature?" in Martin Ruess and Stephen Cutcliffe, eds., *The Illusory Boundary: Environment and Technology in History* (Charlottesville: University of Virginia Press, 2010), 101.

19. Elvin, "The Environmental History of China: An Agenda of Ideas," *Asian Studies Review* 14:2 (1990): 39–53.

20. Murphy, "Man and Nature," 315.

21. George Babcock Cressey, *China's Geographic Foundations: A Survey of Its Land and Its People* (New York: McGraw Hill, 1934), 1, 3.

22. Richard Louis Edmonds, *Patterns of China's Lost Harmony: A Survey of the Country's Environmental Degradation and Protection* (New York: Routledge, 1994), 35.

23. Mark Elvin, "The Environmental Impasse in Late-Imperial China," in Brantly Womach, ed., *China's Rise in Historical Perspective* (Lanham, MD: Rowman and Littlefield, 2010), 155.

24. See Kenneth Pomeranz, "The Transformation of China's Environment, 1500–2000," in Edmund Burke III and Kenneth Pomeranz, eds., *The Environment and World History* (Berkeley: University of California Press, 2009), 118–164.

25. See Pierre-Étienne Will, "Clear Waters versus Muddy Waters: The Zhengbai Irrigation System of Shaanxi Province in the Late Imperial Period," in Elvin and Liu, *Sediments of Time*.

26. Charles Greer, *Water Management in the Yellow River Basin of China* (Austin: University of Texas Press, 1979), 27–28.

27. McNeill, "China's Environment in World Perspective," 36; also see Pomeranz, "The Transformation China's Environment," 121.

28. Needham, *Science and Civilization in China*, vol. 4, pt. 3, 248–250.

29. Greer, *Water Management*, 33–34.

30. Yao, *Huanghe shuili shi yanjiu*, 36.

31. Ibid., 36–37.

32. Yao Hanyuan, *Zhongguo shuili fazhan shi* (Shanghai: Shanghai renmin chubanshe, 2005), 65–71.

33. Xu Jiongxin, "A Study of Long Term Environmental Effects of River Regulation on the Yellow River of China in Historical Perspective," *Geografiska Annaler*, Series A, Physical Geography 12 (1968): 67.

34. Wu Ruobing and Fan Chengtai, "Huaihe xiayou de honglao zaihai dui cetanlu," in Zhongguo shuili xuehui shuilishi yanjiuhui, ed., *Jianghuai shuilishi lunwenji* (Beijing: Zhongguo shuili xuehui shuilishi yanjiuhui, 1993), 16; Jing Cunyi, "Hongzehu de xingcheng yu bianqian," in Huaihe weiyuanhui, ed., *Huaihe shuilishi lunwenji* (Beijing: Shuili dianli bu Huaihe weiyuanhui, 1987), 109; Zheng Zhaojing, "Qian subei shuilishi de wenti," in *Huaihe shuilishi lunwenji*, 26.

35. See Ray Huang, *Dictionary of Ming Bibliography*, vol. 2 (New York: Columbia University Press, 1976), 1107–1115; Edward Vermeer, "P'an Chi-Hsun's Solutions for the Yellow River Problems of the Late 16th Century," *T'oung Pao* 70:3 (1987): 33–67; see also Zheng, "Qian subei," 25–26; Huang, *Huaihe liuyoude*, 58–59; Yao, *Zhongguo shuili fazhanshi*, 562–563.

36. Randall A. Dodgen, "Hydraulic Evolution and Dynastic Decline: The Yellow River Conservancy, 1796–1855," *Late Imperial China* 12:2 (December 1991): 40–41.

37. Hu Ch'ang-tu, "The Yellow River Conservancy in the Ch'ing Dynasty," *Far Eastern Quarterly* 14:4 (August 1955): 511–512; also see Vermeer, "P'an Chi-hsun," 33–62; Wu and Fan, "Huaihe xiayou de honglao," 19; see also Mark Elvin, *The Retreat of the Elephant: An Environmental History of China* (New Haven: Yale University Press, 2006), 137.

38. Charles Davis Jameson, "River, Lake and Land Conservancy in Portions of the Provinces of Anhui and Kiangsu, North of the Yangtze River," *Far Eastern Review* 9:6 (November 1912): 250; see also Wang Zulie, *Huaihe liuyou zhili zongshu* (Bengbu: Shuili dianlibu zhihuai weiyuanhui, 1987), 11–12; Zheng, "Qian subei," 25–26; Dodgen, "Hydraulic Evolution and Dynastic Decline," 41–42; Ma Jun, *China's Water Crisis* (Norwalk, CT: EastBridge, 2004), 6; Xu, "A Study of Long Term Environmental Effects," 67–68; Mark Elvin, "Nature, Technology and Organization in Late-Imperial China," *Nova Acta Leopoldina* NF 98:360 (2009): 151.

39. Elvin, *Retreat of the Elephant*, 130.

40. Ibid., 110–112.

41. Peter Perdue, "A Chinese View of Technology and Nature?" in Martin Ruess and Stephen Cutcliffe, eds., *The Illusory Boundary: Environment and Technology in History* (Charlottesville: University of Virginia Press, 2010), 107.

42. Eduard Vermeer, "Population and Ecology along the Frontier in Qing China," in Elvin and Liu *Sediments of Time*, 35.

43. William Lavely and R. Bin Wong, "Revising the Malthusian Narrative: The Comparative Study of Population Dynamics in Late Imperial China," *Journal of Asian Studies* 57:3 (August 1998): 717.

44. Dwight Perkins, *Agricultural Development in China, 1368–1968* (Edinburgh: Edinburgh University Press, 1969), 16.

45. Peter Perdue, *Exhausting the Earth: State and Peasant in Hunan, 1500–1850* (Cambridge: Harvard University Press, 1987), 93–94.

46. Perdue, "A Chinese View of Technology and Nature?" 106.

47. Will, "The Zheng-Bai Irrigation System," 235.

48. There is some dispute about this conclusion. Vermeer contends that part of the record of increased floods is simply better because the recording of these events was more carefully conducted in the late imperial period. See Vermeer, "Population and Ecology." Data come from Elvin, *Retreat of the Elephant*, 25.

49. Robert Marks, *Tigers, Rice, Silt and Silk: Environment and Economy in Late Imperial South China* (Cambridge: Cambridge University Press, 2006), 338.

50. For more, see Perdue, "A Chinese View of Technology and Nature," 108–110; see also Vermeer, "Population and Ecology," 235–279.

51. See Philip Huang, *The Peasant Economy and Social Change in North China* (Palo Alto: Stanford University Press, 1985).

52. Ge Quansheng et al., "Winter Half-Year Temperature Reconstruction for the Middle and Lower Reaches of the Yellow River and Yangtze River, China, During the Past 200 Years," *Holocene* 13:6 (2003): 933–940; see also Guolin Feng et al., "Abrupt Climate Changes in North China and North Hemisphere," in Q. Luo, ed., *Proceedings of the 2010 Second IITA International Conference on Geoscience and Remote Sensing* (New York: Institute of Electrical and Electronics Engineers, 2010), 335–336.

53. Ge et al., "Winter Half-Year Temperature Reconstruction," 938.

54. Tan Liangshang et al., "Climate Patterns in North Central China during the Last 1800 Years and Their Possible Driving Force," *Climate of the Past* 7 (2011): 685–692.

55. Ibid.; see also Tan Liangcheng et al., "Centennial- to Decadal-Scale Monsoon Precipitation Variability in the Semi-Humid Region, Northern China during the Last 1860 Years: Records from Stalagmites in Huangye Cave," *Holocene* 21:2 (2010): 287–296; Ding Yihui, "Detection, Causes and Projection of Climate Change over China: An Overview of Recent Progress," *Advances in Atmospheric Sciences* 24:6 (2007): 954–971.

56. David D. Zhang et al, "Climate Change and War Frequency in Eastern China over the Last Millennium," *Human Ecology* 35 (2007): 403.

57. See Tan et al., "Centennial- to Decadal-Scale Monsoon Precipitation," 294.

58. Pierre-Étienne Will, "Un cycle hydraulic en Chine: la province du Hubei du XVIe au XIXe siècle," *Bulletin de l'ecole François d'Extrême-Orient* 68 (1980): 261–287.

59. See Vermeer, "Population and Ecology," 31–49.

60. Elvin, "The Environmental Impasse," 151–170.

61. For a summation of these approaches, see David D. Zhang et al., "Global Climate Change, War, and Population Decline in Recent Human History," *Proceedings of the National Academy of Sciences (PNAS)* 104:49 (December 4, 2007): 19214–19219; for a China-centered counter to Malthus, see Lavely and Wong, "Revising the Malthusian Narrative," 717.

62. See Donald Worster, "Transformations of the Earth: Toward an Agroecological Perspective in History," *Journal of American History* 76:4 (March 1990): 1087–1106.

63. Will, "The Zheng-Bai Irrigation System," 309.

64. Dodgen, "Hydraulic Evolution and Dynastic Decline," 37–42; see also Pietz, *Engineering the State;* and Yao, *Zhongguo shuili fazhanshi*, 563–564.

65. Robert Alan Hackman, "The Politics of Water Conservancy in the Huai River Basin, 1851–1911" (Ph.D. dissertation, University of Michigan, 1979); see also Huang Lisheng, "Huaihe liuyoude shuili shiye," (MA thesis, Taiwan Normal University, 1986); Hu, "The Yellow River Administration," 507; On the complicated nature of China's monetary system in the nineteenth century, see Ma Debin, "Money and Monetary System in China in the 19th and 20th Century: An Overview," Working Papers 159/12 41940, Department of Economic History, London School of Economics, January 2012, available at http://eprints.lse.ac.uk/41940/. Dodgen, "Hydraulic Evolution and Dynastic Decline," 47; see also Hackman, *The Politics of Water Conservancy*, 28.

66. Qian Ning, "Fluvial Processes in the Lower Yellow River after Levee Breaching at Tongwaxiang in 1855," *International Journal of Sediment Research* 5:2 (April 1990): 9.

67. Ibid., 2.

68. Tang Bo, "Tongwaxiang gaidao hou Qingting de shizheng jiqi de shi," *Lishi jiaoxue* 8 (2008): 51–53; Wang Lin and Wan Jinfeng, "Huanghe Tongwaxiang juekou yu Qing zhengfu neibu de fudao yu gaidao zhe zheng," *Shandong shifan daxue xuebao* 48:4 (2003): 88–93; Xia Mingfang, "Tongwaxiang gaidao hou Qing zhengfu dui Huanghe de zhili," *Qingshi yanjiu* 4 (1995): 40–51.

69. Wang and Wan, "Huanghe Tongwaxiang juekou," 88–89; Shen Baixian, "Sanshi nian lai Zhongguo zhi shuili shiye," *Number Two Historical Archives* (Nanjing) 320:2 (n.d.): 11.

70. See Kenneth Pomeranz, *The Making of a Hinterland: State, Society, and Economy in Inland North China, 1853–1937* (Berkeley: University of California Press, 1993).

71. Elvin, *Retreat of the Elephant*, 164.

CHAPTER 3 *Transforming the Land of Famine*

1. See Pierre-Étienne Will, "Un cycle hydraulique en China," *Bulletin de l'Ecole Française d'Extrême-Orient* 68 (1980): 261–288.

2. Lillian M. Li, *Fighting Famine in North China: State, Market, and Environmental Decline, 1690s–1990s* (Stanford: Stanford University Press, 2007), 2, 31.

3. I consciously adopt this phrase from William Kirby, "The Internationalization of China," *China Quarterly*, no. 150 (1997): 433–458.

4. During the late imperial period, Zhili Province constituted what is today Beijing, Tianjin, Hebei Province, Inner Mongolia Autonomous Region, western Liaoning Province, and northern Henan Province.

5. William Lockhart, "The Yang-tse-Keang and the Hwang-Ho, or Yellow River," *Proceedings of the Royal Geographical Society of London* 3:5 (1958–1959): 288–289.

6. "Journeys in the Interior of China—Discussion," *Proceedings of the Royal Geographical Society and the Monthly Record of Geography* 2:3 (March 1880): 163.

7. G. James Morrison, "Journeys in the Interior of China," *Proceedings of the Royal Geographical Society and the Monthly Record of Geography* 2:3 (March 1880): 146.

8. Ney Elias, "Notes of a Journey to the New Course of the Yellow River, in 1868," *Journal of the Royal Geographical Society of London* 40 (1870): 5.

9. Ibid., 16.

10. Li, *Fighting Famine in North China*, 272.

11. Timothy Richard, *Forty-Five Years in China: Reminiscences by Timothy Richard* (New York: Frederick A. Stokes, 1916), 130, 133; see also Paul Richard Bohr, *Famine in China and the Missionary: Timothy Richard as Relief Administrator and Advocate of National Reform, 1876–1884* (Cambridge, MA: Harvard East Asian Research Center, 1972). For more on the famine, see He Hanwei, *Guangxu Chunian (1876–1879) Huabei de da hanzai* (Hong Kong: Chinese University Press, 1980).

12. Li, *Fighting Famine in North China*, 247. For other adaptive strategies, see Pierre-Étienne Will, *Bureaucracy and Famine in Eighteenth Century China* (trans. Elborg Forster) (Stanford: Stanford University Press, 1990).

13. Lillian Li, "Introduction: Food, Famine, and the Chinese State," *Journal of Asian Studies* 41:4 (August 1982): 697.

14. Li, *Fighting Famine in North China*, 268.

15. For otherwise superb treatments of the Boxers, see Joseph Esherick, *The Origins of the Boxer Uprising* (Berkeley: University of California Press, 1987); Paul Cohen, *History in Three Keys: The Boxers as Event, Experience, and Myth* (New York: Columbia University Press, 1998).

16. First published in *Xueyi* 3:8 (January 1922); translation by Song Sikang.

17. Elias, "Notes of a Journey," 31; see also Stephen Wheeler, "Obituary: Ney Elias, C.I.E.," *Geographical Journal* 10:1 (July 1897): 102.

18. As conveyed in Morrison, "Journeys in the Interior," 164.

19. J. G. W. Fijnje van Salverda, *Memorandum Relative to the Improvement of the Hwang-Ho or Yellow River in North-China* (The Hague: Martinus Nijhoff, 1891), vii.

20. Ibid., 103.

21. Ibid., 54–78.

22. Li, *Fighting Famine in North China*, 302.

23. See Vannevar Bush, "Biographical Memoir of John Ripley Freeman, 1855–1932," in *Biographical Memoirs*, vol. 8 (New York: National Academy of Sciences, 1935), 170–187; George T. Mazuzan, "Our New Gold Goes Adventuring: The American International Corporation in China," *Pacific Historical Review* 43:2 (May 1974): 212–232; see also David Pietz, *Engineering the State:*

*The Huai River and Reconstruction in Nationalist China* (New York: Routledge, 2002).

24. John R. Freeman, "Flood Problems in China," *Transactions of the American Society of Civil Engineers* 85 (1922): 1417.

25. Ibid., 1436 (emphasis added).

26. Ibid., 1444.

27. "Notes and Comments," *National Review (China)* 15:22 (1914): 684; Zhang Jian, *Zhang Jizi jiulu(er)* (Taipei: Wenhai chubanshe, 1983), 590–591; see also "Current News," *National Review (China)* 18:7 (1915): 40.

28. Song Xishang, *Li Yizhe de shengping* (Taipei: Zhonghua cong shu bian shen weiyuanhui, Taiwan shudian zong jing xiao, 1964), 1–4.

29. Shi Deqing and Kong Ling, "Zhongguo jin xiandai shuili de kaituo—Li Yizhi," *Shuili fazhan yanjiu* (June 2005): 60–61; Tiao Yun and Tang Deyuan, "Wo guo jindai shuili kexue jia Li Yizhi," *Shanxi shida xuebao* 1 (1984): 115–125.

30. See Oliver J. (O. J.) Todd, *The China That I Knew* (Palo Alto: author, 1973), 3–72.

31. Pietz, *Engineering the State*, 41–46.

32. Oliver Edmund Clubb, "Floods of China: A National Disaster," *Journal of Geography* 31:5 (May 1932): 200.

33. "E. C. Lobenstein to T. K. Tseng, Executive Secretary, National Flood Relief Commission," Number Two Historical Archives (Nanjing) 579:1 (1932), as quoted in Pietz, *Engineering the State*, 67.

34. Ibid., 67; for quotation, see "Items of News Received Regarding Conditions in North Anhwei—Extract from a Letter of Mr. A. G. Robinson, Chief Inspector of Pengpu District, to Sir John Hope Simpson (April 13, 1932)," Number Two Historical Archives (Nanjing) 579:31, as quoted in Pietz, *Engineering the State*, 68; see also Lillian Li, "Life and Death in a Chinese Famine: Infanticide as a Demographic Consequence to the 1935 Yellow River Flood," *Comparative Studies in Society and History* 33:3 (1991): 466–510.

35. See William C. Kirby, *Germany and Republican China* (Stanford: Stanford University Press, 1984); Chin Fen, *The National Economic Council: History, Organization, and Activities* (Nanking: National Economic Council, 1935); Margherita Zanasi, *Saving the Nation: Economic Modernity in Republican China* (Chicago: University of Chicago Press, 2006).

36. See Pietz, *Engineering the State*, 99.

37. See Shen Yi, "Huanghe zhili yanjiu zhi mudi ji fanwei," in Shen Yi, ed., *Huanghe wenti taolunji* (Taipei: Taiwan shangwu yinshuguan, 1971), 320–321; see also David Ekbladh, "*Mr. TVA*: Grass-Roots Development, David Lilienthal, and the Rise and Fall of the Tennessee Valley Authority as a Symbol for U.S. Overseas Development, 1933–1973," *Diplomatic History* 26:3 (Summer 2002): 335–374.

38. See Li Yizhi, "Da Weibei gejie huanyinghui yanjiang shuili," in Zhongguo shuili gongcheng xuehui, ed., *Li Yizhi quanji* (Taipei: Zhonghua yeshu weiyuanhui, 1956), 223–228; Li Yizhi, "Shaanxi Weibei shuili gongchengju yinjing di yi qu baogao shu," in Zhongguo shuili gongcheng xuehui, ed., *Li Yizhi quanji* (Taipei: Zhonghua yeshu weiyuanhui, 1956), 228–271; Li Yizhi, "Shaanxi

Weibei shuili gongchengju yinjing di er qu baogao shu," in Zhongguo shuili gongcheng xuehui, ed., *Li Yizhi quanji* (Taipei: Zhonghua yeshu weiyuanhui, 1956), 271–292.

39. See James Scott, *Seeing Like a State: How Various Schemes to Improve the Human Condition Have Failed* (New Haven: Yale University Press, 1998).

40. O. J. Todd, *Development of Water Power at the Yellow River Falls and Elsewhere in South and Central Shansi to Aid in Irrigation and Industrial Expansion—A Report from Surveys Made in 1934 for Shansi Water Conservancy Commission* (Beijing: China International Famine Relief Commission, 1934), 10.

41. Li Yizhi, "Shuili tonglun," in Zhongguo shuili gongcheng xuehui, ed., *Li Yizhi quanji* (Taipei: Zhonghua yeshu weiyuanhui, 1956), 1–12.

42. Charles K. Edmunds, "Taming the Yellow River," *Asia* (June 21, 1921): 538.

43. For more on Freeman and comparative look at development of hydraulics lab in the United States, see Martin Reusse, "The Art of Scientific Precision: River Research in the United States Army Corps of Engineers to 1945," *Technology and Culture* 40:2 (1999): 292–323.

44. Shen Yi, *Huanghe wenti taolun ji* (Taipei: Taiwan shangwu yinshuguan, 1971), 2–3; Iwo Emelung, "Der Gelbe Fluß in Deutschland: Chinesisch-Deutsche Beziehungen aug dem Gebiet des Wasserbaus in den 20er and 30er Jahren des," *Oriens Extremus* Inhaltsverzeichnis OE 38 (1995): 151–182.

45. Emelung does a superb job of narrating these events; see "Der Gelbe Fluß in Deutschland," 162–163; see also the exchanges between Freeman, Engels, and Shen in Shen, "Huanghe wenti taolun ji."

46. O. J. Todd and S. Eliassen, "The Yellow River Problem," *Proceedings of the American Society of Civil Engineers* 66:8, pt. 2 (October 1940): 352. For newspaper reporting on the model, see Fritz Zielesch, "A 'Yellow River' in the Alps," *The Sun* (February 3, 1935): MS16.

47. Quanguo jingji weiyuanhui shuili chu, *Engesi zhidao Huanghe shiyan baogao huibian* (Nanjing: Quanguo jingji weiyuanhui, 1935).

48. Ben Chie Yen, "From Yellow River Models to Modeling of Rivers," *International Journal of Sediment Research* 14:2 (1999): 85.

49. League of Nations' Communication and Transit Organization, "Report by the Committee of Experts on Hydraulic and Road Questions" (Geneva: League of Nations, 1935), 9.

50. For a sampling of this phraseology, see Shuili dianli bu Huan'ge shuili weiyuanhui zhi Huang yanjiuzu, *Huange de zhili yu kaifa* (Shanghai: Shanghai jiaoyu chubanshe, 1984), 1.

51. Wang Hongyu, "Zhuixun dahe yinji gongzhu hexie weilai," in Guo Guoshun, *Huanghe: 1946–2006* (Zhengzhou: Huanghe shuili weiyuanhui, 2006), 3.

52. Robert Bagley, "Shang Archaeology," in Michael Loewe and Edward Shaughnessy, eds., *The Cambridge History of Ancient China* (New York: Cambridge University Press, 1999), 133; see also Lothar van Falkenhausen, "On the Historiographical Orientation of Chinese Archaeology," *Antiquity* 67:257 (1993): 839–849.

53. Chang Kwang-chih, "China on the Eve of the Historical Period," in Loewe and Shaughnessy, *The Cambridge History of Ancient China*, 54–57.

54. For a sampling of the abundant literature on the relationship of archaeology and nationalism in the post-Soviet era, see Philip L. Kohl, "Nationalism and Archaeology: On the Construction of Nations and the Reconstructions of the Remote Past," *Annual Review of Anthropology* 27 (1998): 223–246.

55. James Leibold, "Competing Narratives of Racial Unity in Republican China: From the Yellow Emperor to Peking Man," *Modern China* 32:2 (April 2006): 181–220.

56. Ibid., 186.

57. Ibid., 196.

58. Ibid., 203. For other analyses of the debate, see Frank Dikotter, "Culture, 'Race,' and Nation: The Formation of National Identity in Twentieth Century China," *Journal of International Affairs* 49:2 (1996): 590–605; Prasenjit Duara, *Rescuing History from the Nation: Questioning Narrative of Modern China* (Chicago: University of Chicago Press, 1996).

59. Chiang Kai-shek, *China's Destiny* (New York: MacMillan, 1947), 4–5 (emphasis added).

60. Edward Friedman, "Reconstructing China's National Identity: A Southern Alternative to Mao-Era Anti-Imperialist Nationalism," *Journal of Asian Studies* 53:1 (February 1994): 67–91.

61. See Huanghe shuili weiyuanhui Huanghe zhizong bianjuan shi, *Huanghe da shi ji* (Zhengzhou: Huanghe shuili chubanshe, 2001), 171–172; O. J. Todd, "The Yellow River Dike Breaks of 1935," in O. J. Todd, *Two Decades in China: Comprising Technical Papers, Magazine Articles, Newspaper Stories and Official Reports Connected with Work Under His Own Observation, Association of Chinese and American Engineers, 1938*; Reprint (Taipei: Ch'eng Wen, 1971), 91.

62. Diana Lary, "Drowned Earth: The Strategic Breaching of the Yellow River Dike, 1938," *War in History* 8:2 (2001): 196.

63. It is not entirely clear when these recollections were "discovered." See Wang Zhibin, "1938 nian Huanghe juekou duo Huai luekao," in Shui dian zhi Huaihe weiyuanhui, *Huaihe shuili shi lunwenji* (n.p., 1987), 39–47.

64. Ibid.; see also Xiong Xianyu, "Lujin xin bashi kangzhan shiqi shouwei Huanghe de huigu," *Huanghe shizhi ziliao* 2 (1989): 10–12.

65. Micah Muscolino, "Refugees, Land Reclamation, and Militarized Landscapes in Wartime China: Huanglongshan, Shaanxi, 1937–1945," *Journal of Asian Studies* 69:2 (May 2010): 453–478. For a sampling of flood statistics, see Huanghe shuili weiyuanhui, ed., *Minguo Huanghe dashi ji* (Zhengzhou: Huanghe shuili chubanshe, 2004), 131; Huanghe shuili weiyuanhui, *Shiji Huanghe, 1901–2000* (Zhengzhou: Huanghe shuili chubanshe, 2001), 54; Lary, "Drowned Earth," 202–204.

66. Lary, "Drowned Earth," 205.

67. I thank Jane Sayers for her brilliant article: "China's Mother River Scolds Her Young: Modernization and the National," *Transformations* 5 (December 2002), http://www.transformationsjournal.org/journal/issue_05/pdf/janesayers.pdf, accessed January 19, 2012.

68. Sue Tuohy, "The Sonic Dimensions of Nationalism in Modern China: Musical Representation and Transformation," *Ethnomusicology* 45:1 (Winter 2001): 112.

69. Hung Changtai, "The Politics of Song: Myths and Symbols in the Chinese Communist War Music, 1937–1949," *Modern Asian Studies* 30:4 (October 1996): 904.

70. See Bonnie McDougall, *Mao Zedong's "Talks at the Yan'an Conference on Literature and Art": A Translation of the 1943 Text with Commentary*, Michigan Monographs in Chinese Studies (Book 39) (Ann Arbor: University of Michigan Center for Chinese Studies, 1980).

71. Tuohy, "The Sonic Dimensions of Nationalism in Modern China," 109.

72. Xian Xinghai, *The Yellow River Cantata* (New York: Leeds Music Corporation, 1947).

73. This and all previous references from ibid.; for a brief personal reminiscence, see Hao Jiangtian, *Along the Roaring River: My Wild Ride from Mao to the Met* (Hoboken: John Wiley & Sons, 2008), 13.

74. For more on Xian Xinghai, see Yang Hon-lun, "The Making of a Musical Icon: Xian Xinghai and His Yellow River Cantata," in Annie J. Randall, ed., *Music, Power, and Politics* (New York: Routledge, 2005).

75. Shuilibu Huanghe shuili weiyuanhui, ed., *Huanghe zhi: Huanghe guihua zhi* (Zhengzhou: Henan renmin chubanshe, 1991), 73–76; see also "General Description of the Yellow River Basin," comp. Public Works Commission, 1946, Percy Othus Collection, Pennsylvania State University Special Collections, AX/B40/HCLA/06466, 40–44.

76. Elizabeth Perry, *Rebels and Revolutionaries in North China, 1845–1945* (Stanford: Stanford University Press, 1983).

77. Odoric Y. K. Wou, *Mobilizing the Masses: Building Revolution in Henan* (Stanford: Stanford University Press, 1994), 91–97.

78. Ibid., 166.

79. See ibid., 274–278.

80. J. Franklin Ray, "UNRRA in China: A Case Study of the Interplay of Interests in a Program of International Aid to an Undeveloped Country," paper submitted by the International Secretariat of the Institute of Pacific Relations to the Tenth Conference of the Institute of Pacific Relations, Stratford-upon-Avon, September 1947 (New York: Institute of Pacific Relations, 1947).

81. Ibid., 7–8.

82. George Woodbridge, *UNRRA: The History of the United Nations Relief and Rehabilitation Administration*, vol. 2 (New York: Columbia University Press, 1950), 371; see also Wang Dechun, *Lian he guo shan hou jiu ji zong shu yu Zhongguo,1945–1947* (Beijing: Renmin chubanshe, 2004); Shen Yi, "Huanghe zhukou yu jiuji shanho," *Huanghe wenti taolunji* (Taipei: Taiwan shangwu yinshuguan, 1971), 364–366.

83. Zu Shibao, "Huayuankou: Xianzheng Huanghe bianqian," in Guo Guoshun, ed., *Huanghe: 1946–2006* (Zhengzhou: Huanghe shuili weiyuanhui, 2006), 209–211; Huanghe shuili weiyuanhui, *Shiji Huanghe*, 60.

84. Ray, "UNRRA in China," 23–24; see also O. J. Todd, "The Yellow River Reharnessed," *Geographical Review* 39:1 (January 1949): 41; Woodbridge, *UNRRA*, 433.

85. Ray, "UNRRA in China," 52.

86. Woodbridge, *UNRRA*, 382.

87. For more on Huayuankou, see Diana Lary, *The Chinese People at War: Human Suffering and Social Transformations, 1937–1945* (Cambridge: Cambridge University Press, 2010), 62.

88. Lary, "Drowned Earth," 204.

89. See Zu, "Huayuankou," 210.

90. "Construction: The Earth Mover," *Time* (May 3, 1954), 86.

91. Shen Yi, "Du Huanghe zhiben jihua gaiyaoshu hou," *Huanghe wenti taolunji* (Taipei: Taiwan shangwu yinshuguan, 1971), 292–293.

92. Republic of China, Supreme Economic Council, Public Works Commission, "Preliminary Report on the Yellow River Project by the Yellow River Consulting Board" (1948), Percy Othus Collection, Pennsylvania State University Special Collections, 1/08, 7.

93. Shen Yi, "Huanghe zhili yanjiu zhi mudi ji fanwei," *Huanghe wenti taolunji* (Taipei: Taiwan shangwu yinshuguan, 1971), 316–325.

94. For brief biographical profiles, see Abel Wolman and W. H. Lyles, "John Lucian Savage, 1879–1967, Biographical Memoir," National Academy of Sciences, 1978, available at http://books.nap.edu/html/biomems/jsavage.pdf; "Captain James P. Growdon, Army," n.d., http://projects.militarytimes.com /citations-medals-awards/recipient.php?recipientid=12225; Arlington National Cemetery, "Eugene Reybold," 2008 available at http://www.arlingtonceme tery.net/eugene-reybold.htm; for more on the history of the Three Gorges Dam, see Yin Liangwu, "The Long Quest for Greatness: China's Decision to Launch the Three Gorges Project" (Ph.D. dissertation, Washington University, 1996).

95. Republic of China, Supreme Economic Council, Public Works Commission, "Preliminary Report, 11–14.

96. Shuilibu Huanghe shuili weiyuanhui, "Huanghe zhi: Huanghe guihua zhi," 78–79; Charles Greer, *Water Management in the Yellow River Basin* (Austin: University of Texas Press, 1979), 47–48.

97. Li, *Fighting Famine in North China*, 2007, as quoted in Kenneth Pomeranz, "Development with Chinese Characteristics? Convergence and Divergence in Long-Run Comparative Perspective," Max Weber Lecture Series, June 2011.

98. Arnold Pacey, *Technology in World Civilization: A Thousand-Year History* (Cambridge: MIT Press, 1991).

CHAPTER 4 *Making the Water Run Clear*

1. David Nye, *America as Second Creation: Technology and Narratives of New Beginnings* (Cambridge: MIT Press, 2003), 3.

2. Shuili bu Huanghe shuili weiyuanhui, *Renmin zhili Huanghe liushi nian* (Zhengzhou: Huanghu shuili chubanshe, 2006), 82–83.

3. Fu Zuoyi, "Mao zhuxi de lingdao juedingle zhi Huai gongcheng de shengli," *Zhi Huai huikan* 1 (1951): 25–29; see also Subei zhi Huai zong zhihuibu gongwu chu, "Subei guan'gai zongqu tufang gongcheng chubu zongjie," 1954, Jiangsu Provincial Archives, 4:4.

4. Fuyang zhuanqu zhi Huai zhihui bu, "Fuyang zhuanqu guanyu 1953 nian dongji zhixing jigan mingong zhi de jingyan," *Zhi Huai huikan* 3 (1952): 434.

5. Jiangsu zhi Huai zong zhihui bu, "Subei zhi Huai gongcheng baogao," May 3, 1952, Jiangsu Provincial Archives, Nanjing, 15:15.

6. Fu, "Mao zhuxi de lingdao," 26.

7. Zhang Chongyi, "Jianguo chuqi shuili gongcheng jianshe de qiji—fuzeling shuiku xiujian jishi," *Jiang Huai wenshi—Dangdai xiezhen* 4 (1997): 79–80.

8. Ibid.

9. See Xu Suzhi, "Renzhen guanche zhi Huai fangzhen zuohao minli dongyuan yu gongzi gaige gongzuo," *Zhi Huai weikan* 3 (1953): 429–439.

10. Wang Jihe, "Huanghe ren 'jia,'" in Luo Xiangxin, ed., *Huanghe: Wangshi* (Zhengzhou: Huanghu shuili chubanshe, 2006), 2; Huanghe shuili weiyuanhui, *Shiji Huanghe* (Zhengzhou: Huanghe shuili chubanshe, 2001), 90.

11. Lily Xiao Hong Lee et al., *Biographical Dictionary of Chinese Women*, vol. 2 (Armonk, NY: M. E. Sharpe, 2003), 430–432; for Qian's reflections, see "Qian Zhengying zishu," available at http://www.waterinfo.net.cn/qianzhy/34.htm.

12. Shuili bu Huanghe shuili weiyuanhui, *Renmin zhili Huanghe liushi nian*, 98–100.

13. Ibid., 101–102.

14. "Qingnian tuanyuan Xiao Zhenxing qiang du juekou," *Renmin ribao* (August 15, 1952): 2.

15. Jun Qian, "Zhi Huang baozheng shengchan de yingmo men," *Xin Huanghe* 3 (1950): 33.

16. Jun Qian, "Women keyi anju Huanghe bei'an le: Pingyuan sheng yingmo dahui Huanghe zhanlan ji," *Xin Huanghe* 3 (1950): 35–36.

17. Ibid.

18. Ibid.

19. Ibid.

20. Fu Zuoyi, "Zhi shui wu nian," *Renmin ribao* (October 8, 1954): 2.

21. "Wu nian lai de shuili jianshe," *Renmin ribao* (October 8, 1954): 2; note that 1954 data go only through June.

22. Wang Huayun, "Jiu nian lai zhi Huang gongzuo de chengjiu," *Renmin ribao* (August 11, 1955): 2.

23. Wang Huayun, "Shenqie de huai nian: Huiyi Mao Zedong tongzhi dui zhi Huanghe shiye de guanhuai," *Huanghe shizhi ziliao* 14 (October 1986): 2–4.

24. Shuili bu Huanghe shuili weiyuanhui, *Renmin zhili Huanghe liushi nian*, 92.

25. Ibid., 116.

26. Ibid.

27. Ibid., 117; see also Gong Shiyang, "Renmin Shengli qu zai women shouzhong shengli jiancheng," in Luo, *Huanghe: Wangshi*, 42–44.

28. Henan sheng yin Huang guan'gai ji Wei guanli ju, "Yin Huang guan'gai ji Wei shuili gongcheng fazhan shi," Institute for Water and Hydropower Research, Beijing (May 1959), 51267.

29. Ibid., 6.

30. Ibid., 7.

31. Ibid., 7; for figures on the entire project published in 1954, see Zhang Hanying, "Renmin shengli qu," *Renmin ribao* (October 8, 1954): 2.

32. "Renmin shengli qu yijing zhengqu fangshui," *Renmin ribao* (April 17, 1952): 2.

33. Ding Man, "Yin Huang guan'gai qu zong de yige nongcun—Wangguanying," *Renmin ribao* (April 17, 1952): 2.

34. Ibid.

35. Ibid.

36. Zhang, "Renmin shengli qu," 2.

37. Ding, "Yin Huang guan'gai qu zong de yige nongcun—Wangguanying," 2.

38. For more, see Paul R. Josephson, "'Projects of the Century' in Soviet History: Large-Scale Technologies from Lenin to Gorbachev," *Technology and Culture* 36:3 (July 1995): 519.

39. "Sulian shuli jianshe fazhan jianshi," *Renmin shuili* (May 1951): 57–61.

40. Josephson, "'Projects of the Century,'" 519.

41. Ibid., 520; see also Paul Josephson, *Industrialized Nature: Brute Force Technology and the Transformation of the Natural World* (Washington, DC: Island Press, 2002); Xu Liangying and Fan Dianian, *Science and Socialist Construction in China* (Armonk, NY: M.E. Sharpe, 1982).

42. Zhang Hanying, "Xuexi Sulian xianjin jingyan tigao women de sixiang yu jishu shuiping," *Renmin shuili* (December 1952): 6–9.

43. Fu, "Zhi shui wu nian," 2.

44. See *China News Analysis* "Water Conservancy," (March 5, 1954): 6–7.

45. Fu Zuoyi, "1953 nian quanguo shuili huiyi kaicaoci," in Zhonghua renmin gongheguo shuilibu bangongting, ed., *1949–1957 nian li ci quanguo shuili huiyi baogao wenjian* (Beijing: n.p., 1957), 120–122; also see Chi Wenshun, "Water Conservancy in Communist China," *China Quarterly*, no. 23 (1965): 39.

46. Deng Zihui, *Report on the Multi-Purpose Plan for Permanently Controlling the Yellow River and Exploiting Its Water Resources* (Peking: Foreign Languages Press, 1955), 9; Chinese-language original in Deng Zihui, *Guanyu gen zhi Huanghe shuihai he kaifa Huanghe shuili de zonghe guihua de baogao* (Shanghai: Renmin chubanshe, 1955).

47. Ibid., 14.

48. Ibid., 15.

49. Ibid., 16.

50. Ibid., 22.

51. Ibid.

52. Ibid.

53. Ibid., 23.

54. Ibid., 24.

55. Ibid., 10–11.

56. Ibid., 32.

57. Ibid., 33.

58. Ibid.

59. Ibid., 34.

60. Ibid., 47.

61. Ibid., 28

62. Ibid., 28.

63. Ibid., 42.

64. Ibid., 40.

65. Ibid.

66. Ibid., 41.

67. Ibid., 32.

68. Ibid., 39.

69. Ibid., 29.

70. Huanghe shuili weiyuanhui, *Renmin zhili Huanghe liushi nian*, 122.

71. Joseph Stalin, "Economic Problems of the USSR: Character of Economic Laws Under Socialism," available at http://www.marxists.org/reference/archive/stalin/works/1951/economic-problems/cho2.htm.

72. Deng, *Report on the Multi-Purpose Plan*, 21.

73. David J. M. Hooson, "The Middle Volga: An Emerging Focal Region in the Soviet Union," *Geographical Journal* 126:2 (June 1960): 184–85; see also Philip Micklin, "Environmental Costs of the Volga-Kama Cascade of Power Stations," *Water Resources Bulletin* 10:3 (June 1974): 565–572; and A. A. Grigoryev, "Soviet Plans for Irrigation and Power: A Geographical Assessment," *Geographical Journal* 118:2 (June 1953): 170–171.

74. M. F. Grin, "Communism and the Transformation of Nature," *Priroda* 1 (1962): 25–36 (in Russian), quoted in P. M. Kelly et al., "Large-Scale Water Transfers in the USSR," *GeoJournal* 7:3 (1983): 208.

75. As quoted in Boris Komarov, *The Destruction of Nature in the Soviet Union* (Armonk, NY: M. E. Sharpe, 1980), 60–61.

76. Ibid., 61.

77. Philip Pryde, "The Quest for Environmental Quality in the USSR," *American Scientist* 60:6 (November–December 1972): 739–745; and David F. Duke, "Seizing Favors from Nature: The Rise and Fall of Siberian River Diversion," in T. Tvedt and E. Jakobsson, eds., *A History of Water, Volume A: Water Control and River Biographies* (London: IB Taurus, 2006), 3–34.

78. Deborah A. Kaple, *Dream of a Red Factory: The Legacy of High Stalinism in China* (Cambridge: Cambridge University Press, 1994), 6.

79. Cheng Xuemin, "Sulian zhuanjia duiyu Huanghe guihua de juda bangzhu," in Zhonghua renmin gongheguo shuilibu bangongting, ed., *Genzhi Huanghe shuihai kaifa Huanghe shuili* (Beijing: Caizheng jingji chubanshe, 1955), 74–84.

80. Shuili bu Huanghe shuili weiyuanhui, *Renmin zhili Huanghe liushi nian*, 156–167.

81. Ibid., 155.

82. Ibid.

83. "Yige zhansheng ziran de wei da jihua," in Zhonghua renmin gongheguo shuilibu bangongting, *Genzhi Huanghe shuihai kaifa Huanghe shuili*, 40.

84. Ibid.

85. Ibid., 42.

86. Zhang Hanying, "Zhili Huanghe de xin de lichengbei," in Zhonghua renmin gongheguo shuilibu bangongting, *Genzhi Huanghe shuihai kaifa Huanghe shuili*, 43–49.

87. Liu Hao, "Qian nian mengxian yao shixian," in Zhonghua renmin gongheguo shuilibu bangongting, ed., *Genzhi Huanghe shuihai kaifa Huanghe shuili*, 90.

88. Ibid., 93.

89. Hua Shan, "Sanmenxia," in Zhonghua renmin gongheguo shuilibu bangongting, *Genzhi Huanghe shuihai kaifa Huanghe Sshuili*, 85.

90. Jun Quan, "Bian Huanghe wei li he—Zhi Huang zhanlan guanji," Xinhua she xinwengao, April 25, 1955.

91. Xiao He, "Renmin you liliang zhengfu Huanghe," *Renmin ribao* (October 9, 1955): 2.

92. Ibid.

93. Ibid.

94. Ibid.

95. Ibid.

96. Ibid.

97. Ibid.

98. Jun, "Bian Huanghe wei li he," 2.

99. Wang Jianping, "Cong zhi Huanghe zhanlan he dao Huanghe bowu-guan," in Guo Guoshun, ed., *Huanghe: 1946–2006* (Zhengzhou: Huanghe shuili weiyuanhui, 2006), 268–277; for additional but brief account of the exhibition, see "Zhi li Huanghe zhanlan zai Beijing zhanchu," *Renmin ribao* (October 17, 1955), 2.

100. Ibid.

101. Odd Arne Westad, "Introduction," in Odd Arne Westad, ed., *Brothers in Arms: The Rise and Fall of the Sino-Soviet Alliance, 1945–1963* (Stanford: Stanford University Press, 2000), 3.

102. J. A. Allan, "Natural Resources as National Fantasies," *Geoforum* 14:3 (1983): 243–247.

103. Mary Zaccone, "Some Aspects of Surplus Labor, Water Control, and Planning in China, 1949–1960" (Ph.D. dissertation, University of North Carolina, 1963), 232.

104. Ibid., 232–233; see also Charles Greer, *Water Management in the Yellow River Basin of China* (Austin: University of Texas Press, 1979), 119.

105. Michel Oksenberg, "Policy Formation in Communist China: The Case of Mass Irrigation Campaign, 1957–1958" (Ph.D. dissertation, Columbia University, 1970), 29; see also Mei Chengrui and Harold E Dregne, "Review Article: Silt and the Future Development of the Yellow River," *Geographic Journal* 358 (2003): 12; and Zaccone, "Some Aspects of Surplus Labor," 120.

106. Robert Carin, *Irrigation Schemes in Communist China* (Hong Kong: Union Research Institute, 1963), 36–38.

107. Quoted in ibid., 41.

108. Quoted in ibid., 2.

109. Oksenberg, "Policy Formation in Communist China," 32. For more on collectivization, see Nicholas Lardy, "Economic Recovery and the First Five-Year Plan," in Robert MacFarquhar and John K. Fairbank, eds., *Cambridge History of China: The People's Republic of China (Part I)—The Emergence of Revolutionary China, 1949–1965* (Cambridge: Cambridge University Press, 1987), 144–184.

110. "1955 nian quanguo shuili huiyi zongjie baogao," *Xin Huanghe* 2 (1955): 19–23; also "1955 nian quanguo shuili huiyi zongjie baogao," in Zhonghua renmin gongheguo shuili bu bangongting, ed., *1949–1957 lici quanguo shuili huiyi baogao wenjian*, Institute for Water and Hydropower Research, Beijing, 1957, 145: 49–57 [1], 218–228.

111. Central Intelligence Agency, *The Program for Water Conservancy in Communist China, 1949–1961* (Washington, DC, 1962), 16; also Oksenberg, "Policy Formation in Communist China," 32; and Carin, *Irrigation Schemes in Communist China*, 43.

112. Andrew Mertha, *China's Water Warriors: Citizen Action and Policy Change* (Ithaca: Cornell University Press, 2008), 104.

113. Stuart Schram, "Mao Tse-tung's Thought from 1949–1976," in Roderick MacFarquhar and John K. Fairbank, eds., *The Cambridge History of China—The People's Republic, Part 2: Revolutions with the Chinese Revolution, 1966–1982* (Cambridge: Cambridge University Press, 1991), 16–19.

114. Kulaks were relatively affluent farmers in the Russian imperial and early Soviet periods who were deemed class enemies of the peasantry and largely eliminated as they were charged with obstructing collectivization.

115. Frederick Teiwes, "Establishment and Consolidation of the New Regime," in MacFarquhar and Fairbank, *The Cambridge History of China—The People's Republic, Part 2: Revolutions with the Chinese Revolution, 1966–1982*, 110–113.

116. Mao Tse-tung, "The Question of Agricultural Cooperation (July 31, 1955)," in *Communist China, 1955–1959: Policy Documents with Analysis* (Cambridge, MA: Harvard University Press, 1962), 94.

117. Ibid., 95–96.

118. Mao Tse-tung, "Preface to 'Socialist Upsurge in China's Countryside,'" in *Communist China, 1955–1959*, 117–118; "Draft Program for Agricultural Development, 1956–1967," in *Communist China, 1955–1959*, 119–126.

119. Fu Zuoyi, "Wei tiqian he chao'e wancheng wu nian jihua he zai qi nian zhi shi'er nian nei xiaomie putong shuihan zaihai er douzheng," in Zhonghua renmin gongheguo shuilibu bangongting, *1949–1957 nian li ci quanguo shuili huiyi baogao wenjian*, 249–250.

120. Ibid., 256.

121. Ibid., 253.

122. Oksenberg gives a sampling of this data in his "Policy Formation in Communist China," 23.

123. Fu Zuoyi, "Yi jiu wu liu nian shuili jianshe de zhanjia he yi jiu wu qi nian shuili jianshe de fangzhen yu renwu," in Zhonghua renmin gongheguo shuilibu bangongting, *1949–1957 nian li ci quanguo shuili huiyi baogao wenjian*, 293–295.

124. On retreat from the first "leap forward," see Robert MacFarquhar, *The Origins of the Cultural Revolution: Contradictions among the People* (London: Oxford University Press, 1974), 86–89.

125. Huanghe shuili weiyuanhui, *Renmin zhili Huanghe liushi nian*, 157–158.

126. Huanghe zhi bianzhuan weiyuanhui, *Huanghe zhi* (Di liu) (Zhengzhou: Huanghe shuili chubanshe, 2001), 154.

127. Huanghe shuili weiyuanhui, *Renmin zhili Huanghe liushi nian*, 159.

128. Ibid., 159–162; for the range of critiques of the Sanmenxia project, see the articles in *Zhongguo shuili* 7, 8 (1957).

129. Xu Liangying and Fan Dainian, *Science and Socialist Construction in China* (Armonk, NY: M.E. Sharpe, 1982), 184; translation of Xu Liangying and Fan Dainian, *Kexue he wo guo shehui zhuyi jianshe* (Beijing: Renmin chubanshe, 1957).

130. Oksenberg, "Policy Formation in Communist China," 281.

131. Stuart Schram, "Mao Tse-tung's Thought from 1949–1976," 23–25; Teiwes, "Establishment and Consolidation of the New Regime," 122–125.

132. [Mao Zedong?] "Shenme hua," *Renmin ribao* (August 2, 1957): 4.

133. Lu Zheng, "Jiu dian Huanghe wan li sha zhe zhan Sanmenxia shuiku de jiaoxun," *Wenshi cankao* 24 (2010): 32–36.

134. Mao Zedong, "The Situation in Summer 1957" (July 1957), as quoted in Schram, "Mao Tse-tung's Thought from 1949–1976," 32.

135. Teiwes, "Establishment and Consolidation of the New Regime," 139–142.

136. "Xianqi yige xingxiu nongtian shuili de rechao," *Renmin ribao* (September 25, 1957): 1; see "Guanyu jin zhong ming chun da guimo de kazhan xingxiu nongtian shuili he jifei yundong de jueding," *Renmin ribao* (September 25, 1957): 1.

137. "Xianqi yige xingxiu nongtian shuili de rechao," 1.

138. Chi Wen-shun, "Water Conservancy in Communist China," *China Quarterly*, no. 23 (1965): 39.

139. "Da xing shuili bi xu yikao qunzhong," *Renmin ribao* (December 15, 1957): 1; see also Yu Ming, "Dapo da xing shuili de sixiang zhang'ai," *Renmin ribao* (November 5, 1957): 4; "Shuili shi nongye de mingmai," *Renmin ribao* (December 22, 1957): 2; Yu Ming, "Xingxiu shuili zhong de san ge wenti," *Renmin ribao* (December 12, 1957): 2.

140. Frank Dikotter, *Mao's Great Famine: The History of China's Most Devastating Catastrophe, 1958–1962* (New York: Walker, 2010), 27.

141. Robert MacFarquhar, *The Origins of the Cultural Revolution: The Great Leap Forward, 1958–1960* (New York: Columbia University Press, 1983), 77.

142. Ibid., 34.

143. Mao Zedong, "Sixty Points on Working Methods—A Draft Resolution from the Office of the Centre of the CPC," in *Selected Works of Mao Zedong*, vol. 8 (Hyderabad: Kranti, n.d.), available at http://www.marxists.org/reference/archive/mao/selected-works/volume-8/index.htm, accessed March 19, 2012).

144. As quoted in MacFarquhar, *The Origins of the Cultural Revolution*, vol. 1, 186.

CHAPTER 5 *Creating a Garden on the North China Plain*

1. Frank Dikotter, *Mao's Great Famine: The History of China's Most Devastating Catastrophe, 1958–1962* (New York: Walker, 2010), 47–48.

2. As quoted in Robert MacFarquhar, *The Origins of the Cultural Revolution: The Great Leap Forward, 1958–1960* (New York: Columbia University Press, 1983), 25.

3. Michel Oksenberg, "Policy Formation in Communist China: The Case of Mass Irrigation Campaign, 1957–1958" (Ph.D. dissertation, Columbia University, 1970), 14–15.

4. Liu Shaoqi, "The Present Situation, the Party's General Line for Socialist Construction and Its Future," *Communist China, 1955–1959: Policy Documents with Analysis* (Cambridge: Harvard University Press, 1962), 420–438.

5. MacFarquhar, *The Origins of the Cultural Revolution*, 75.

6. Stuart Schram, "Mao Tse-tung's Thought from 1949–1976," in Roderick MacFarquhar and John K. Fairbank, eds., *The Cambridge History of China—The People's Republic, Part 2: Revolutions with the Chinese Revolution, 1966–1982* (Cambridge: Cambridge University Press, 1991), 34.

7. Nicholas Lardy, "The Chinese Economy under Stress, 1958–1965," in Roderick MacFarquhar and John K. Fairbank, eds., *Cambridge History of China: The People's Republic of China, Part I—The Emergence of Revolutionary China, 1949–1965* (Cambridge: Cambridge University Press, 1991), 362.

8. Schram, "Mao Tse-tung's Thought from 1949–1976," 36.

9. For a succinct description, see Lardy, "The Chinese Economy under Stress," 365–367.

10. Chi Wen-shun, "Water Conservancy in Communist China," *China Quarterly*, no. 23 (1965): 39.

11. Li Baohua, "Shuili yundong de xin xingshi," *Renmin ribao* (June 23, 1958): 2.

12. Ibid.

13. Ibid.

14. Ibid.

15. Ibid.

16. Ibid.

17. Wei Weilian and Yao Liwen, "Shui gan yi dong chun, sheng guo ji qian nian," *Renmin ribao* (March 13, 1958): 2.

18. Ibid.

19. Ibid.

20. Ibid. See also "Xu shui wei zhu, xiao xing wei zhu, she ban wei zhu," *Renmin ribao* (March 21, 1958): 1. The Dragon King was the god of rain in Chinese mythology.

21. "Xianqi geng da de nongtian shuili gaochao," *Renmin ribao* (October 14, 1958): 3.

22. Quoted in Mary Zaccone, "Some Aspects of Surplus Labor, Water Control, and Planning in China, 1949–1960" (Ph.D. dissertation, University of North Carolina, 1963), 92.

23. "Zhonggong zhongyang guanyu shuili zuo de zhishi," *Renmin ribao* (September 11, 1958): 1.

24. Ibid.

25. "Xianqi geng da de nongtian shuili gaochao," 3.

26. Ibid.

27. Ibid.

28. Ibid.

29. "Nanshui beidiao—Jiang He xieshou," *Renmin ribao* (October 19, 1958): 2.

30. Chi, "Water Conservancy in Communist China," 40.

31. "Zhonggong zhongyang guanyu shuili zuo de zhishi," 1; on rural labor shortages, see Zaccone, "Some Aspects of Surplus Labor," 127–128.

32. James Nickum, "A Collective Approach to Water Resource Development: The Chinese Commune System" (Ph.D. dissertation, University of California, Berkeley, 1974), 8.

33. Central Intelligence Agency, *The Program for Water Conservancy in Communist China, 1949–1961* (Washington, DC, 1962), 12.

34. Ibid., 11.

35. *Current Background* 530 (Hong Kong: American Consulate General, 1958), 1. For additional context, see MacFarquhar, *The Origins of the Cultural Revolution: The Great Leap*, 102. 1 catty = 1.1023 pounds (500 grams).

36. Chan Ying-keung, *Mass Mobilization for Development: Water Conservancy in China* (Hong Kong: Chinese University of Hong Kong Social Research Center, 1979), 16.

37. David Nye, *American Technological Sublime* (Cambridge: MIT Press, 1994), xiv, quoting David Kertzer, *Ritual, Politics, and Power* (New Haven: Yale University Press, 1989), 67.

38. Jun Qian and Chen Jian, "Sanmenxia shi de danshen," *Renmin ribao* (April 5, 1957): 2.

39. "Huanying Sanmenxia shuili shuniu gongcheng kaigong," *Renmin ribao* (March 29, 1957): 2.

40. "Dajia lai zhiyuan Sanmenxia!" *Renmin ribao* (April 14, 1957): 1.

41. "Zhengfu Huanghe di kai—Juguo zhumu de Sanmenxia shuili shuniu gongcheng zhengshi kaigong," *Renmin ribao* (April 14, 1957): 2; see also Wang Huayun, "Huanghe lishi de xin de yiye," *Renmin ribao* (April 14, 1957): 2.

42. Huanghe shuili weiyuanhui, *Renmin zhili Huanghe liushi nian* (Zhengzhou: Huanghu shuili chubanshe, 2006), 159–160.

43. Ibid., 163.

44. "Zhandou ba tian zhanduan Huanghe," *Renmin ribao* (November 27, 1958): 1.

45. Ibid.

46. "Huanghe xin liu shi ke kaishile," *Renmin ribao* (November 22, 1958): 2.

47. Zhang Lihun, "Yiqie weile jieliu," *Renmin ribao* (November 22, 1958): 2.

48. Ibid.

49. Ibid.; see also "Zhunbei tong Huanghe hong shui zui hou bodou," *Renmin ribao* (May 5, 1959): 1.

50. "Beijing guoqing zhi ye," *Renmin ribao* (October 2, 1958): 2.

51. Guang Weiran, "Sanmenxia de he chang," *She kan* 7 (1958): 68–73; Zhou Zhuquan, "Ping 'Sanmenxia da he chang,'" *Renmin yinyue* 12 (1958): 5–6, 2.

52. "Beijing guoqing zhi ye," 2.

53. Lardy, "The Chinese Economy under Stress," 381.

54. Peter Ho, "Mao's War against Nature? The Environmental Impact of the Grain-First Campaign," *China Journal* 50 (July 2003): 54; see also Dikotter, *Mao's Great Famine*, 135; Audrey Ronning Topping, "Foreword: The River Dragon Has Come," in Dai Qing, ed., *The River Dragon Has Come!* (Armonk, NY: M.E. Sharpe, 1998), xxii; Zaccone, "Some Aspects of Surplus Labor," 181–182.

55. See Chi, "Water Conservancy in Communist China," 40.

56. Rhoads Murphy, "Man and Nature in China," *Modern Asian Studies* 4:1 (Fall 1967): 329.

57. Vaclav Smil, *China's Environmental Crisis: An Inquiry into the Limits of National Development* (Armonk, NY: M.E. Sharpe, 1993), 59; see also Murphy, "Man and Nature in China," 330–331.

58. Central Intelligence Agency, *The Program for Water Conservancy in Communist China*, 14.

59. Robert Carin, *Irrigation Schemes in Communist China* (Hong Kong: Union Research Institute, 1963), 106, 118–119.

60. Dong Bin et al., "Rice Impact in Henan Irrigation Districts along the Lower Yellow River Reaches," International Water Management Institute, 2003, available at http://www.iwmi.cgiar.org/Assessment/files_new/publica tions/Workshop%20Papers/IYRF_2003_Dong-Bin.pdf.

61. Carin, *Irrigation Schemes in Communist China*, 109.

62. Audrey Donnithorne, *China's Economic System* (New York: Praeger, 1967), 130.

63. Bruce Stone, "The Chang Jiang Diversion Project: An Overview of Economic and Environmental Issues," in Asit K. Biswas et al., eds., *Long-Distance Water Transfer: A Chinese Case Study and International Experiences* (New York: Tycooly, 1983), 202–203, 204.

64. Zaccone, "Some Aspects of Surplus Labor," 128.

65. Dong et al. "Rice Impact in Henan Irrigation Districts," 130.

66. Fred Pearce, *When the Rivers Run Dry* (Boston: Beacon Press, 2007), 110.

67. Zaccone, "Some Aspects of Surplus Labor," 128–129.

68. Oksenberg, "Policy Formation in Communist China," 651.

69. See Carin, *Irrigation Schemes in Communist China*, 80; and Zaccone, "Some Aspects of Surplus Labor," 130–132.

70. Huanghe shuili weiyuanhui, *Renmin zhili Huanghe liushi nian*, 185.

71. Ma Jun, *China's Water Crisis* (Norwalk, CT: EastBridge, 2004), 10.

72. Vaclav Smil, "Controlling the River," *Geographical Review* 45:1 (Winter 2001): 266; 186, 200–203.

73. Huanghe shuili weiyuanhui, *Renmin zhili Huanghe liushi nian,*, 191–192.

74. Ibid., 187–189.

75. As quoted in Carin, *Irrigation Schemes in Communist China*, 2–3.

76. As quoted in ibid. 116.

77. James Nickum, ed., *Water Management Organization in the People's Republic of China* (Armonk, NY: M.E. Sharpe, 1981), 57.

78. On "high modernism," see James C. Scott, *Seeing Like a State: How Various Schemes to Improve the Human Condition Have Failed* (New Haven: Yale University Press, 1998); on China, see Peter Perdue, "A Chinese View of Technology and Nature?" in Martin Ruess and Stephen Cutcliffe, eds., *The Illusory Boundary: Environment and Technology in History* (Charlottesville: University of Virginia Press, 2010), 115.

79. Paul Josephson, *Industrialized Nature: Brute Force Technology and the Transformation of the Natural World* (Washington, DC: Island Press, 2002).

80. Lillian Li, *Fighting Famine in North China* (Palo Alto: Stanford University Press, 2007), 4.

81. See Murphy, "Man and Nature in China," 325–328; and Robert Weller, *Discovering Nature: Globalization and Environmental Culture in China and Taiwan* (New York: Cambridge University Press, 2006), 59.

82. Nye, *American Technological Sublime*, 3–5.

83. Zhao Jijun and Jan Woudstra, "In Agriculture Learn from Dazhai: Mao Zedong's Revolutionary Village and the Battle against Nature," *Landscape Design* 32:2 (April 2007): 171; see also Mitch Meisner, "Dazhai: The Mass Line in Practice," *Modern China* 4:1 (January 1978): 27–62.

84. Shuili bu Huanghe shuili weiyuanhui, *Renmin zhili Huanghe liushi nian*, 200–208; for other perspectives, see Shuili dianli bu Huanghe shuili weiyuanhui, "Huanghe xiayou fanghong ji hedao zhengli chakan," Institute for Water and Hydropower Research 59333, February 1960; Huanghe shuili weiyuanhui, "Huanghe xiayou zhili wenti xueshu taolunhui," Institute for Water and Hydropower Research 59708, November 1962; "Huanghe Sanmenxia shuiku yunyong fanshi de yanjiu yu jianyi," Institute for Geographic Sciences and Natural Resources, Chinese Academy of Sciences, Beijing, 21544, n.d.

85. Shuili bu Huanghe shuili weiyuanhui, *Renmin zhili Huanghe liushi nian*, 211–214.

86. Ibid., 260; for detailed investigations on consequences of mass irrigation development on North China Plain, see Zhongguo kexueyuan and Shuili dianli bu, "Huanghe xiayou yin Huang guan chu cisheng yanjianhu yubao he yufang de jige wenti," Institute for Geographic Sciences and Natural Resources, Chinese Academy of Sciences Z029019, February 1962; Zhongguo kexueyuan and Shuili dianli bu, "Huabei pingyuan yixie diqu turang yanjianhua bianhua qingkuang de diacha fenxi," Institute for Geographic Sciences and Natural Resources, Chinese Academy of Sciences Z031657, July 1964.

87. Henan sheng dizhi ju, "Taolun Henan sheng yin Huang Renmin shengli qu guanqu yanjiantu de xingcheng ji fangzhi wenti," Institute for Geographic Sciences and Natural Resources, Chinese Academy of Sciences D016719 Z032163, October 1964, 1; see also "Renmin shengli qu guanqu tu rang yanfen dongtai de chubu fenxi," Institute for Geographic Sciences and Natural Resources, Chinese Academy of Sciences D013547 Z030383, 1964. Shuili bu Huanghe shuili weiyuanhui, *Renmin zhili Huanghe liushi nian*, 260;

Lou Puling and Ho Linxiang, "The Rehabilitation of an Irrigation System along the Yellow River," *Irrigation and Drainage Systems* 2 (1988): 12–13.

88. Ibid., 262; see also "Henan fan xian yin Huang guan'gai gai jian zhong dao de shiji," *Renmin ribao* (June 15, 1969): 3; "Henan Fengqui yin Huang guan'gai jian zhong dao shiji," *Renmin ribao* (April 18, 1970): 4; "Shandong Huanghe xiayou yin Huan yuguan fazhan nongye de jingyan," *Renmin ribao* (May 6, 1970): 3.

89. Ibid., 263.

90. See Donnithorne, *China's Economic System*, 136–137; and James Nickum, *Hydraulic Engineering and Water Resources in the People's Republic of China: Report of the U.S. Water Resources Delegation, August–September 1974* (Stanford: U.S.-China Relations Program, 1977), 27–29.

91. Ibid., 27; see also Ministry of Water Conservancy, *Small Hydropower Development in China* (Beijing: China Water Resources and Electric Power Press, 1980).

92. Charles Greer, *Water Management in the Yellow River Basin of China* (Austin: University of Texas Press, 1979), 113–114.

93. Long Zhentao, "Dixia shui," *Renmin ribao* (August 8, 1961): 6.

94. See He Cheng, "Ji jing he dixia shui de liyong," *Renmin ribao* (September 8, 1961): 2; and "Tigong fengfu dixia shui ziyuan—Baozheng da jing zhunque you heli," *Renmin ribao* (November 8, 1964): 5; "Wei ji shi gi liangshi he jingji zuowu zhuyao yan di tigong shuiwen dizhi ziliao," *Renmin ribao* (August 27, 1964): 1; see also articles in Shuili dianlibu shuili si and Shuili dianlibu kexue yanjiusuo, *Qunzhong zhaoshui* (Beijing: Shuili dianli chubanshe, 1973).

95. "Kaifa dixia shui dee jianbing—ji Hebei sheng Mancheng Xian zhuanjing dui de xianjin shiji," *Renmin ribao* (December 3, 1972): 2.

96. Ibid.

97. Ibid.; see also Nickum, *Hydraulic Engineering*, 51; and Smil, *China's Environmental Crisis*, 44.

98. "Mao Zedong sixiang zhiyin Lin xian renmin xiucheng le Hongqi qu," *Renmin ribao* (April 21, 1966): 2.

99. Ibid.

100. Ibid.

101. Ibid.; see also "Renmin qunzhong you wuxian de chuangzao li," *Renmin ribao* (April 21, 1966): 2.

102. See "Lin xian renmin shi nian jianku fendou," *Renmin ribao* (July 9, 1969): 1; "Hongqi qu pan," *Renmin ribao* (December 12, 1973): 3; "Yuenan hu zhimin laodong qingnian tuan youhao fang hua daibiaotuan canguan Lin xian Hongqi qu," *Renmin ribao* (March 30, 1973): 4.

103. "Hongqi qu (ge ci)," *Renmin ribao* (October 9, 1975): 4.

104. Du Jingyuan and Max Woodworth, "Irrigation Society in China's Northern Frontier, 1860s–1920s," *Cross Currents: East Asian History and Culture Review* 1 (December 2011), available at http://cross-currents.berkeley.edu.

105. James Nickum, "Dam Lies and Other Statistics: Taking the Measure of Irrigation in China, 1931–1991," East-West Center Occasional Papers, Environment Series 18 (January 1995), 55–57; see also Takashi Kume, Kiyoshi Torii, and Toru Mitsuno, "Approach to Land-Use Analysis in the Hetao Irri-

gation Project of Inner Mongolia, China, Based on Satellite Image Data," available at http://www.geospatialworld.net/Paper/Application/ArticleView .aspx?aid=266, 2000.

106. Bruce Stone, "The Chang Jiang Diversion Project: An Overview of Economic and Environmental Issues," in Asit K. Biswas et al., eds., *Long-Distance Water Transfer: A Chinese Case Study and International Experiences* (New York: Tycooly, 1983), 204.

107. J. M. Gonçalves, L. S. Pereira et al., "Modelling and Multicriteria Analysis of Water Saving Scenarios for an Irrigation District in the Upper Yellow River Basin," *Agriculture Water Management* 94 (2007): 93–95.

108. For an excellent overview, see Barry Naughton, *The Chinese Economy: Transitions and Growth* (Cambridge: MIT Press, 2006).

109. Jan-Erik Gustafsson, "Four Decades of Water Management in China," *Stockholm Journal of East Asia Studies* 44 (2003): 82; see also Vaclav Smil, *The Bad Earth Environmental Degradation in China* (Armonk, NY: M.E. Sharpe, 1984), 80.

110. Nickum, for example, makes this argument. See his *Hydraulic Engineering*, 37.

111. Wang Rusong et al., *China Water Vision* (Beijing: China Meteorological Press, 2000), 15–16.

112. Lillian Li, "Introduction: Food, Famine, and the Chinese State," *Journal of Asian Studies* 41:4 (August 1982): 701.

113. Robert Weller, *Discovering Nature: Globalization and Environmental Culture in China and Taiwan* (New York: Cambridge University Press, 2006), 164.

114. Zhu Shouquan et al., "Effect of Diverting Water from South to North on the Ecosystem of the Huang-Huai-Hai Plain," in Biswas et al., *Long-Distance Water Transfer.*

115. Eloise Kennedy et al., "Combining Urban and Rural Water Use for a Sustainable North China Plain," 2003, available at http://www.iwmi.cgiar.org /assessment/files_new/publications/Workshop%20Papers/IYRF_2003 _Kendy.pdf.

116. Shui Fu, "A Profile of Dams in China," in Dai, *The River Dragon Has Come!*, 22–23.

CHAPTER 6  *Managing Legacies, Managing Growth*

1. "Hongqi qu bei zha an de qianqian houhou," *Shanxi shuili* 3 (1993): 8–9.

2. Michael Eng and Ma Jun, "Building Sustainable Solutions to Water Conflicts in the United States and China," *China Environmental Series* 8 (Washington, DC: Woodrow Wilson Center, 2011): 173–175.

3. "Hongqi qu bei zha an de qianqian houhou," 9; for more, see Wang Jinyou, "Hongqi qu gaosu le women shenme?" *Sixiang zhengzhi gongzuo yanjiu* 12 (2004): 26; Wang Wenjun and Gu Hua, "Zhanghe shui shi jiufen de chengyin ji qi dui ce tanxi," *Haihe shuili* 2 (2011): 14–15.

4. For engaging comparisons, see Mark Elvin, "The Environmental Impact of Economic Development," in Brantly Womach, ed., *China's Rise in Historical Perspective* (Lanham, MD: Rowman and Littlefield, 2010), 151–170.

5. For data sources, see Zhang Guanghui et al., "Huabei pingyuan shui ziyuan jique qinshi yu yinyuan," *Dili kexue yu huanjing bao* 33:2 (June 2011): 172–176; Wang Shiqin et al., "Huabei pingyuan qianceng dixiashui shuiwei dongtai bianhua," *Dili xubao* 63:5 (May 2008): 462–472.

6. For water quality classification, see Ministry of Water Resources of the People's Republic of China, "Dibiao shui huanjing zhiliao biaozhung," available at http://english.mep.gov.cn/pv_obj_cache/pv_obj_id_A36B86CCCE5B37CA4 FE3F0594931C59C8C700800/filename/W020061027509896672057.pdf; see also Ministry of Water Resources of the People's Republic of China, "China's Glaciers in Danger of Drastic Shrinkage: Report," available at http://www.mwr .gov.cn/english/news/201010/t20101009_238173.html; Ministry of Water Resources of the People's Republic of China, "Ground Water Tapping Out in North China," available at http://www.mwr.gov.cn/english/news/201004/t20100429 _201821.html); Ministry of Water Resources of the People's Republic of China, "Report on the State of the Environment," available at http://english.mep.gov .cn/pv_obj_cache/pv_obj_id_D7893A5849565C06D7D90DD66B7DFA 21671D3D00/filename/P020110411532104009882.pdf.

7. See "Shuilibu zhang tan Zhongguo de shui wenti," *Renmin ribao* (March 22, 1996): 10; "Huanghe duanliu: Zhi Huang xin keti," *Renmin ribao* (May 24, 1997): 5. For an excellent overview of this research, see Tang Qiuhong et al., "Hydrological Cycles Change in the Yellow River Basin during the Last Half of the Twentieth Century," *Journal of Climate* 21 (2008): 1790–1806.

8. Judith Bannister, "Population, Public Health and the Environment in China," *China Quarterly*, no. 156 (1998): 1014.

9. Tang et al., "Hydrological Cycles," 1791.

10. Jan-Erik Gustafsson, "Four Decades of Water Management in China," *Stockholm Journal of East Asia Studies* 44 (2003): 106–111. For examples of research on Yellow River drainage, see Fu Guoyin et al., "Hydro-Climatic Trends of the Yellow River Basin for the Last 50 Years," *Climatic Change* 65 (2004): 149–178; Yang Dawen et al., "Analysis of Water Resources Variability in the Yellow River of China during the Last Half Century Using Historical Data," *Water Resources Research* 40:6 (June 2004), W06502; J. Xia et al., "The Renewability of Water Resources and Its Quantification in the Yellow River Basin, China," *Hydrological Processes* 18 (2004): 2327–2336; Xu Jiongxin, "Temporal Variation of River Flow Renewability in the Middle Yellow River and the Influencing Factors," *Hydrological Processes* 35 (2005): 620–631; Luliu. Liu et al., "Estimation of Water Renewal Times for the Middle and Lower Sections of the Yellow River," *Hydrological Processes* 17 (2003): 1941–1950; G. Wang, "Eco-Environmental Degradation and Causal Analysis in the Source Region of the Yellow River," *Environmental Geology* 40 (2001): 884–890.

11. Cai Ximing, "Water Stress, Water Transfer and Social Equity in Northern China—Implications for Policy Reforms," *Journal of Environmental Management* 87 (2008): 15.

12. Robert Ash and Richard Louis Edmonds, "China's Land Resources, Environment and Agricultural Production," *China Quarterly*, 156 (1998), 836–880; Richard Louis Edmonds, *Patterns of China's Lost Harmony: A Survey of the*

*Country's Environmental Degradation and Protection* (New York: Routledge, 1994), 70–71.

13. See Cao Jinqing, *China along the Yellow River: Reflection on Rural Society* (New York: Routledge, 2005); also Richard Sanders, "The Political Economy of Chinese Environmental Protection: Lessons of the Mao and Deng Years," *Third World Quarterly–Journal of Emerging Areas* 222 (1997): 1201–1214.

14. Michael Webber et al., "The Yellow River in Transition," *Environmental Science and Policy* 11:5 (2008): 425; idem, "Managing the Yellow River: Questions of Borders, Boundaries, and Access," *Transforming Cultures eJournal* 46:6 (December 2001): 907–920, available at http://epress.lib.uts.edu.au/journals/TfC/.

15. Webber et al., "The Yellow River in Transition," 426.

16. Ibid.

17. Liu Changming and Jun Xia, "Water Problems and Hydrological Research on the Yellow River and the Hai and Huai River Basins in China," *Hydrological Processes* 18:12 (2004): 2200.

18. Webber et al., "The Yellow River in Transition," 426.

19. See, e.g., Ash and Edmonds, "China's Land Resources," 837–843.

20. Gavin McCormack, "Water Margins: Competing Paradigms in China," *Critical Asian Studies* 33:1 (2001): 7. McCormack cites chapter 2 of "Land and Food," United Nations Environment Programme, GEO-2000 (Global Environmental Outlook, 2000), available at http://www.cger.nies.go.jp/geo2000/english/0064.htm.

21. Jerry McBeath and Jenifer Huang McBeath, "Environmental Stressors and Food Security in China," *Journal of Chinese Political Sciences* 14:49 (2009): 68. For a wider perspective, see Jenifer Huang McBeath and Jerry McBeath, *Environmental Change and Food Security in China* (New York: Springer, 2010).

22. Shu Dufa et al., "A New Stage of Nutrition Transition in China," *Public Health Nutrition* 5:1A (2002): 171.

23. Ibid. On nutritional changes, see also B. M. Popkin et al., "The Nutrition Transition in China: A Cross-Sectional Analysis," *European Journal of Clinical Nutrition* 47:5 (1993): 333–346.

24. Liu Jongguo et al., "China's Move to Higher-Meat Diet Hits Water Security" (correspondence), *Nature* 454 (July 24, 2008): 397; see also Liu Jongguo and H. H. G. Savenijie, "Food Consumption Patterns and Their Effect on Water Requirements in China," *Hydrology and Earth System Sciences* 12 (2008): 887–898.

25. For data on yields per hectare, see Gustafsson, "Four Decades of Water Management in China," 80.

26. Cheng Hefa et al., "Meeting China's Water Shortage Crisis: Current Practices and Challenges," *Environmental Science and Technology* 43 (2009): 241; 10,000 yuan roughly equal to US$80,000 at end-of-year 2006 exchange rate.

27. M. Wang et al., "China's Puzzle Game: Four Spatial Shifts in Development," in Michael Weber et al., eds., *China's Transition to a Global Economy* (New York: Palgrave MacMillan, 2002), as quoted in Webber et al., "The Yellow River in Transition," 424–425; see also United Nations Development

Programme, *Human Development Report 2004* (New York: Oxford University Press, 2004).

28. Webber et al., "The Yellow River in Transition," 425.

29. Ibid., citing Elizabeth Economy, *The River Runs Black* (Ithaca: Cornell University Press, 2005).

30. Cheng et al., "Meeting China's Water Shortage Crisis," 240.

31. First articulated in the immediate post–Great Leap Forward retrenchment period by Zhou Enlai as a strategy to revive economic growth, the "Four Modernizations" was a set of policies promoted by Deng Xiaoping after 1978 designed to modernize agriculture, defense, industry, and science and technology by increased investment in modern sectors and opening access to foreign trade and technology transfers.

32. Cheng et al., "Meeting China's Water Shortage Crisis," 241.

33. McBeath and McBeath, "Environmental Stressors," 55; and Fu Xiaolan and V. N. Balasubramanyan, "Township and Village Enterprises in China," *Journal of Development Studies* 39:4 (2003): 27–46.

34. Mark Wang et al., "Rural Industries and Water Pollution in China," *Journal of Environmental Management* 79 (2006): 649.

35. Fu and Balasubramanyan, "Township and Village Enterprises in China," 30.

36. Z. L. Zhu and D. L. Chen, "Nitrogen Fertilizer Use in China—Contributions to Food Production, Impacts on the Environment and Best Management Strategies," *Nutrient Cycling in Agroecosystems* 63 (2002): 117.

37. Webber et al., "Managing the Yellow River," 117.

38. Lin Zhen et al., "Three Dimensions of Sustainability of Farming Practices in the North China Plain: A Case Study from Ningjin County of Shandong Province," *Agriculture Ecosystems and Environment* 105 (2005): 515. For a slightly different version, see Lin Zhen et al., "Sustainability of Farmers' Soil Fertility Management Practices: A Case Study in the North China Plain," *Journal of Environmental Management* 79 (2006): 409–419.

39. Zhu and Chen, "Nitrogen Fertilizer Use in China," 120–121.

40. Lin et al., "Three Dimensions of Sustainability," 520; see also Chen Jianyao et al., "Nitrate Pollution from Agriculture in Different Hydrogeological Zones of the Regional Groundwater Flow System in the North China Plain," *Hydrogeology Journal* 13 (2005): 481–492.

41. Bai Xuemei and Shi Peijun, "Pollution Control in China's Huai River Basin," *Environment* 48:7 (September 2006): 25.

42. Wang Xun et al., "Huaihe wuran qianxi," *Zhongguo huanjing jiance* 22:1 (2006): 96–98.

43. Bai and Shi, "Pollution Control in China's Huai River Basin," 25.

44. Wang et al., "Rural Industries and Water Pollution in China," 651.

45. As quoted in ibid.

46. Bai and Shi, "Pollution Control in China's Huai River Basin," 36.

47. Ibid., 25.

48. Ibid., 28.

49. Debbie Yan Lee, "Child Mortality and Water Pollution in China: Achieving Millennium Development Goal 4," July 2007, available at http://

www.wilsoncenter.org/sites/default/files/child_mortality_jul07.pdf; see also Christine Boyle, "Water-Borne Illness in China," available at http://www.wilsoncenter.org/topics/docs/waterborne_Aug07.pdf.

50. Li Jie, "Huanghe duanliu: Zhi Huang xin keti," *Renmin ribao* (May 24, 1997): 5.

51. Liu Zhenying, "Jiang Chunyun yaoqing zhuanjia zuotan shi qiangdiao ba Huanghe duanliu zuowei yige zhongda keti yanjiu jiejue," *Renmin ribao* (September 30, 1997): 2.

52. Bai Jianfeng, "Muqin he: Ni de ruzhi hai neng tang duojiu?" *Renmin ribao* (June 13, 2007): 10.

53. Xu Jiongxin, "A Study of Anthropogenic Seasonal Rivers in China," *Catena* 15:1 (January 2004): 26; see also Li Dong et al., "Huanghe xia you hedao duanliu qingkuang ji tedian," *Renmin Huanghe* 10:10 (1997): 10-12.

54. Xu Jiongxin, "High-Frequency Zones of River Desiccation Disasters in China and Influencing Factors," *Environmental Management* 149:81 (1997): 101-102; see also Liu Dewen, "Huanghe yousi lu," *Shandong nongye* 9 (1997): 8-9; Yang Zhaofei, "Huanghe duanliu de shengtai sikao," *Zhongguo huanjing guanli* 5 (October 1997): 4-7; He Naiwei et al., "Wan li Huanghe wan gu liu," *Shengtai jingji* 5 (October 1997): 9-15.

55. Xu, "High-Frequency Zones," 109-110. For similar conclusions, see Fu Guobin et al., "Hydro-Climatic Trends of the Yellow River Basin for the Last 50 Years," *Climatic Change* 65 (2004): 149-178; Liu Changming and Zhang Shifeng, "Drying Up of the Yellow River: Its Impacts and Counter-Measures," *Mitigation and Adaptation Strategies for Global Change* 26:2 (June 2001): 265-272. For an example of greater stress on climatic factors, see Qian Weihong and Zhu Yafen, "Climate Change in China from 1889 to 1998 and Its Impact on the Environmental Condition," *Climate Change* 50:4 (2001): 419-444.

56. Xu , "High-Frequency Zones," 109-110; see also Liu Changming and Jun Xia, "Water Problems and Hydrological Research on the Yellow River and the Hai and Huai River Basins in China," *Hydrological Processes* 18:12 (2004): 2199-2200.

57. Fu et al., "Hydro-Climatic Trends of the Yellow River Basin," 170-171.

58. See Ma Jun, *China's Water Crisis* (Norwalk, CT: EastBridge, 2004), 28; and Xu, "A Study of Anthropogenic Seasonal Rivers in China," 23.

59. Ren Liliang, "Impacts of Human Activity on River Runoff in the Northern Area of China," *Journal of Hydrology* 261:1-4 (2002): 204-217.

60. Liu and Zhang, "Drying Up of the Yellow River," 209-211.

61. Elizabeth Economy, "Asia's Water Security Crisis: China, India, and the United States," in Ashley Tellis et al., eds., *Strategic Asia 2008-2009: Challenges and Choices* (Washington, DC: National Bureau of Asian Research, 2008): 365-389.

62. See http://www.time.com/time/specials/packages/0,28757,1975813,00.html.

63. Dai Qing, *Yangtze! Yangtze!* (London and Toronto: Earthscan Canada, 1994); Dai Qing, ed., *The River Dragon Has Come!: The Three Gorges Dam and the Fate of China's Yangtze River and Its People* (Armonk, NY: M.E. Sharpe, 1997).

64. Yin Deyong, "China's Attitude toward Foreign NGOs," available at http://law.wustl.edu/WUGSLR/Issues/Volume8_3/Yin.pdf.

65. *Huang tudi* (Guangxi Film Studio, 1984), released with subtitles in the United States in 1985.

66. Bonnie McDougall, *The Yellow Earth* (Hong Kong: Chinese University of Hong Kong, 1991).

67. As translated in Stephen Field, "*He shang* and the Plateau of Ultrastability," *Bulletin of Concerned Asian Scholars* 23:3 (July–September 1991): 11. This and other articles are included in a "Symposium on *He shang*," *Bulletin of Concerned Asian Scholars* 23:3 (July–September 1991).

68. Ibid., 10–11. For a slightly different translation, see Su Xiaokang and Wang Luxiang, *Deathsong of the River: A Reader's Guide to the Chinese TV Series Heshang* (Ithaca: Cornell East Asia Series, 1991), 190–192.

69. Su and Wang, *Deathsong of the River,* 217–218.

70. Field, "*He shang*," 4.

71. Wang Jing, "*He shang* and the Paradoxes of Chinese Enlightenment," *Bulletin of Concerned Asian Scholars* 23:3 (July–September, 1991), 25.

72. "Records of Comrade Deng Xiaoping's Shenzhen Tour (II)," People's Daily Online, available at http://english.people.com.cn/200201/25/eng20020125_89333.shtml.

73. Edward Friedman, "Reconstructing China's National Identity," *Journal of Asian Studies* 53:1 (February 1994): 70; see also Jane Sayers, "China's Mother River Scolds Her Young: Modernization and the National Landscape," *Transformations* 5 (December 2002), available at http://www.transformationsjournal.org/journal/issue_05/pdf/janesayers.pdf.

74. For related discussion, see Wang Hui, *The End of the Revolution* (New York: Verso, 2010).

75. For recent examples that immediately embed the Yellow River in this nationalist discourse, see Bai, "Muqin he," 10; Wang, "Huanghe hui biancheng nei liu he ma?" 2; He et al., "Wan li Huanghe wan gu liu," 9; Yang, "Huanghe duanliu de shengtai sikao," 4.

76. For examples of the latter, see Patrick Tyler, "A Tide of Pollution Threatens China's Prosperity," *New York Times* (September 25, 1994): 3; and Marilyn Beach, "Beijing: Water, Pollution, and Public Health in China," *Lancet* 358:9283 (September 1, 2001): 735.

77. Lester Brown, *Who Will Feed China? Wake-Up Call for a Small Planet* (New York: W. W. Norton, 1995).

78. See Vaclav Smil, *Feeding the World: A Challenge for the 21st Century* (Cambridge: MIT Press, 2000); see also Robert L. Paarlberg, "Feeding China: A Confident View," *Food Policy* 22:3 (1997): 269–279; Hong Yan and Li Xiubin, "Cultivated Land and Food Supply in China," *Land Use Policy* 17 (2000): 73–88; Wang Huixiao et al., "Problems, Challenges, and Strategic Options of Grain Security in China," *Advances in Agronomy* 103 (2000): 101–147; Chen Xikang, "Feeding One Billion: A Study on China's Grain Problem in the Twenty-First Century," *Chinese Economy* 29:1 (January–February 1996): 22–41; Liu Yanhua et al., "Food Security of China and Lester Brown's Point of View," *Journal of Chinese Geography* 7:4 (1997): 1–8; Wu Xiaofeng, "Zhongguo de liangshi wenti: Yuanjing yu qiujie celue," *Jingji yanjiu cankao* 67 (1997):

2–11; Li Chenggui, "Liangshi yanjiu zhong de ruogan wu qu," *Dangdai jingji kexue* 5 (1998): 64–68; Liu Jianhua, "21 shiji shui lai yanghuo Zhongguo," *Keji wencui* 6 (1995): 122–125; Guowuyuan xinwen bangongshi, "Zhongguo de liangshi wenti," *Renmin ribao* (October 25, 1996): 2.

79. Robert Ash and Richard Edmonds, "China's Land Resources, Environment and Agricultural Production," *China Quarterly*, no. 156 (1998): 836–879; and Vaclav Smil, "China's Agricultural Land," *China Quarterly*, no. 158 (1999): 414–429.

80. Paarlberg, "Feeding China: A Confident View."

81. Chen, "Feeding One Billion," 26–27 (emphasis added).

82. Ibid., 25.

83. Ibid., 23.

84. Hong and Li, "Cultivated Land and Food Supply in China," 73.

85. Liu et al., "Food Security of China and Lester Brown's Point of View," 5–6. The original article is in rough English. I have taken the liberty of providing a more succinct rendering.

86. Chen, "Feeding One Billion," 36–38.

87. See Andrew Mertha, *China's Water Warriors: Citizen Action and Policy Change* (Ithaca: Cornell University Press, 2008).

88. For a profile of "muddling along" in the energy sector, see David Pietz, "The Past, Present, and Future of China's Energy Sector," in Gabriel Collins et al., eds., *China's Energy Strategy* (Newport, RI: Naval Institute Press, 2008).

89. Ibid., 211.

90. Wang Yahua, "Water Dispute in the Yellow River Basin: Challenges to a Centralized System," *China Environment Series* 6 (2004): 94–98.

91. Ma, *China's Water Crisis*, 39–40.

92. Wang Houjie et al., "Interannual and Seasonal Variation of the Huanghe (Yellow) River Water Discharge over the Last Fifty Years: Connections to ENSO Events and Dams," *Global and Planetary Change* 18 (2004): 2197–2210; see also Mei Chengrui and Harold E. Dregne, "Review Article: Silt and the Future Development of the Yellow River," *Geographic Journal* 358 (2003): 7–22.

93. Wang, "Water Dispute in the Yellow River Basin," 95–96.

94. For a representative set of proposals, see Yang Zhaofei, "Huanghe duanliu de shengtai sikao," *Zhongguo huanjing guanli* 5 (October 1997): 4–7.

95. See Justin Yifu Lin and Liu Zhiqiang, "Fiscal Decentralization and Economic Growth in China," *Economic Development and Cultural Change* 49:1 (October 2000): 1–21.

96. For a dissenting view, see Cai Hongbin and Daniel Treisman, "Did Government Decentralization Cause China's Economic Miracle?" *World Politics* 58:4 (2006): 505–535.

97. Bai and Shi, "Pollution Control in China's Huai River Basin," 28–29.

98. Ibid., 29; for the most comprehensive English-language treatment of the Huai plan, see Economy, *The River Runs Black*.

99. Richard Sanders, "The Political Economy of Chinese Environmental Protection: Lessons of the Mao and Deng Years," *Third World Quarterly-Journal of Emerging Areas* 222 (1997): 1206.

100. Ibid.

101. As quoted in ibid.

102. Ibid., 1207; see also Richard Louis Edmonds, "The Environment in the People's Republic of China: 50 Years On," *China Quarterly*, no. 156 (1998): 725–732.

103. Economy, "Asia's Water Security Crisis," 377.

104. Bernadette McDonald and Douglas Jehl, *Whose Water Is It?* (Washington, DC: National Geographic Society, 2003), 46.

105. For a superb exploration, see Patricia Wouters et al., "The New Development of Water Law in China," *Water Law Review* 7:2 (Spring 2004): 243–308.

106. Chang Li and Lynn White, "The Fifteenth Central Committee of the Chinese Communist Party," *Asian Survey* 38:3 (March 1998): 231.

107. Zhongguo kexueyuan dili yanjiusuo, "Guanyu nanshui beidiao diqu de ziran tiaojian yu zonghe kaocha wenti" (March 2, 1959), Institute for Geographic Sciences and Natural Resources, Chinese Academy of Sciences Z028114, Beijing; see also Liu Changming and Laurence Ma, "Interbasin Water Transfer in China," *Geographical Review* 73:3 (July 1983): 253–270; Vaclav Smil, *China's Environmental Crisis: An Inquiry into the Limits of National Development* (Armonk, NY: M. E. Sharpe, 1993), 48; Edmonds, "Patterns of China's Lost Harmony," 117.

108. Philip P. Micklin, "Soviet Water Diversion Plans: Implications for Kazakhstan and Central Asia," *Central Asian Survey* 1:4 (1983): 9.

109. For other thoughts on the Grand Canal and the South-to-North Water Diversion project, see Steven Solomon, *Water: The Epic Struggle for Wealth, Power, and Civilization* (New York: HarperCollins, 2010), 431, 445.

110. Chen Zhikai, "Nanshui beidiao shi zaofu renmin de qianqiu weiye," *Zhongguo shuili bao* (November 30, 2002): 2; for additional reports, see "Nanshui beidiao: Cong guihua da shishi," *Renmin zhengxie* (December 3, 2002): B1; Pan Jiazheng, "Guanyu nanshui beidiao de jiu dian kanfa," *Guangming Ribao* (March 29, 2002): B1; Jiang Xia, "Nanshui beidiao gongcheng zongti guihua queding san tiao shui luxian," *Renmin ribao* (November 11, 2002): 2.

111. "Nanshui beidiao shang la!" *Zhongguo caijing bao* (November 26, 2002): 1.

112. For additional biographical information, see Judith Shapiro, *Mao's War against Nature* (New York: Cambridge University Press, 2001), 51–62; see also Zhao Cheng, *Chang he gu lu: Huang Wanli jiushi nian rensheng cangsang* (Wuhan: Changjiang wenyi chubanshe, 2004).

113. Energy Probe Research Foundation, "Zhang Guangdou's Interview on Beijing TV" (January 29, 2004), available at http://eprf.probeinternational .org/node/3797/. For a recent biography of Zhang, see Guo Mei and Zhou Zhangyu, *Zhang Guangdou zhuan* (Nanjing: Jiangsu renmin chubanshe, 2011).

114. "Waterworks Evoke Debate," *China Daily*, December 20, 2003, available at http://www.chinadaily.com.cn/en/doc/2003-12/20/content_291989.htm.

115. The National Environmental Protection Agency was upgraded in administrative rank and renamed the State Environmental Protection Agency (SEPA) in March 1998. In 2008 SEPA was upgraded to ministerial rank and renamed the Ministry of Environmental Protection.

116. South-to-North Water Diversion Project Stepped Up," *China Daily* (December 15, 2008), available at http://www.chinadaily.com.cn/china/2008-12/15/content_7306531.htm; "South-North Water Diversion Project Helps Relieve Drought," *China Daily* (February 19, 2011), available at http://www.chinadaily.com.cn/business/2011-02/19/content_12044260.htm; "Water Diversion Project Operates Next Year," *China Daily* (February 4, 2012), available at http://www.chinadaily.com.cn/china/2012-02/04/content_14538267.htm; "Water Diversion Project Back on Fast Track," *China Daily* (February 11, 2011), available at http:// europe.chinadaily.com.cn/china/2011-02/25/content_12077829/.

117. Edward Wong, "Plan for China's Water Crisis Spurs Concern," *New York Times* (June 2, 2011): A1.

118. For an excellent overview, see Richard Sanders, "The Political Economy of Chinese Environmental Protection: Lessons of the Mao and Deng Years," *Third World Quarterly: Journal of Emerging Areas* 20:6 (1999): 1208–1209.

119. Hong Yang et al., "Water Scarcity, Pricing Mechanism and Institutional Water Scarcity in Northern China Irrigated Agriculture," *Agricultural Water Management* 61 (2003): 143–161; see also Ximing Cai, "Water Stress, Water Transfer and Social Equity in Northern China—Implications for Policy Reforms," *Journal of Environmental Management* 87:1 (April 2008): 14–25; Qu Futian et al., "Sustainable Natural Resource Use in Rural China: Recent Trends and Policies," *China Economic Review* 22:4 (December 2011): 444–460.

120. Fred Pearce, *When the Rivers Run Dry* (Boston: Beacon Press, 2007 ), 111.

121. Economy, "Asia's Water Security Crisis," 373.

122. Nathan Nankivell, "The National Security Implications of China's Emerging Water Crisis," Jamestown Foundation, September 1, 2005, available at http://www.asianresearch.org/articles/2694.html.

123. Report cited in "China Needs to Cut Use of Chemical Fertilizer: Research," Reuters, January 14, 2000, available at http://www.reuters.com/article/idUSTRE60D20T20100114/.

124. Bin Yang and Lu Yanpin, "The Promise of Cellulosic Ethanol Production in China," *Journal of Chemical Technology and Biotechnology* 82:1 (January 2007): 6.

125. Ibid.

126. See E. Gnansounou, A. Dauriat, and C. E. Wyman, "Refining Sweet Sorghum to Ethanol and Sugar: Economic Trade-offs in the Context of North China," *Bioresource Technology* 96 (2005): 985–1002.

127. Economy, "Asia's Water Security Crisis," 370.

128. See Benjamin Haas, "China Misses Output Targets as It Envies U.S. Shale Gas Success (February, 20, 2014), available at http://www.bloomberg.com/news/2014-02-20/china-misses-output-targets-as-it-envies-u-s-shale-gas-success.html.

129. National Intelligence Council, *Global Trends 2025: A Transformed World* (Washington, DC, 2008), 52.

130. Javier Blas, "China Rules Out Pursuit of African Farmland," Financial Times.com (April 20, 2009), available at http://www.ft.com/cms/s/0/9d2cdee8-2dcf-11de-9eba-00144feabdc0.html.

131. Alexandra Spiedloch and Sophia Murphy, "Agricultural Land Acquisitions: Implications for Food Security and Poverty Alleviation," in Michael Kugelman and Susan Levenstein, eds., *Land Grab? The Race for the World's Farmland* (Washington, DC: Woodrow Wilson International Center for Scholars, 2009), 42.

132. Michael Kugelman, "Introduction," in Kugelman and Levenstein, *Land Grab?*, 1.

133. Ibid.

134. National Intelligence Council, *Global Trends 2025*, 75.

135. Joshua Brown, "When Rivers Run Dry," University of Vermont, Spring 2010, available at http://www.alumni.uvm.edu/vq/spring2010/knowledge.asp.

136. For these and other data, see "Water Pollution in China," available at http://prezi.com/tlajqcl7qwbb/water-pollution-in-china/.

137. See Ma, *China's Water Crisis*.

138. Kenneth Pomeranz, "The Great Himalayan Watershed: Water Shortages, Mega-Projects and Environmental Politics in China, India, and Southeast Asia," *The Asia Pacific Journal: Japan Focus*, available at http://japanfocus.org/-Kenneth-Pomeranz/3195, 1.

139. "Climate Change Takes Toll on Grain Harvest," November 5, 2010, ChinaDaily.com, available at http://www.chinadaily.com.cn/china/2010-11/05/content_11505107.htm.

140. Ibid.

141. Pomeranz, "The Great Himalayan Watershed."

142. Jennifer Turner and Linden Ellis, "China's Growing Ecological Footprint," *China Monitor* 7 (March 2007): 9.

143. Xu Jianchu, "The Highlands: As Shared Water Tower in a Changing Climate and a Changing Asia," Working Paper, World Agroforestry Centre, Nairobi, 2007, 1.

144. National Intelligence Council, *Global Trends 2025*, 66.

145. "Southeast Asia," International Rivers Network Berkeley, available at http://www.internationalrivers.org/southeast-asia/.

146. Economy, "Asia's Water Security Crisis," 379–380.

# Acknowledgments

A HOST OF INSTITUTIONS and individuals were vital to this project. First, I thank the National Science Foundation, the National Endowment for the Humanities, the Mellon Foundation, and the American Philosophical Society for providing support for fieldwork in China. Second, my sincerest gratitude goes to the Institute for Water and Hydropower Research (Beijing), the Institute of Geographic Sciences and Natural Resources Research of China's Academy of Sciences (Beijing), and the Yellow River Conservancy Commission (Zhengzhou). The wonderful administration and staff of these organizations were extraordinarily helpful in facilitating my research. In the UK, my sincerest thanks go to the staff of the Needham Research Institute for a remarkably rewarding research and writing experience during the spring of 2006. Much of the writing of this manuscript was conducted at the School of Historical Studies at the Institute for Advanced Study (Princeton). It is truly a special place, and my tenure there in 2011–2012 as the Will S. Doney Fellow was incredibly productive. Third, I thank Washington State University, particularly the College of Arts and Sciences, which provided research support that indeed made this entire project possible. Last, there are seemingly innumerable individuals who contributed their expertise and assistance to this project. I will attempt to thank them all but will inevitably omit some. For this, I apologize. My deepest thanks go to Iwo Amelung, Chen Maoshan, Nicola DiCosmo, Fu Guobing, Hou Qianliang, Jung (Molly) Huang, Jiang Dejuan, John Kicza, William Kirby, Isabelle Lewis, Li Lijuan, John Moffet, Melody Negron, Wu Qiang, Roger Schlesinger, Song Sikang, Ray Sun, Pat Thorsten, Tan Xuming, Ezra Vogel, Ai Wang,

Paul Whitney, Yu Jingjie, and Zhou Xin. And to several of my admired and respected colleagues—Ken Pomeranz, Robert Marks, Micah Muscolino, Mark Elvin, and Don Worster—many thanks for reading and commenting on all or part of the manuscript. All my gratitude to Kathleen McDermott and Andrew Kinney at Harvard University Press for a smooth and productive relationship. Finally, all my love to Valeria and Olivia, who endured my long absences for research and writing. Despite all the help I received, there will undoubtedly be errors of fact and interpretation. For these, I take sole responsibility.

# Index